CZECHOSLOVAK ACADEMY OF SCIENCES

# BIOLOGY OF COCCINELLIDAE

CZECHOSLOVAK ACADEMY OF SCIENCES

Scientific Editor Doc. Dr. Josef Mařan
Scientific Adviser Doc. Dr. Karel Hůrka, CSc.

Graphic Design by Petr Drbal

# BIOLOGY OF COCCINELLIDAE

Ivo Hodek

with keys for identification
of larvae by co - authors

1973

Dr. W. JUNK N. V.,
Publishers
The Hague

ACADEMIA,
Publishing House
of the Czechoslovak
Academy of Sciences,
Prague

Distribution throughout the world with
the exception of Socialist countries:

Dr. W. Junk N. V., Publishers, 13, van Stolkweg
The Hague, Holland

To all those scientists
who have now passed on
and who laid the foundation
of our present knowledge of Coccinellidae

# Preface

This book provides the first monograph of *Coccinellidae*. Although the group finds inclusion both in Clausen's (1940) "Entomophagous Insects" and in Balduf's (1935) "Entomophagous Coleoptera", reference in these works is limited to three and twenty pages respectively. Moreover, the last thirty years since these books appeared have seen a great deal of work on the group.

The use of insecticides largely destroyed the early attempts at biological control and interest remained low for as long as insecticides appeared quite successful. However, the problems of insecticides soon became apparent, and in the last decade there have been tremendous developments in biological control, particularly in combination with other measures to form integrated control. In these developments aphids and mites, which perhaps include the most widespread crop pests, have received special attention and therefore there has been aroused great interest in coccinellids as potential control agents. Moreover, coccinellids have long proved valuable in the biological control of coccids. Therefore, coccinellids have become of very considerable interest among the entomophagous insects of the world.

It is now quite impossible to compress all the information that exists about the group into a brief volume; to attempt this would even be undesirable, for the important ecological features of coccinellids would then become lost in a morass of less useful details. The underlying aim of the writer has been to survey the group in a way which is instructive and especially includes the type of information likely to be needed by those working on coccinellids. In particular, those who start work on coccinellids will find the book will save many hours of searching the literature and of trying to interpret conflicting statements.

Although it is hoped the book is sufficiently comprehensive for this purpose, it has also been designed to be sufficiently short and readable for use by the unspecialized applied entomologists as well as to interest the layman, who merely has a liking for this group.

To achieve these aims, it has been necessary to make a selection of the available material about coccinellids. This selection has been governed by the following principles:

— papers which identify causal relationships have been given priority over those which are purely descriptive, because the former are more useful in connection with the utilisation of coccinellids in biological or integrated control;

— most stress has been laid on more recent work because this normally gives references to older work; recent reviews which give a good summary of earlier work receive special mention;

— as far as possible, papers are also mentioned which tend to be "hidden" from many workers because either they are in difficult languages or are in inaccessible journals.

In spite of the ecological emphasis of the book, it has been felt necessary to devote the first three chapters (co-written by appropriate specialists) to other topics. Thus Chapter 1 concerns the taxonomy, morphology and anatomy of adults necessary to understand the terminology of the group and the structural features which are relevant to ecology (e. g. variation in mouthparts of groups with different food habits).

While the identification of adult coccinellids presents no special problems because good standard keys have long been available particularly for the Palaearctic fauna, the identification of larvae is still difficult. The only key, even for Palaearctic species, is the key to English species by F. I. van Emden (1949). Only recently has a key to central European species been produced by Klausnitzer (1970), the co-author of Chapter 2. It therefore is worthwhile to include keys to larvae in this chapter of the book. To make their identification as simple as possible, the book features the first colour plates of whole 4th instar larvae of the most common Palaearctic species. There is further a simple pictorial field key.

Chapter 3 discusses the notorious variability of coccinellids on a genetic and zoogeographic basis.

The book has been produced in English because most scientists either have a working knowledge of this language or have easy access to translation facilities. It is hoped that a style of English has been maintained which will be as clear as possible to those for whom English is not a native language. It is hoped that the subject matter has been subdivided in a helpful way but, in any case, overlap between sections has been retained without repetition by the extensive use of cross-references.

The names of authors of coccinellid species are not used in the text (with exception of identification keys) but may be found in the subject index — pars *Coccinellidae.* Also the synonymy of aphids is confined to the index — pars *Aphidoidea.*

July 1971 *I. Hodek*

# Acknowledgements

It is a pleasure to express my indebtedness to Dr. H. F. van Emden who made my English more understandable and for his valuable comments. Thanks are also due to Prof. M. J. Way who read and criticized the last chapter.

Grateful acknowledgement is hereby made to many authors who permitted me to use their unpublished data or "in press" manuscripts, esp. to Drs H. D. Brown, G. Iperti, M. Kehat, Lim Sook Ming, Ya. Lusis, G. Moter, H. Sasaji, B. C. Smith, U. Tadmor. Specialists in all parts of the world have greatly helped by sending me reprints of their papers.

For permission to reproduce re-drawn illustrations, thanks are due to the Editors of Annals of applied Biology, Annual Review of Entomology, British Journal of Animal Behaviour, Bulletin of entomological Research, Bulletin de la Société entomologique d'Egypte, Canadian Entomologist, Citrus Magazin, Entomologia experimentalis et applicata, Entomophaga, Fruits, Japanese Journal of applied Entomology and Zoology, Journal of Animal Ecology, Kontyu, Transactions of the Royal entomological Society of London, Zeitschrift für Morphologie und Ökologie der Tiere, Zoologische Jahrbücher. I am indebted to the Director of "Station de Recherches de Lutte Biologique et de Zoologie agricole", I. N. R. A., Antibes, to F. E. Skinner and B. Matsumoto for permission to use their photos.

# Contents

# ERRATA

P. 30, 8th line from above: for canel read canal;
p. 155, 15th line from below: for fallsot read falls to;
p. 163, 28th line from above: for Picea. A. bipunctata read Picea, A. bipunctata;
p. 181, 21st line from below: for condititons read conditions;
p. 193, last line above the figure: for marked with 2 read marked with *;
p. 201, 2nd line from above: for coccon read cocoon;
5th line from above: for cykle read cycle;
p. 224, left half of Fig. 9.02 is reversed;
p. 256, right column: after E. chrysomelina add 62, 175,
after H. arcuata add 200;
p. 257, right column: after P. auritus add 122, 124,
after R. lophantae add 95, 101.
Photo 33: for smallr read smaller;
photo 39: for situatedin read situated in; after „no aggregations" read „are ...";
photo 41: for 220 m read 1200 m;
photo 49, 50: for divised read devised;
photo 57, 58: for ni Sierra Nevada read in Sierra N.
Plate XVII, Symbols: for left tlank read left flank.

# 1 TAXONOMY AND MORPHOLOGY OF ADULTS

I. KOVÁŘ

## 1.1 Morphology

Adult coccinellids are of minute to medium size, 0.8—18 mm large. The body (Pl. I, 1—8) is most often oval, sometimes almost 3 times as long as wide. The dorsal surface is convex; the ventral surface is always flat. The body surface is either bare or covered with short recumbent hairs and is more or less shiny. In the *Sukunahikonini*, there are also longer erect hairs in addition to the short recumbent ones. Sculpturing is in the form of fine or coarse punctures, densely arranged.

**Head** (Pl. II, 1—5). — The individual parts of the cranium (epicranium, frons, genae, clypeus) are fused. The anterior part of the cranium is simple and projects forwards, in the *Chilocorinae* it is strongly widened laterally. The antennae are inserted in front of the eyes in the dorsolateral corners of the frons, but ventrally in the *Chilocorinae*. The lateral part of the cranium bears anteriorly large compound eyes with a finely, in the *Coccidulinae* and the *Lithophilinae* rather coarsely, faceted surface. Anteriorly an indistinct suture separates the gula from the basal part of the labium (submentum), which forms a medial projection. Between this projection and the inner margin of the genae are deep grooves which form the insertions for the basal parts of the maxillae. The inner skeleton of the head — the tentorium — is formed by narrow tentorial arms, the tentorial bridge is missing.

The mouth-parts are of biting type. The labrum is always much narrower than the anterior margin of the cranium and is clearly visible from above except in the *Chilocorinae*.

The massive, wide, sickle-shaped **mandible** has incisor and molar parts on its inner side. The incisor part is formed by two teeth to give the mandible a bifid apex. The molar part is formed by the molar projection (mola) which usually consists of two teeth situated obliquely beside each other. Ventrally a membraneous prostheca is present, growing between the incisor and molar areas and bearing a continuous row of bristles. The outer side of the mandible is simple, strongly curved and concave at the base. The base of the mandible is wide and has two condyles, of which the ventral one is stronger. Both ventral and dorsal sides of the mandible bear both setae and sensillae. According to the food of the beetle, various types of mandible can be distinguished; these differ particularly in the construction of the apex (incisor part) and the mola (molar part) (Strouhal 1926a, b, Pradhan 1938).

The phytophagous type (Pl. III, 1—3) can be characterized as highly specialized, adapted to biting-off and roughly preparing plant tissue, including the tough epidermis. Typically the incisor part consists of four (three) large, apically round or blunt teeth carrying larger or lesser accessory teeth, including the teeth of the "molar region". The distinct molar projection is replaced by a row of coarse teeth which form the molar part of the mandible (*Henosepilachna*, *Subcoccinella*). From this typical phytophagous mandible, other mandibles can be derived (*Affidentula*) in which a gradual reduction of accessory teeth (including the teeth of the "molar region" up to the reduction of the subapical tooth) and a strong shortening of apical teeth take place.

The **carnivorous** type is the basic and widely represented type of mandible in the family *Coccinellidae* and corresponds to the general description given earlier. Within this type there are several trends. In aphidophagous coccinellids of the tribe *Coccinellini* (Pl. III, 4—8) both the terminal teeth are comparatively small and split off each other near the top; the ventral tooth is also somewhat smaller. The inner part of the mandible is entirely smooth between the top and the mola; only in some genera (Strouhal 1926b) it is covered with minute, blunt teeth. This dentation is very conspicuous in *Tytthaspis sedecimpunctata* (Pl. III, 9—10), (mycophagous species), and *Bulaea lichatschovi* (phytophagous species). In contrast, such dentation is missing in *Coleomegilla maculata* (polyphagous species, feeding on pollen and aphids). Butt (1951) found the molar region of *C. maculata* to differ from that of aphidophagous species.

The ventral terminal tooth is relatively well-developed in the *Scymnini* (Pl. IV, 11—12), *Coccidulini* (Pl. IV, 5—6), and *Stethorini*. In coccidophagous species the ventral tooth is of approximately the same size as the dorsal one (*Noviini*) (Pl. IV, 3—4), or is reduced (*Telsimiini*), or is quite absent (some *Chilocorini*) (Pl. IV, 1—2). In the subfamily *Sticholotinae* the apex of the mandible is not divided and comprises a single tooth; the molar projection is either reduced or absent.

The **mycophagous** type (Pl. IV, 7—10) can be derived from the basal carnivorous type with a mola and occurs in part of the tribe *Psylloborini*. The apex of the mandible bears two teeth, of which the ventral one divides into a row of further teeth, the size of which diminishes towards the base. The number of teeth in this row is different in different species.

The maxilla (Pl. V, 1—6) consists of cardo, stipes, lacinia, galea and the maxillary palpus. The area called stipes consists of three sclerites which form a tube-shaped structure. The galea articulates with the distal part of the stipes; in the *Epilachninae* it is covered with many relatively short setae and is much longer than in carnivorous coccinellids. The lacinia is oblong and flat and bears setae similar to those on the galea on the ventral side. In carnivorous coccinellids these setae are rather long and thin, in the *Psylloborini* stronger, curved setae are present in addition. The maxillary palpus is four-segmented; the apical segment is securiform, weakly divergent anteriorly or nearly parallel-sided; only in some *Sticholotinae* it is much elongated and conical.

Two parts can be distinguished on the labium (Pl. VI, 1—3): the partly movable prelabium (prementum) and postlabium (mentum + submentum) which are connected

---

**Plate I**

Shape of the body. 1 — *Hippodamia tredecimpunctata;* 2 — *Paranaemia vittigera;* 3 — *Semiadalia undecimnotata;* 4 — *Subcoccinella vigintiquatuorpunctata;* 5 — *Coccidula scutellata;* 6 — *Hyperaspis reppensis;* 7 — *Henosepilachna vigintioctomaculata;* 8 — *Afissa mystica.* (I. Kovář del.)

**Plate II**

1, 2, 4, 5 — Head capsules, dorsal aspect; 3 — head capsule ventral aspect. 1 — *Adalia bipunctata;* 2 — *Scymnus frontalis;* 3 — *Propylaea quatuordecimpunctata* (cd — cardo, g — gula, mt — mentum, plb — prelabium, of — occipital foramen, smt — submentum, st — stipes, t — tentorial arm); 4 — *Chilocorus renipustulatus;* 5 — *Ortalia sp.* (I. Kovář del.)

**Plate III**

Mandibles. 1, 2 — *Henosepilachna elaterii;* 3 — *Subcoccinella vigintiquatuorpunctata;* 4, 5 — *Adalia bipunctata;* 6, 7, 8 — *Anisosticta novemdecimpunctata;* 9, 10 — *Tytthaspis sedecimpunctata* (I. Kovář del.)

PLATE I

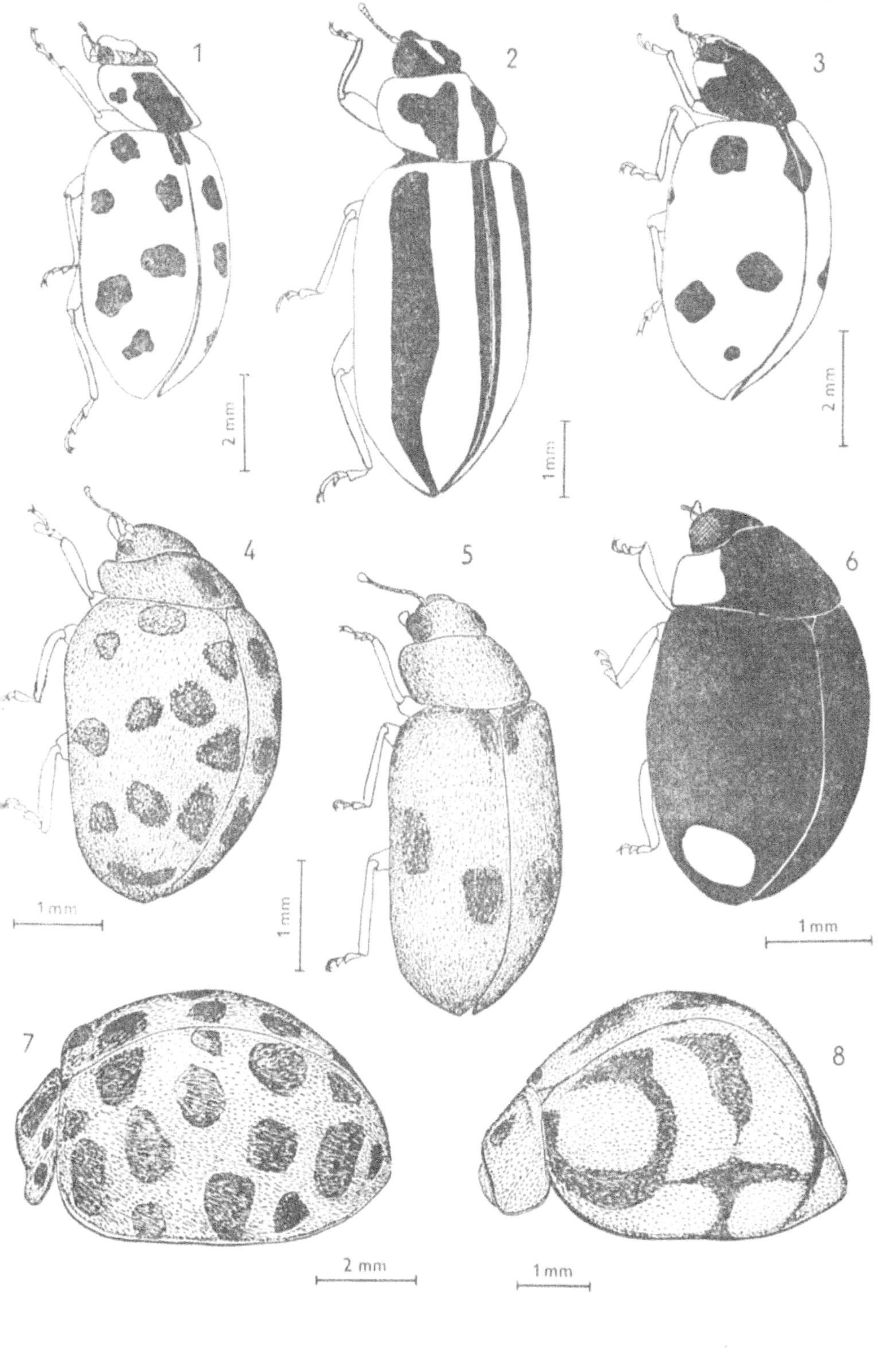

PLATE II

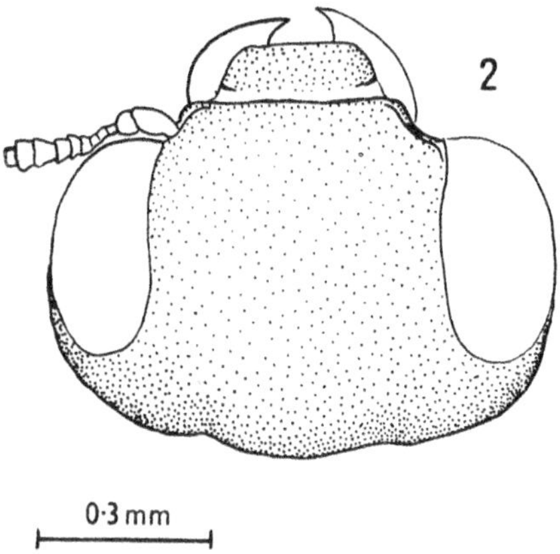


2

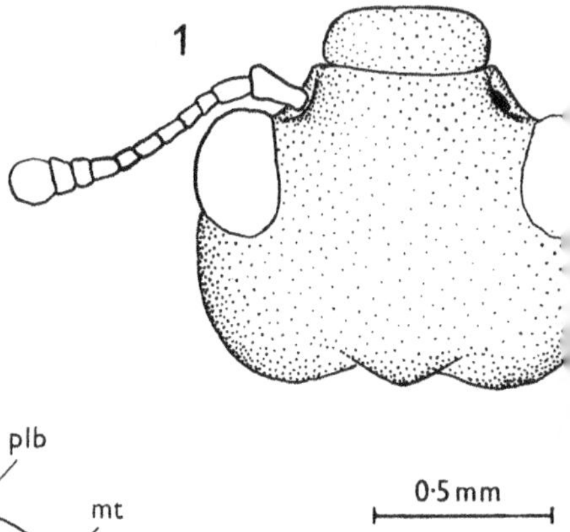


1

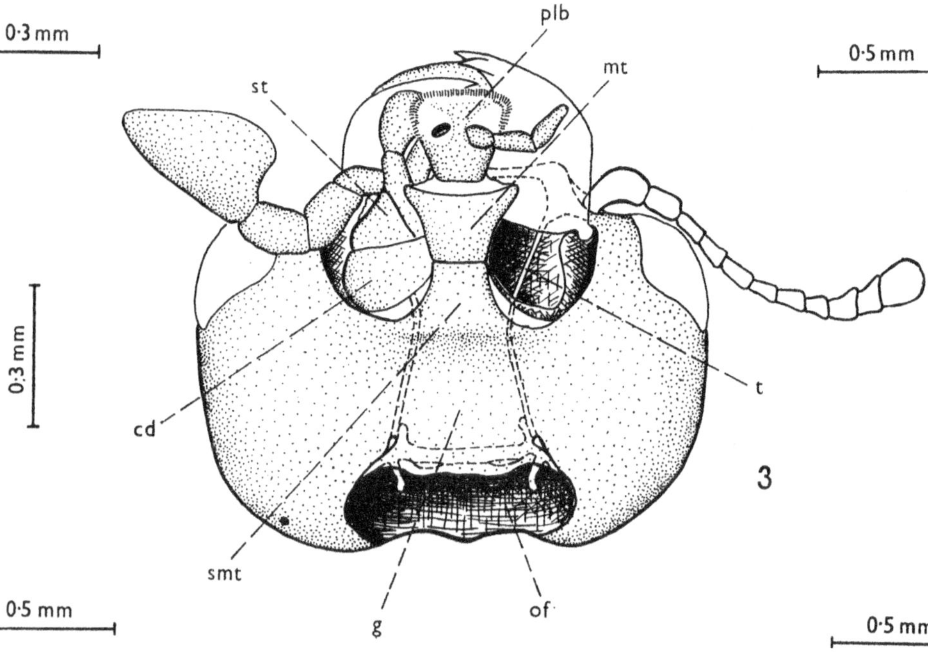


3

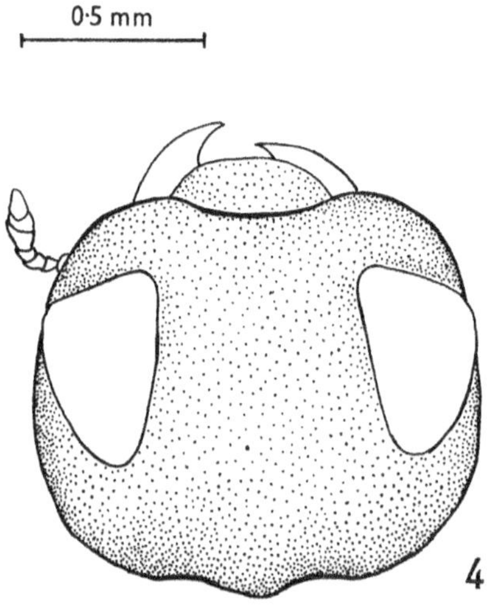


4

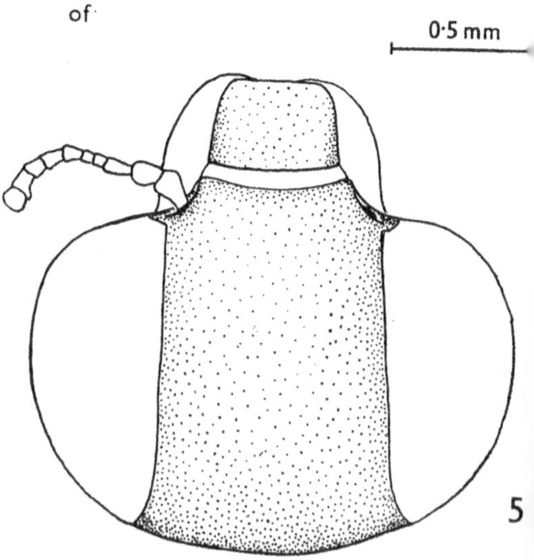


5

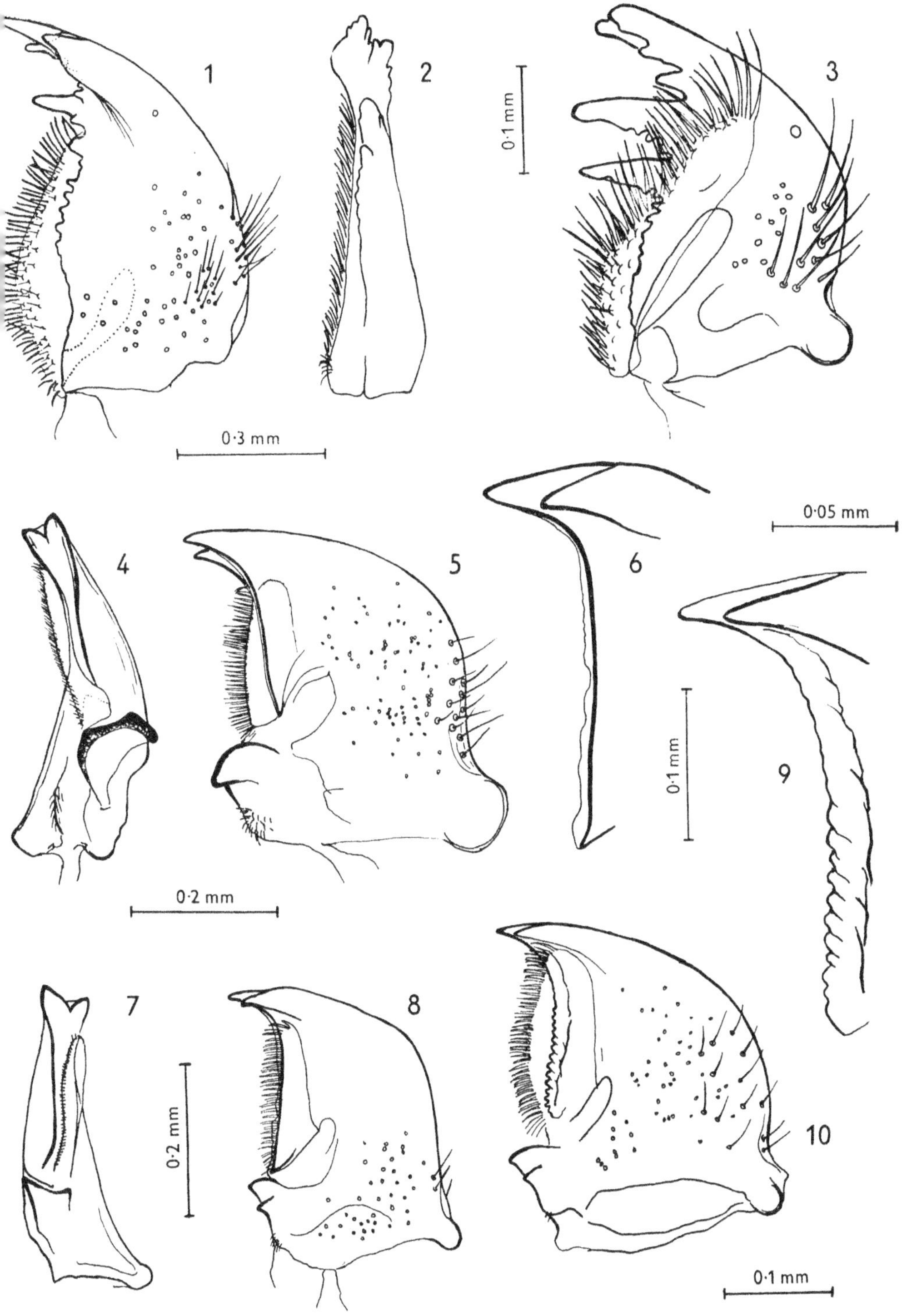
1
2
3
0·1 mm
0·3 mm
0·05 mm
4
5
6
9
0·1 mm
0·2 mm
7
8
0·2 mm
10
0·1 mm

by a membrane. The anterior margin of the prementum is densely covered with fine, spine-like hairs. In most small species the prelabium is nearly quadrate, but in the *Coccinellinae* it projects strongly forwards and in the *Epilachninae*, by contrast, it is narrowed in front. The labial palp is three-segmented with a small basal segment, which is completely missing in the *Noviini*. The medial projection of submentum is more or less wide and is only narrow in the *Sticholotinae*. The basic shape of the antennae (Pl. VI, 4) is eleven-segmented and weakly clavate (*Coccinellinae*, *Epilachninae*). In many groups the number of segments and their size are variable (Pl. VI, 5—7).

**Thorax.** — The prothorax is bent forwards, the pronotum is convex and sharply separated along its whole length by a distinct ridge from the hypomeron. The anterior margin is most often widely emarginated to cover the basal parts of the head. The prosternum (Pl. VII, 2) most often has the shape of the letter T. Its posterior, the intercoxal process, often bears distinct "prosternal carinae". The front transverse part of the sternum is well-developed, and only rarely very narrow and reduced (*Noviini*). The front coxal cavities are partly closed at the back by a postcoxal process of the hypomeron, which joins medially on to the furcasternum.

Most of the meso- and metanotum is covered by the elytra. The mesothoracic scutellum is the only visible part of the notum. The ventral side of the meso- and metathorax contains both sternal and pleural sclerites (Pl. VII, 1). The mesosternum is small. The mesothoracic pleural region is divided by a pleural suture into two sclerites, the episternum and epimeron. Basally, both these sclerites reach the lateral part of the middle coxal cavity. The metasternum forms the largest part of the ventral side of the metathorax. Medially, small sclerites between the posterior coxal cavities form the exterior part of furcasternum. Almost the whole metasternum is divided by a longitudinal medial suture which is related to the well-developed inner skeleton (metendosternite) (Pl. VII, 3). The metapleuron is divided into the larger episternum and the smaller epimeron.

**Legs** (Pl. VIII, 1—7) are well-developed and of the running type. The construction of the coxae is regular; the front and the larger hind coxae are transversely oval, the middle coxa is almost round. The femur is elongated, slender (*Coccinellinae*), rather stout, unflattened in most small species, or very stout and flat (*Platynaspini*, *Aspidimerini*). In some groups it has a shallow groove for the reception of the tibia. The tibia is slender or with a spine-like projection on its outer side. The tarsus is of trimerous (*Noviini* and a part of *Scymnini*), tetramerous (*Lithophilinae*) or more usually cryptotetramerous (pseudo-trimerous) type; the tarsal formula 3, 3, 3 or 4, 4, 4 is the same for both sexes.

The **elytra** (Pl. IX. 4—6) are convex to a varying degree and are never truncated

---

**Plate IV**

Mandibles. 1, 2 — *Chilocorus renipustulatus;* 3, 4 — *Novius cruentatus;* 5, 6 — *Coccidula scutellata;* 7, 8 — *Thea vigintiduopunctata;* 9, 10 — *Halyzia sedecimguttata;* 11, 12 — *Scymnus abietis*. (I. Kovář del.)

**Plate V**

1, 3, 4 — maxillae; 2, 5, 6 — galea and lacinia. 1, 2 — *Henosepilachna elaterii (*cd — cardo, g — galea, lc — lacinia, mxp — maxillary palpus, st — stipes); 3 — *Exochomus quadripustulatus;* 4, 5 — *Thea vigintiduopunctata;* 6 — *Adalia bipunctata*. (I. Kovář del.)

**Plate VI**

1, 2, 3 — labia; 4, 5, 6, 7 — antennae. 1, 7 — *Novius cruentatus;* 2 — *Lithophilus sp*. (lbp — labial palpus, m — mentum, pmt — prementum); 3 — *Henosepilachna elaterii;* 4 — *Adalia bipunctata;* 5 — *Chilocorus renipustulatus;* 6 — *Scymnus frontalis*. (I. Kovář del.)

PLATE IV

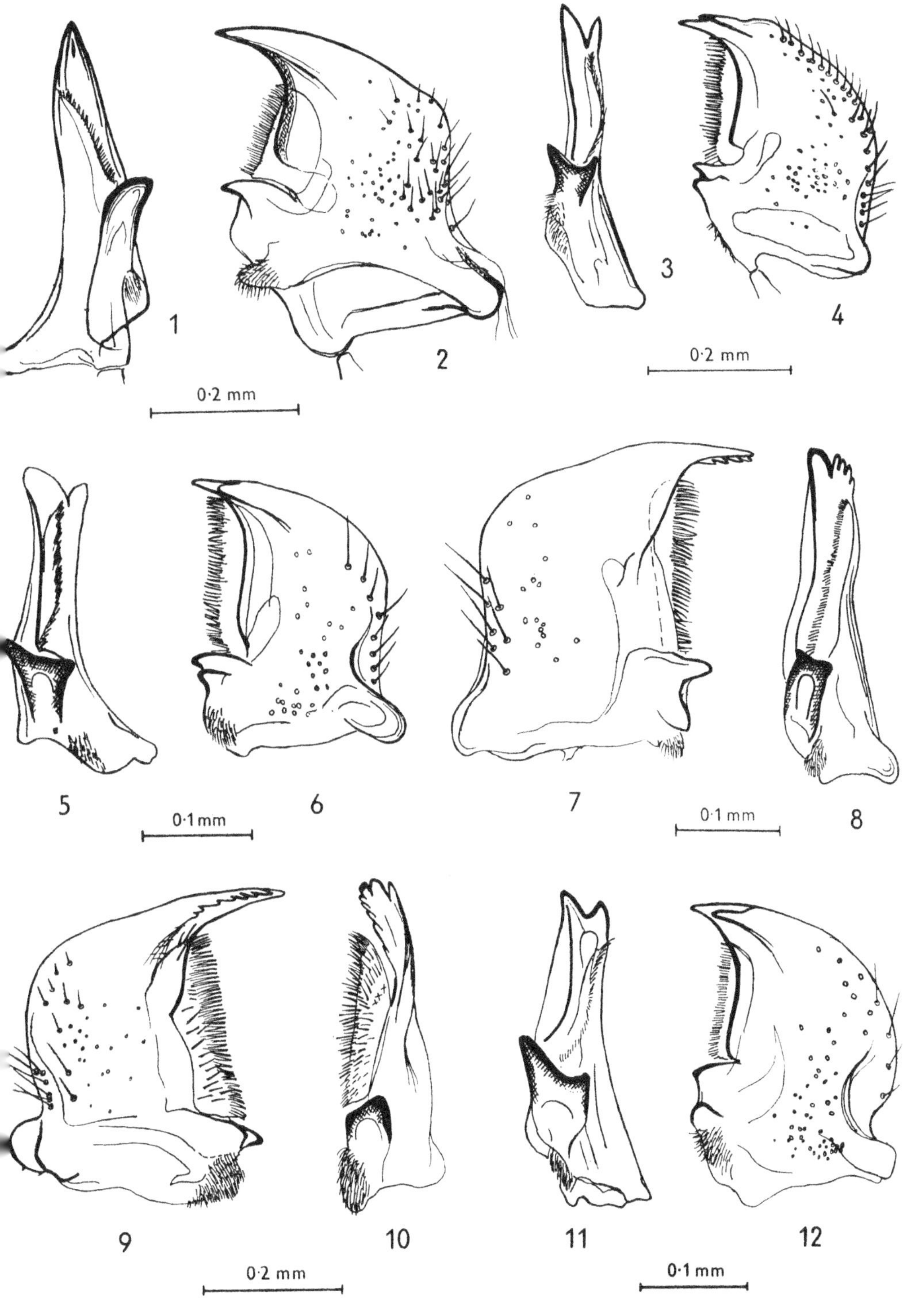

PLATE V

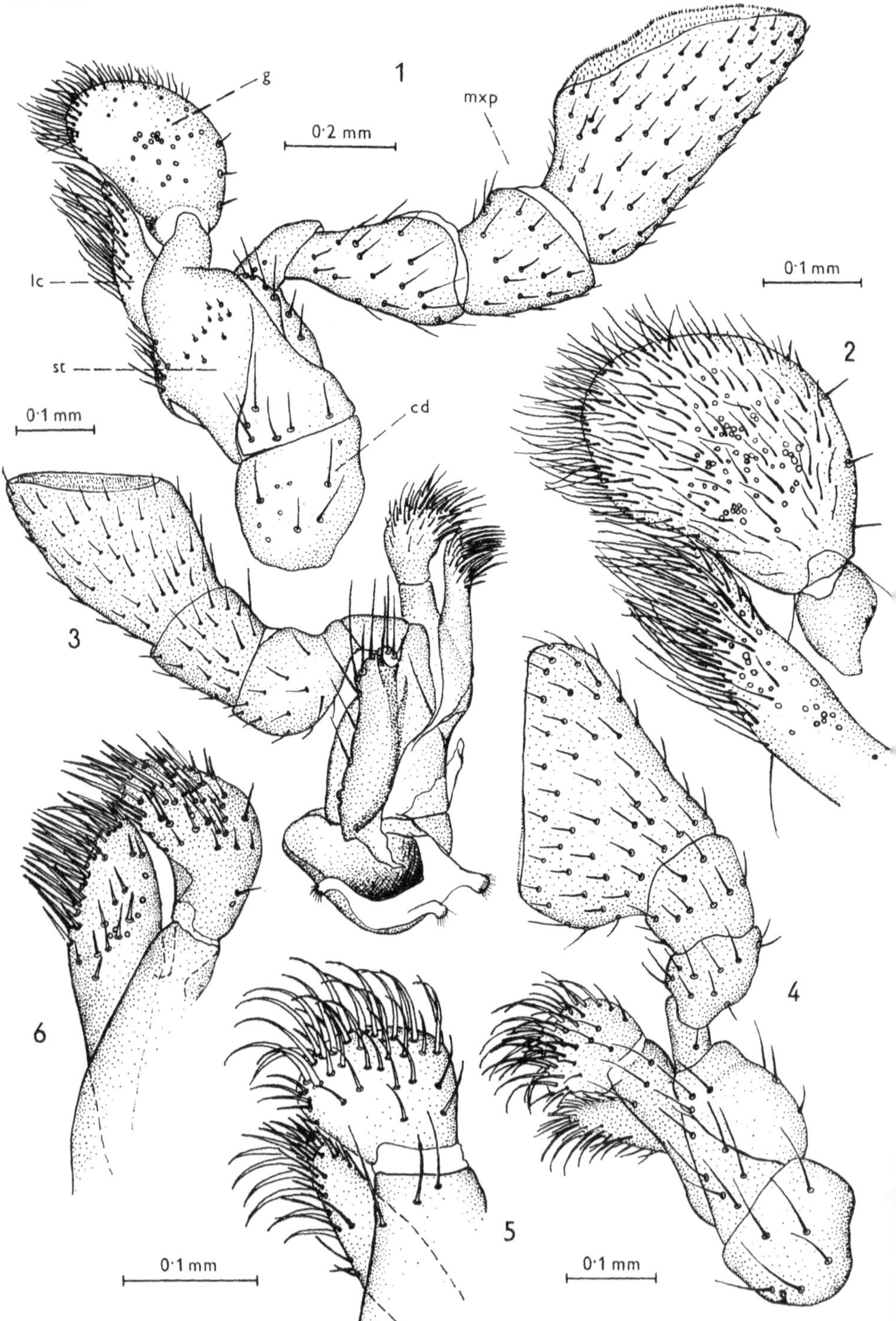

PLATE VI

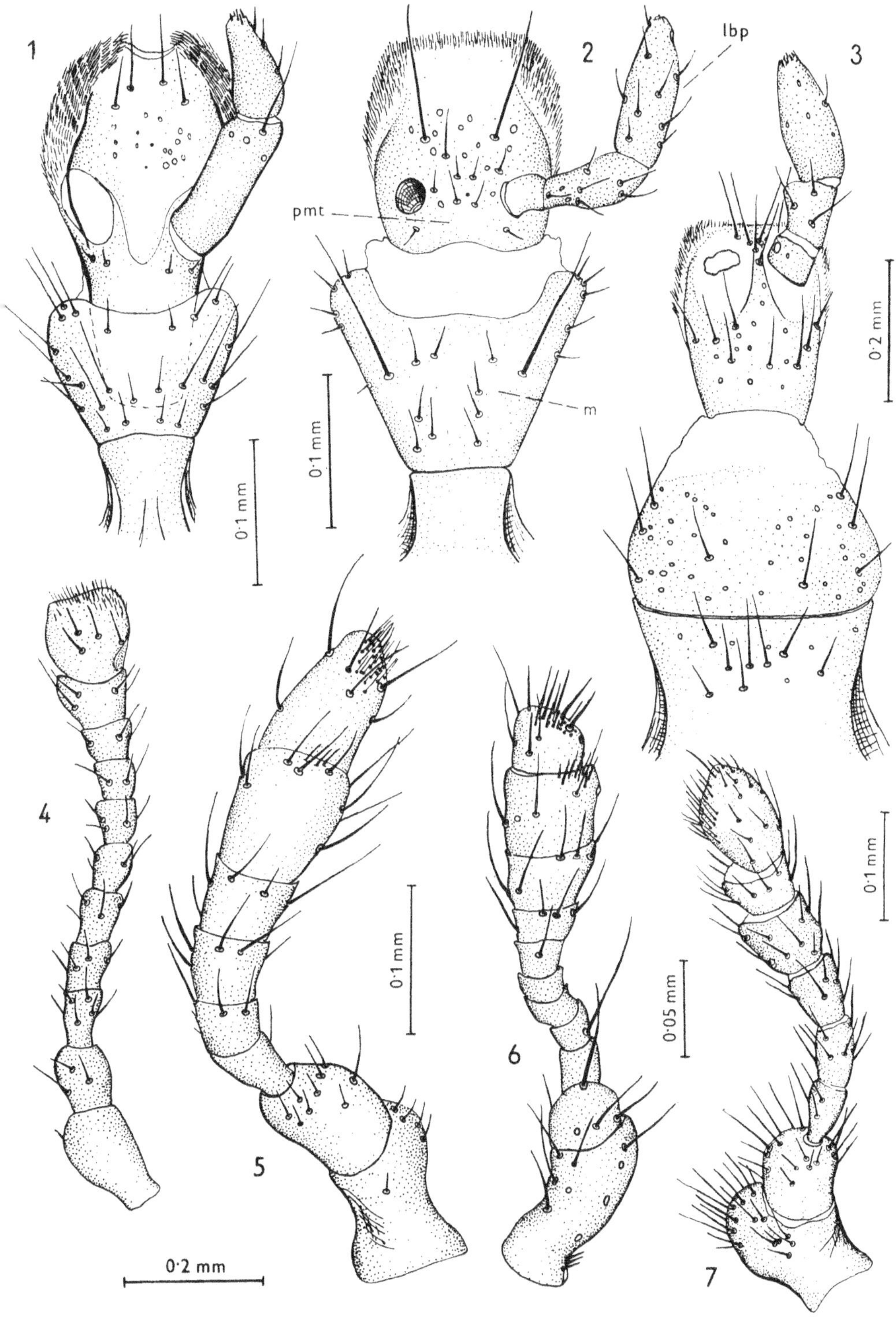

at the apex. The lateral margin is sometimes expanded outwardly, the ventral flat part of the elytron (epipleuron) is narrowed apically.

The **hind wings** (Pl. IX, 1—3) are functional and only rarely reduced (*Lithophilus, Cynegetis, Rhizobius*). The nervature is of cantharoid type, with rather feebly developed veins.

The **abdomen** is ten segmented, the tergal part is represented by eight visible tergites of which only the last visible is well sclerotised. The ventral side includes only five or six visible sternites. The first and second sternites are fused together with the base of the third morphological sternite (the first visible sternite). The first visible sternite is the largest of all the abdominal sternites and bears (except only *Hippodamia*) the femoral line. The length of the seventh sternite ( = fifth visible one) is often greater than the preceding sternites. The posterior margin is either round, completely covering the next (= eighth) sternite (*Lithophilini, Serangiini, Telsimiini* and some other tribes), or is similar in shape to the previous sternite and does not completely cover the eighth (*Coccinellinae, Epilachninae*).

Usually the parts of the ninth and tenth abdominal segments are designated as **genitalia.** The female genitalia (Pl. XI, 4, 7, 8) consist of the paired ninth pleurites and a pair of coxites (ninth sternites) and the tenth tergite. The coxites usually carry a stylus. The shape of the coxites is variable and various types can be distinguished. In many coccinellid groups the coxites are very elongate triangular structures and sometimes they function as an ovipositor.

The genital sclerites of the males (Pl. X, 3) are formed by the ninth and tenth tergites and by the elongated, thin ninth sternite (apodema). The male genitalia proper (Pl. X, 1—3) have two parts, tegmen and sipho. The tegmen is variably thick, symmetrical or asymmetrical and consists of a basal piece, paired paramera (= lateral lobes), and a distinct median piece, sometimes erroneously called the "penis" or "aedeagus". The hypomera (= trabes) articulate with the ventral side of the basal piece of the tegmen. The sipho (= penis, aedeagus) is very long, tubular and ventrally curved. Its base is most often broadened and forms a siphonal capsule. The apex of the sipho bears the gonopore, and the shapes of both sipho and tegmen are unique features in species and often used in taxonomy.

Sex can only be determined with some difficulty in coccinellids; the characters for distinguishing males and females cannot be generalized. Males are usually smaller than females and often have lighter colouration on the front part of the head, sometimes even on the front part of the pronotum and also other parts of the body may differ in colour. The length of the antennae is slightly greater in males than in females. The number

**Plate VII**

1 — *Adalia bipunctata*, thorax and abdomen, ventral aspect (fl — femoral line, $em_2$, $em_3$ — meso- and metaepimeron, $es_2$, $es_3$ — meso-and metaepisternum, $s_2$, $s_3$ — meso- and metasternum); 2 — *Coccidula scutellata*, prothorax, ventral aspect (h — hypomeron, hp — postcoxal process of the hypomeron, pc — prosternal carina, $s_1$ — prosternum); 3 — *Propylaea quatuordecimpunctata*, endoskeleton of meso- and metathorax (ms — metendosternite). (I. Kovář del.)

**Plate VIII**

1, 2, 3, 4 — legs; 5, 6, 7 — tarsal claws. 1, 6 — *Lithophilus sp.* (f — femur, t — tibia, ta — tarsus, tr — trochanter); 2 — *Adalia bipunctata;* 3, 5 — *Cryptogonus orbiculus;* 4 — *Chilocorus renipustulatus;* 7 — *Thea vigintiduopunctata.* (I. Kovář del.)

**Plate IX**

1, 2, 3 — hind wings; 4, 5, 6 — elytra, ventral aspect; 7, 8 — dorsal surface of elytra. 1 — *Scymnus abietis;* 2 — *Semiadalia undecimnotata;* 3 — *Anatis ocellata* (C — Costa, Sc — Subcosta, R — Radius, M — Media, A — Anal vein); 4 — *Chilocorus renipustulatus;* 5, 7 — *Henosepilachna elaterii;* 6 — *Ortalia sp.;* 8 — *Thea vigintiduopunctata.* (I. Kovář del.)

2

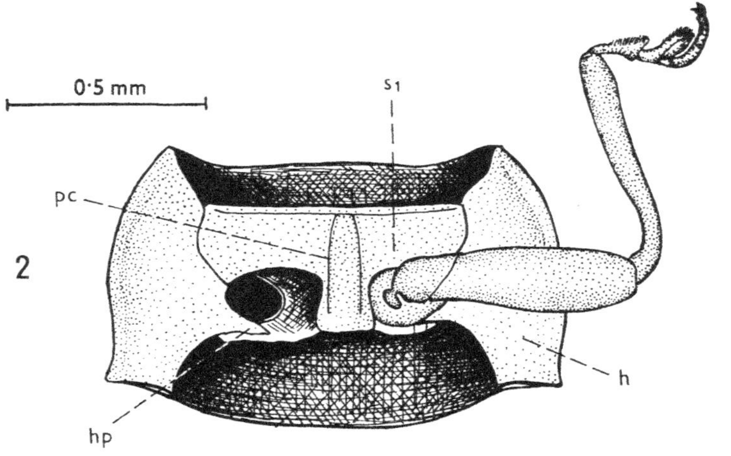


3

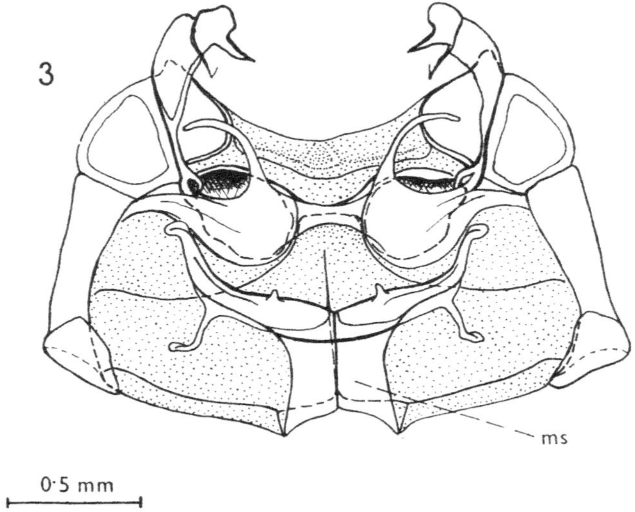


1

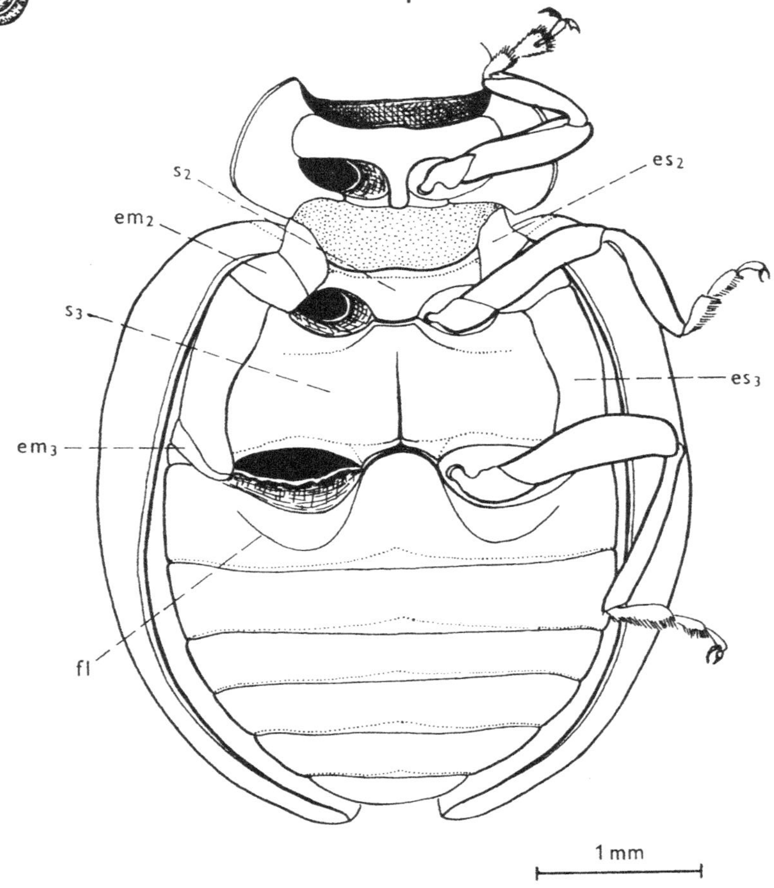

PLATE VIII

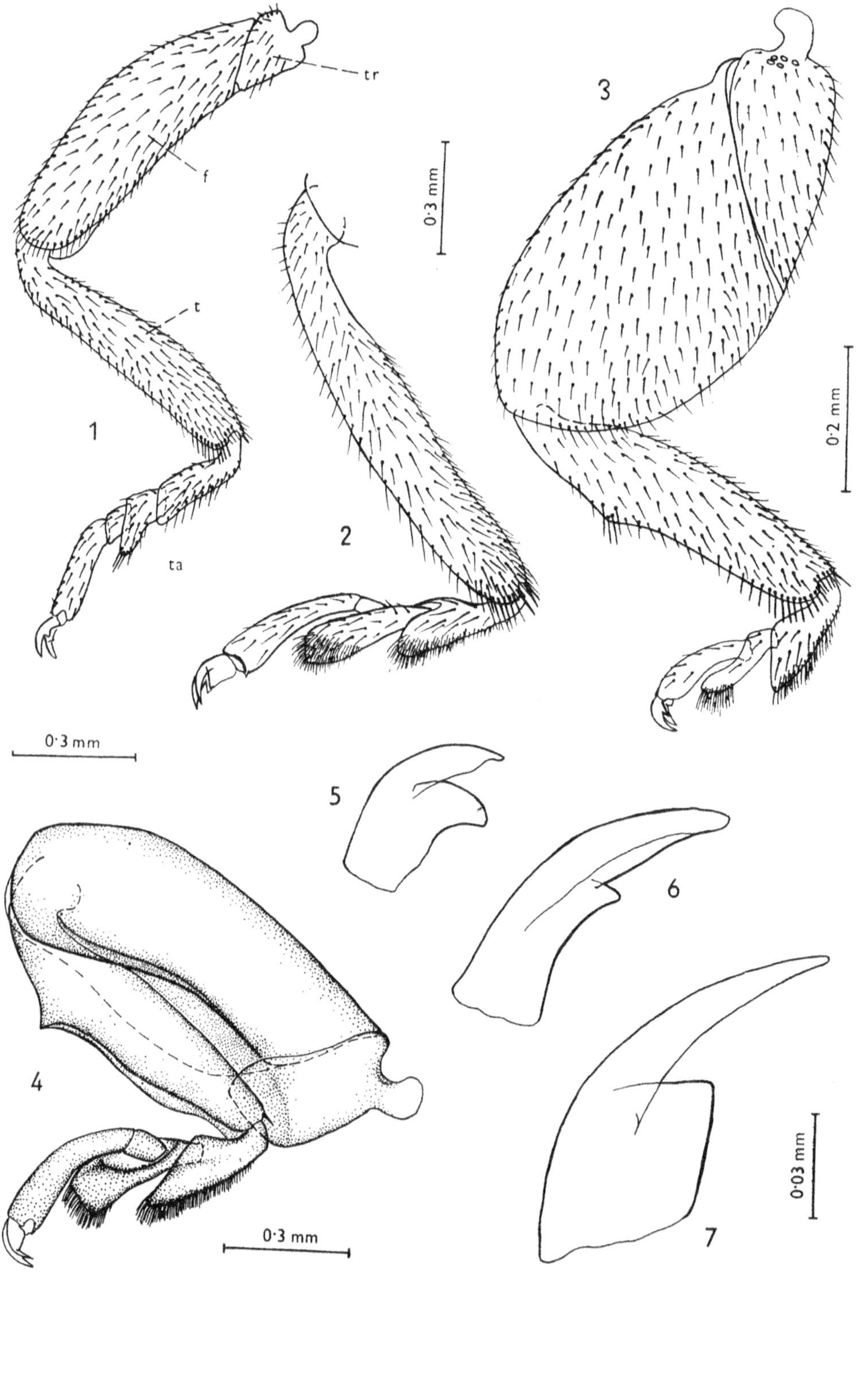

PLATE IX

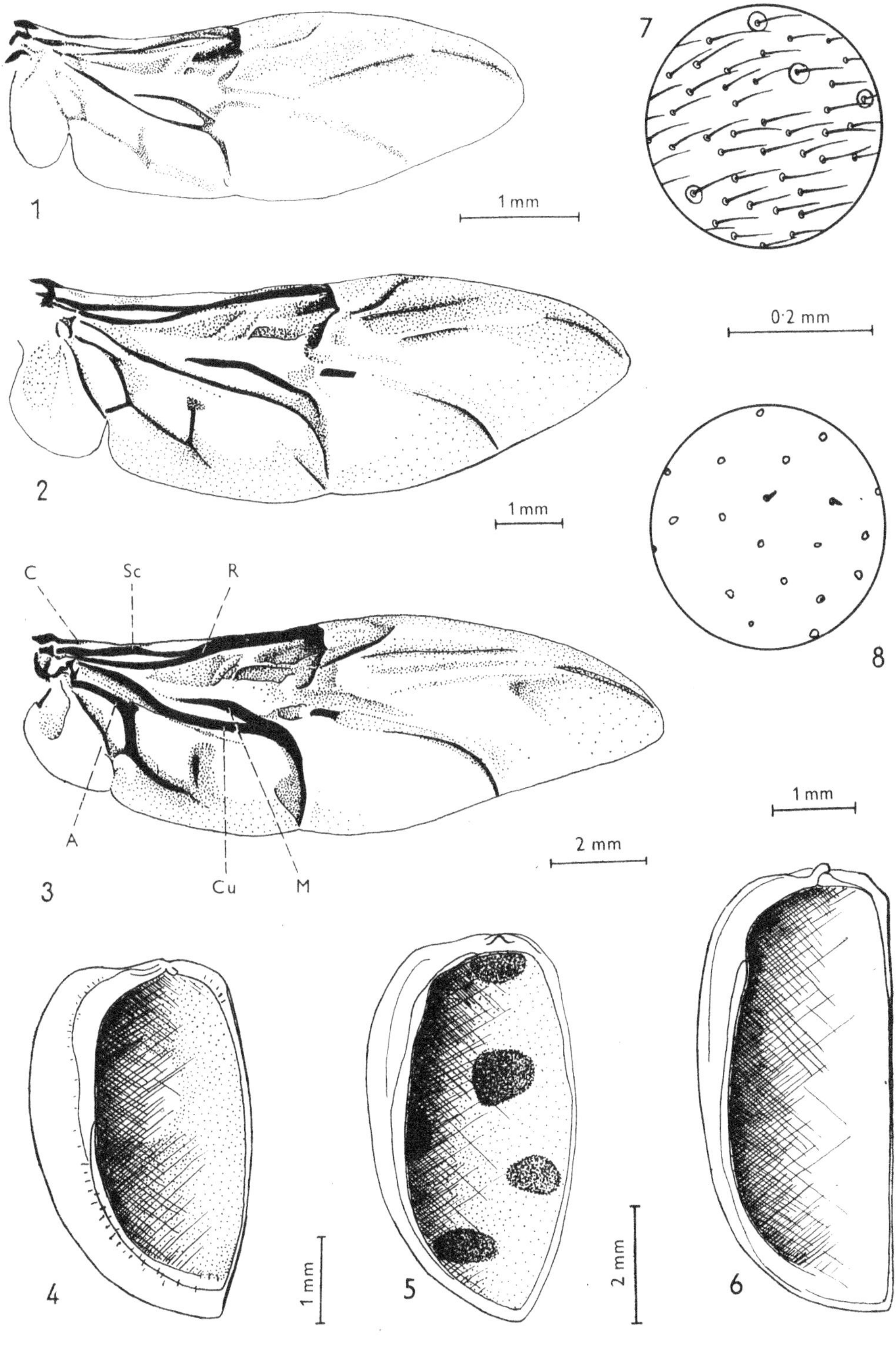

of visible sternites is usually the same in both sexes. In *Chilocorini*, the sixth sternite of the females is hardly visible, and only slightly visible in the males. To distinguish the sexes, it is possible in most species to use the appearance of the last abdominal sternites, and particularly the construction of the eighth abdominal sternite (= the sixth visible sternite). In many species there are differences in the form of the sixth visible sternite, which, more often in males than in females, is emarginated to a different degree at the middle of the hind margin (Pl. XI, 1—6).

## 1.2 Anatomy

The **female reproductive organs** consist of a pair of ovaries, two lateral oviducts, a median oviduct, a vaginal part, a bursa copulatrix and a spermatheca. The nubmer of ovarioles in one ovary is not constant and often varies even within one species. Robertson (1961) records an average number of 17 ovarioles in one ovary with the range of 2 (*Stethorus punctillum*) to 51 (*Coccinella septempunctata*) for 54 species and 28 genera of *Coccinellidae*. The ovarioles are of the meroistic and telotrophic type; nutrition of the oocytes is provided by the nutritive cells of the germarium. The terminal filaments of individual ovarioles are attached together and fix both ovaria in the abdominal cavity. At the base of the terminal filament, the ovariole is extended into a cylindrical germarium, which contains oogonia and trophocytes or nutritive cells. Ermolenko (1963) compares the construction of "mature" and "immature" ovaries in *Cryptolaemus montrouzieri*. In ovipositing females, the vitellarium contains increasing egg follicles at different stages of maturation, which are separated by distinct notches. Ovarioles lead into the upper expanded parts of lateral oviducts by pedicels. The lateral oviducts are relatively short, simple tubes, united to form a somewhat broader medial oviduct. Williams (1945) and Tanasijević (1958) described a pair of collateral glands extending from the median oviduct in some *Epilachninae*. Posteriorly the median oviduct opens into the vagina by a gonopore. Ventrally there protrudes a pouch-like rather large bursa copulatrix and in some species this has at the top a sclerotized funnel-shaped part called the infundibulum. The bursa is connected with the spermatheca (receptaculum seminis) (Pl. X, 4—6) by a short or longer narrow sperm duct which most often arises from the distal part of the infundibulum or in the *Scymnini*, *Chilocorini* and some *Coccinellini* from the distal part of bursa. In *Menochilus* the sperm duct is extremely long and contorted. The following parts of the spermatheca can be recognized: the distal arched part called "cornu" and two projections at the base of the spermatheca. One of these projections ("nodulus") communicates with the sperm duct and the other ("ramus") with the spermathecal (accessory) gland. Often one (usually "ramus") or both projections are missing, and in *Stethorus punctillum* not only the whole spermatheca, but also the bursa copulatrix is missing.

The **male reproductive organs** lie under the alimentary canal and consist of the testes, the vasa deferentia and the ejaculatory duct. The testis is polyfollicular and the number of testicular follicles is variable like the number of the female ovarioles. The follicles are spherical or oval. The vas deferens is a curved tube which at its

---

Plate X

1, 3 — tegmen; 2 — sipho; 4, 5, 6 — spermatheca. 1 — *Adalia decempunctata;* 2 — *Adalia decempunctata* (sc — siphonal capsule); 3 — *Adalia bipunctata* (bp — basal piece, ml — median lobus, p — paramera, t — trabes); 4 — *Adalia conglomerata* (c — cornu, i — infundibulum, n — nodulus, r — ramus, sd — sperm duct); 5 — *Scymnus frontalis;* 6 — *Chilocorus renipustulatus.* (I. Kovář del.)

PLATE X

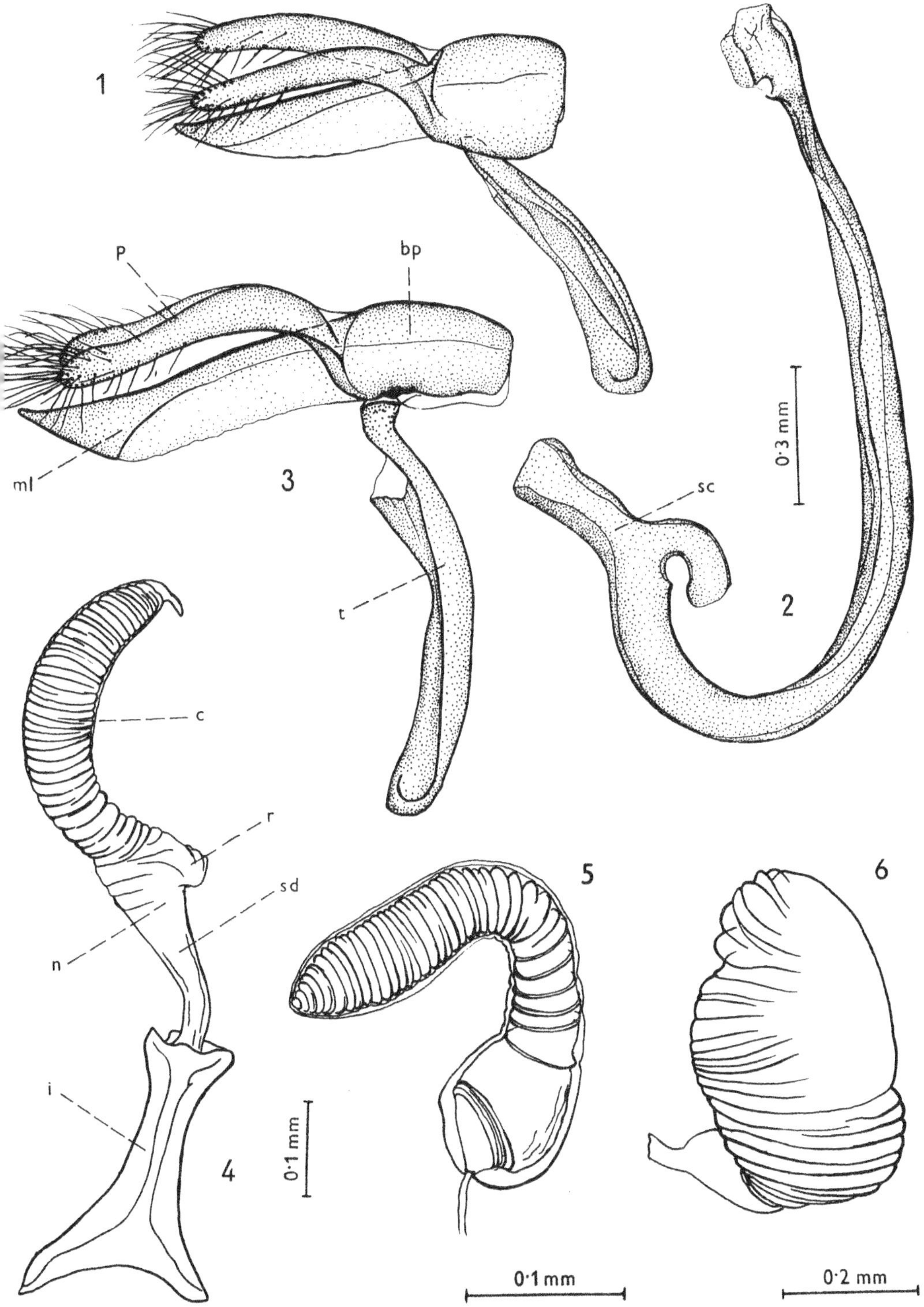

base, near the testis, is swollen and forms the vesicula seminalis. The paired vasa deferentia unite to form a somewhat wider ductus ejaculatorius which issues at the base of the sipho and is approximately of the same length or usually longer than the vas deferens. Two pairs of accessory glands arise from junction of the vasa deferentia. In *Subcoccinella vigintiquatuorpunctata*, there has been observed, in addition to these two pairs, a third pair of accessory glands situated under the vesicula seminalis (Tanasijević 1958).

The **alimentary canel** consists of three histologically distinguishable sections: the stomodeum or fore-gut, the mesenteron or mid-gut, and the proctodeum or hind-gut. The fore-gut is always the shortest section and is composed of the pharynx, oesophagus and crop. The wall of the fore-gut has large inner folds and is composed of cuticular intima, epithelium, basement membrane and muscularis. The oesophagus is much narrower than the pharynx and its intima is thinner and is sometimes provided with teeth. The oesophagus extends into a small conical or pear-shaped chamber — the crop, which lies in the front half of the prothorax. The muscularis of the crop is the most complicated such tissue in the pre-gut. Between the crop and mesenteron there is a constriction which marks externally the position of the oesophageal valve.

The largest section of the alimentary canal is the mid-gut which reaches to about the fifth abdominal sternite. It is about 5—6 times as wide as the oesophagus, and much wider even than this when filled with food. In carnivorous coccinellids, the mesenteron is simple. In phytophagous coccinellids, however, it has two parts: the wide front part continues into a tube which loops forward to form several spiral loops around it. The wall of the mid-gut is thicker than elsewhere in the alimentary canal, and also has regenerative cells (nidi). In *Hyperaspis vinciguerrae* Hafez and El-Ziady (1952) have observed that the whole epithelial lining is shed after every meal and is immediately replaced by a new one from the nidi. Pradhan (1939) also pointed out that the digestive cells are monophasic and are generally used up completely during one period of secretion. Two types of secretion have been observed in coccinellids, merocrine and holocrine. The continuous or, in *Epilachnaindica*, discontinuous (Pradhan 1936) peritrophic membrane is produced by a small group of cells at the base of the pyloric valves. The mesenteron is closed posteriorly by the pyloric valves, which consist of small folds and are provided with the pyloric sphincter. The intima over the epithelium is already very thin. The six Malpighian tubes arise from the pyloric region and form loose loops in the body cavity; they reach the anterior part of the proctodeum. In cross-section they consist of 2—3 epithelial cells, the cell walls of which often disappear. Their inner border, which closes the central cavity, bears cilia. The outer cover of the tubes is formed by a basal membrane.

The hind-gut consists of the anterior intestine (ileum and colon) and the posterior intestine (rectum) and is usually curved for the first third. The ileum is short and usually straight. Six longitudinal folds are visible in cross-section, the intima is very thin and only the circular muscle is developed. The distal parts of the Malpighian tubes become closely attached to the junction of ileum and colon, and become enclosed

Plate XI
1, 2 — abdomen, ventral aspect; 3 — male genital sclerites; 4, 7, 8 — female genital sclerites; 5, 6 — seventh and eighth sternite of abdomen. 1 — *Propylaea quatuordecimpunctata*, male; 2 — *Propylaea quatuordecimpunctata*, female; 3, 4 — *Henosepilachna elaterii* (7S — seventh sternite, 8S — eighth sternite, 9S — ninth sternite, 8T — eighth tergite, 10T — tenth tergite); 5 — *Anisosticta novemdecimpunctata*, male; 6 — *Anisosticta novemdecimpunctata*, female; 7 — *Thea vigintiduopunctata* (cx — coxite, 9P — ninth pleurite, 10T — tenth tergite); 8 — *Scymnus abietis*. (I. Kovář del.)

PLATE XI

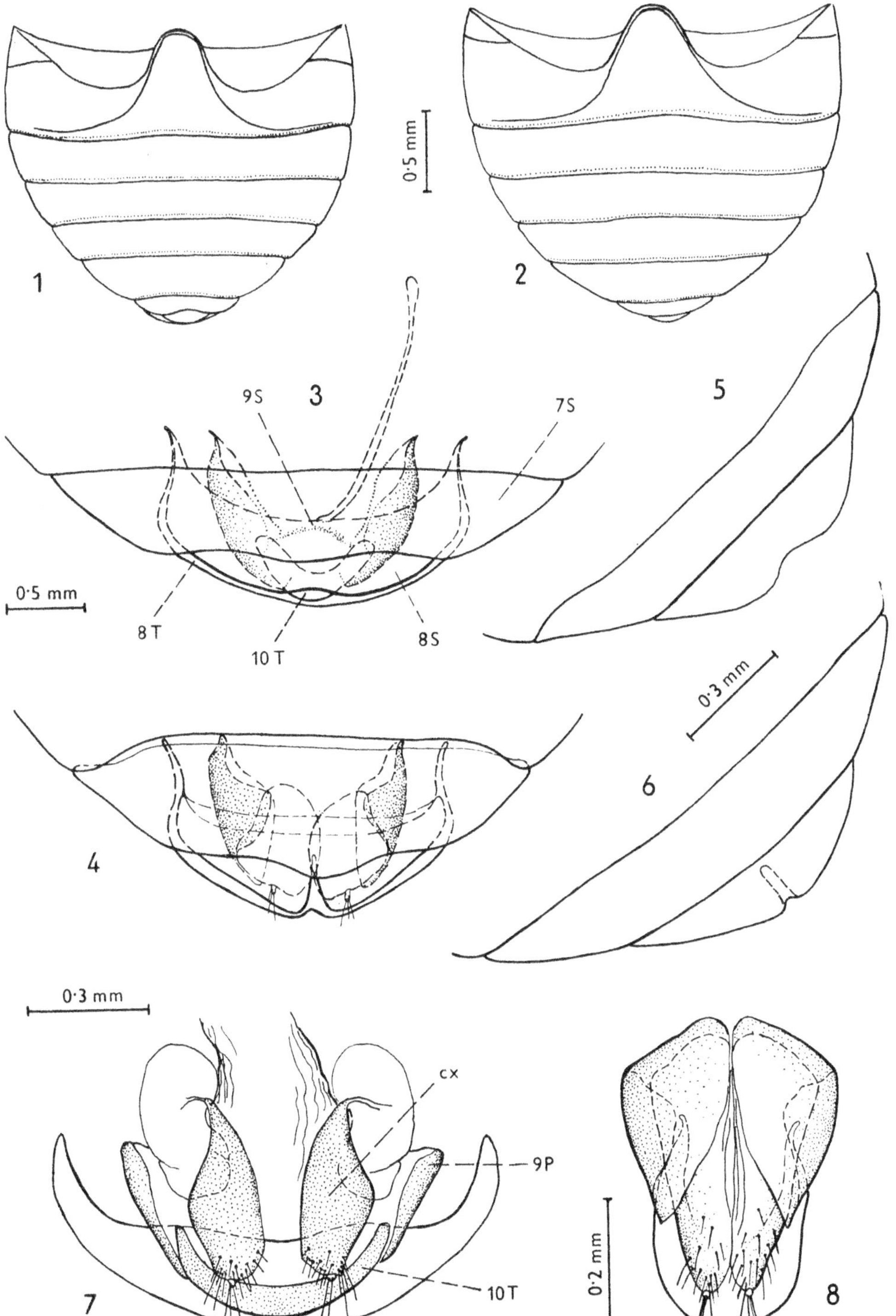

by a thin peritoneal sheath or fascia. The colon is longer than the ileum; its posterior part widens into a conical chamber. The colon wall is thicker than in the ileum and the intima is heavily chitinised. Between the circular muscle layer and fascia there is a space called the fascial chamber. The Malpighian tubes terminate in this chamber, which is filled with loose connective tissue and blood. Near where the colon leads into the rectum, the fascial chamber communicates with the gut cavity by minute pores or canaliculi. The rectum has six longitudinal folds, a thick intima and robust circular musculature which consists of several layers. The rectum opens to the exterior by the anus.

There is no great difference between carnivorous and phytophagous species in the length of fore- and hind-gut. In phytophagous coccinellids, however, the mid-gut and consequently the whole alimentary canal is considerably longer than in carnivorous species; also the Malpighian tubes are longer in phytophagous than in carnivorous species.

## 1.3 Phylogeny and taxonomy

The family *Coccinellidae* belongs to the superfamily *Cucujoidea*, section *Clavicornia*, and is related to the families *Corylophidae*, *Cerylonidae* and *Discolomidae* (Crowson, 1955). A historical survey of the higher classification of coccinellids has been given by Sasaji (1968b). Of the earlier authors, Korscheffsky proposed a system which is included in Junk's "Coleopterorum catalogus" (1930—1931) and which gained general acceptance. According to this system, coccinellids were divided into three subfamilies: *Epilachninae*, *Coccinellinae* (*s. lat.*) and *Lithophilinae*. The phytophagous subfamily *Epilachninae* was considered the most primitive by many authors and its relationship to the family *Chrysomelidae* was discussed. The genus *Lithophilus* was classified into a separate subfamily, mainly on account of the different structure of the tarsus. In the older systems the subfamily *Coccinellinae* (*s. lat.*) was very heterogeneous and contained several developmental branches.

Sasaji (1968b) has presented a more modern system based on the comparative morphology of the larvae (Kamiya 1965) and adults, which is in line with recent views on the phylogeny of the order Coleoptera. The phylogenetic development of coccinellids can be represented in a simplified way as follows (see also the dendrogram) (Pl. XII).

The most primitive subfamily of the family *Coccinellidae*, related to the *Clavicornia* groups mentioned earlier, is the new subfamily *Sticholotinae*. This subfamily has been established by Sasaji (1968) to include the tribes *Sukunahikonini*, *Serangiini*, *Sticholotini* and *Shirozuellini*, which have many common characters. The *Sukunahikonini* are comparatively more primitive than the other tribes. The *Shirozuellini*, established by Sasaji (1967), are a specialized tribe of this subfamily, apparently similar to the *Telsimiini* (*Chilocorinae*). The external structure of the tribe *Sticholotini*, which apparently includes the *Pharini*, *Coelopterini*, *Clanini* and *Microweiseini*, is rather variable and this tribe may eventually be split into several tribes.

Very early on, a branch developed from the *Sticholotinae* — ancestor, leading to two recent subfamilies, *Chilocorinae* and *Scymninae*. A strongly expanded clypeus is characteristic for the *Chilocorinae*, and it appears in the *Telsimiini*, *Platynaspini* and *Chilocorini* [also in the genus *Shirozuella* (*Shirozuellini*)]. The tribe *Telsimiini* has the greatest affinity to *Sticholotinae*, whereas the *Platynaspini* and especially the *Chilocorini* are much more derived groups.

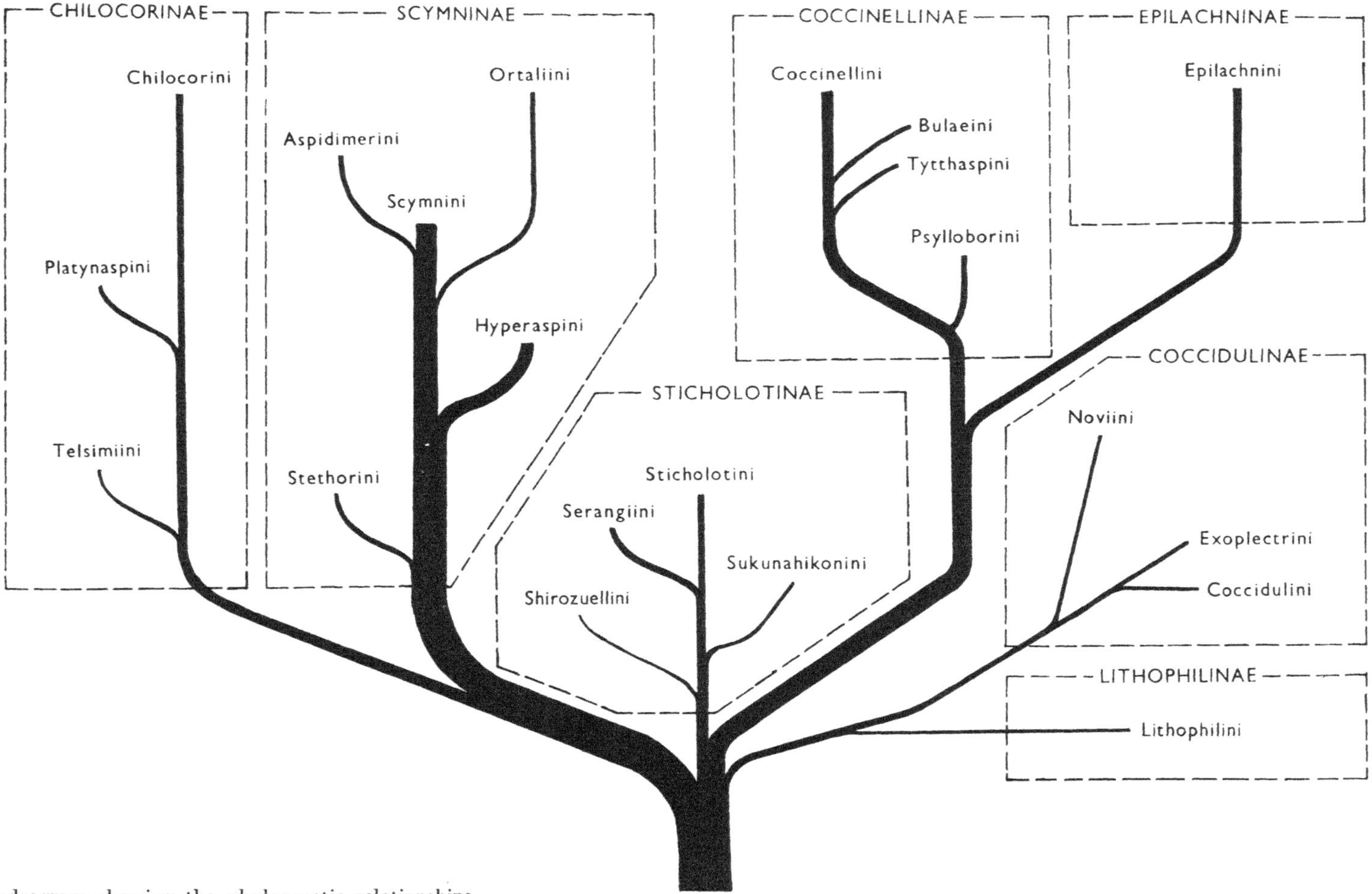


Dendrogram showing the phylogenetic relationships.

The subfamily *Scymninae*, which includes the majority of small species, includes the tribes *Stethorini*, *Scymnini*, *Aspidimerini*, *Hyperaspini* and *Ortaliini*. These tribes are characterized by narrow elytral epipleura; often trimerous tarsi, and a thickened femur. The *Aspidimerini* are in close affinity to the tribe *Scymnini* which forms the root of this developmental branch and includes a great number of species. The *Stethorini* very soon separated from the *Scymnini* — ancestor; they are considered as a derived tribe. The tribe *Hyperaspini* is related to the *Scymnini*. The characters of their larvae would suggest that they are the most primitive group of coccinellids; the adults, however, have many derived characters. The tribe *Ortaliini* occupies the most advanced position in the subfamily *Scymninae* (Sasaji 1968b).

The other developmental branch from the *Sticholotinae* — ancestor evidently arises later and independently of the first branch. It leads, on the one hand, to the carnivorous or mycophagous tribes of the subfamily *Coccinellinae*, and on the other hand to the subfamily *Epilachninae*. The development of both groups has been strongly divergent as a result of the different types of nutrition; however, many characters show that they had a common ancestor.

A complete or partial transition to vegetative food has taken place several times within the originally carnivorous subfamily *Coccinellinae*. The tribe *Psylloborini*, which belongs to this subfamily, has some characters related to the *Epilachninae*. This fact is explained by a secondarily convergent development. A similar tendency, although at another phylogenetic level, is manifested in the tribes *Tytthaspini* and *Bulaeini*, further tribes established by Savoïskaya (1969b).

The subfamily *Coccidulinae* and the related subfamily *Lithophilinae* also have a slight affinity to *Sticholotinae*. Sasaji (1968) included the *Lithophilini* into the subfamily *Coccidulinae* on the basis of their adult character. Klausnitzer (1971a) has divided Sasaji's *Coccidulinae* into two subfamilies on the basis of their larval characters. The subfamily *Coccidulinae* comprises the tribes *Coccidulini* and *Exoplectrini*, which have relatively close affinity. Also the tribe *Noviini* clearly falls with this subfamily, in spite of its having some characters similar to those of the *Scymnini*. The subfamily *Liphophilinae* is most likely the more primitive in this branch, which, judging by the analogy of many characters, had a phylogenetic development similar to the *Epilachninae* — *Coccinellinae* branch.

Some other tribes, such as the *Scymnillini*, *Cranophorini*, *Oeneini*, *Pentiliini*, *Chnoodini* and *Discotomini* have as yet not been sufficiently studied.

A summary of the system of higher categories of coccinellids is given in Tab. 1.01.

Tab. 1.01 Summarized system of higher categories of coccinellids

| | |
|---|---|
| Subfamily: | *Sticholotinae* (*Pseudococcinellidae*)[1]) |
| tribe | *Sukunahikonini* |
| tribe | *Serangiini* |
| tribe | *Sticholotini* (*Pharini*, *Coelopterini*, *Clanini*, *Microweiseini*) |
| tribe | *Shirozuellini* |
| Subfamily: | *Scymninae* |
| tribe | *Stethorini* |
| tribe | *Scymnini* |
| tribe | *Aspidimerini* |
| tribe | *Hyperaspini* |
| tribe | *Ortaliini* |

[1]) Names in parentheses are synonyms.

Subfamily: *Chilocorinae*
tribe *Telsimiini*
tribe *Platynaspini*
tribe *Chilocorini*

Subfamily: *Lithophilinae* (*Tetrabrachinae*)
tribe *Lithophilini*

Subfamily: *Coccidulinae*
tribe *Coccidulini* (*Rhizobiini*)
tribe *Exoplectrini*
tribe *Noviini*

Subfamily: *Coccinellinae*
tribe *Coccinellini* (*Hippodamiini*, *Anisostictini*, *Synonichini*)
tribe *Psylloborini*
tribe *Tytthaspini*
tribe *Bulaeini*

Subfamily: *Epilachninae*
tribe *Epilachnini*

---

# 2 MORPHOLOGY AND TAXONOMY OF THE LARVAE WITH KEYS FOR THEIR IDENTIFICATION[1]

G. I. SAVOÏSKAYA and B. KLAUSNITZER

The morphology and taxonomy of coccinellid larvae have as yet been incompletely studied. The subject was first tackled by Böving (1917), and soon afterwards Gage (1920) described 14 species of American coccinellids. A paper by Strouhal (1926) is devoted to larvae of *Coccinellini* and of *Psylloborini*, one by Binaghi (1941a, b) to larvae of *Chilocorini* and some species of *Scymnini* and one by Kapur (1950) to larvae of *Epilachninae*. van Emden (1949) published a key for the larvae of 36 species of British coccinellids. Three larvae of particular taxonomic interest have been described by Capra (1947) (*Bulaea lichatschovi*), Kanervo (1941) (*Calvia quinquedecimguttata*) and Korschefsky (1934) (*Platynaspis luteorubra*).

The work by Kamiya (1965) is especially interesting in that, on the basis of the larvae of Japanese *Coccinellidae*, the tribal phylogeny of the family is discussed. This same author also described 36 species of Japanese *Coccinellidae* and published a key to them (Sasaji 1968, formerly Kamiya). Klausnitzer (1969) described the larva of *Lithophilus connatus* (Lithophilinae); the larvae of this subfamily had not been previously studied. The same author has produced a monograph with a key to 59 species on the mid-European coccinellid larvae (Klausnitzer 1970a) and also other papers concerning coccinellid larvae (Klausnitzer 1970b, c, 1971a, b, c).

Savoïskaya has published a series of works devoted to the morphology and taxonomy (with keys) for 81 species of coccinellid larvae in the Soviet Union; larvae of 47 species were described for the first time (Savoïskaya 1955, 1957, 1960a, b, 1961, 1962a, b, c, 1963, 1964a, b, 1965, 1966, 1968, 1969a, b, 1970a, b, 1971a, b). The same author's detailed study of the morphology of adults and especially of larvae of coccinellids has suggested possibilities for revising the taxonomy of the family *Coccinellidae*.

Two new tribes (*Tytthaspini* and *Bulaeini*) and a new subgenus (*Neococcinella*) have been erected; also the taxonomy of the genus *Exochomus* and the taxonomic status of some other groups of beetles have been changed (Savoiskaya 1969b, 1971a).

[1])This chapter is mostly written by G. I. Savoïskaya. B. Klausnitzer contributed the diagnoses of twelve species (*A. obliterata*, *S. vigintiguttata*, *R. chrysomeloides*, *N. limonii*, *S. interruptus*' *S. alpina*, *A. conglomerata*, *S. lyncea*, *S. impustulata*, *T. trilineata*, *C. decemguttata*, *B. oblongus*), produced the key for the genera *Rhizobius* and *Calvia* and completed the key for the genera *Synharmonia*, *Scymnus* and *Brumus* and gives other additions to the text and references.

## 2.1 Comparative morphology of the larvae

Larvae of coccinellids in most of the tribes have an elongate, slender body *(Coccinellini, Psylloborini, Scymnini* and others). In the *Hyperaspini* and *Noviini* it is of elongate-ellipsoidal shape and hemispherical in the *Platynaspini.* Usually, the body of the larvae is arched on top and flat below. In the *Coelopterini,* however, the dorsum is not uniformly arched: the head and thorax are flatter than the abdomen, which is closer to hemispherical in shape.

The head of the larvae is frequently square with rounded corners, i. e. of about equal length and width, but in the *Hyperaspini* and *Platynaspini* it is strongly transverse. In the *Serangiini,* the head is elongate (Kamiya 1965). The sides of the head are usually rounded to a greater *(Psylloborini)* or lesser extent *(Coccinellini)* or are straight *(Platynaspini,* particularly the *Chilocorini).*

In most species the head is either completely sclerotized *(Coccinellini, Chilocorini,* etc.) or only in some areas *(Hyperaspini, Stethorini, Platynaspini),* but in the *Scymnini* the head is sclerotized only very slightly. The frontal suture is nearly always well-developed; it is V-shaped or Y-shaped, usually with an elongate epicranial suture *(Epilachnini, Chilocorini partim),* though this may be missing *(Coccinellini, Coccidulini, Psylloborini,* etc.). In the 3rd and 4th instar larvae of the *Scymnini, Platynaspini* and *Hyperaspini* the frontal suture itself is absent. A fronto-clypeal suture is well-developed only in the *Epilachnini.*

The antennae are one, two or three segmented. As a rule the first segment is short and wide; the second is usually rather longer, and sometimes even 2—3 times as long as the first and then appears cylindrical. In most coccinellid larvae the third antennal segment is small, slightly sclerotized and hardly noticeable; in some *Coccinellini (Spiladelpha, Hippodamia,* etc.) the *Tytthaspini* and the *Coccidulini* it is cupola (dome)-like. The second segment always has a long and thick seta or a shorter spine.

The mandibles are triangular in shape. In the *Epilachnini* they are equipped with four or five large teeth at the apex. In other coccinellid larvae the mandibles have only one *(Hyperaspini, Platynaspini, Stethorini, Scymnini, Exochomus, Brumus)* or two teeth *(Coccinellini, Chilocorus, Noviini, Coelopterini)* at the apex. However, there may be a row of smaller teeth *(Psylloborini, Anisosticta* from *Coccinellini)* or thick setae *(Tytthaspini)* below the apical tooth. The mandibles in the *Bulaeini* are very unusual; they have a third small tooth in addition to two well-developed apical teeth. The *Bulaeini* are a specialized tribe, whose larvae, like the adults, feed on pollen (see 6. 111). This type of feeding, as well as the structure of the mandibles and of the frontal suture place this tribe near the *Epilachnini.* Moreover, the behaviour of the larvae in the *Epilachnini* and *Bulaeini* is also similar; they are of limited mobility and collect in large groups, both characteristics of phytophagous insects.

The maxillary palpi are three segmented except in the *Noviini* where they are two segmented. The terminal segment of the maxillary palpi is usually robust and wide, although it narrows towards the apex. Only in the *Coelopterini* is it extremely long and narrow, stick-like, and 5—7 times al long as wide. The structure of the galea is rather variable; Kamiya (1965) distinguishes three principal types.

The armature of the galea is also important in taxonomy. The galea in the *Tytthaspini* is covered with dense, short, thin setae, in the *Psylloborini* by dense, short, thick setae and in other *Coccinellinae* with sparse long setae.

The labial palps are one to two segmented. The submentum is of variable shape, but is often cylindrical. In the *Psylloborini* and *Coelopterini* it is narrowed at the base;

in the *Hyperaspini* and *Tytthaspini* it is wider and transverse. In the *Psylloborini* it is covered with dense, stick-like setae, in the *Tytthaspini* with dense, thin, short setae and in most other larvae of *Coccinellidae* with sparse long setae.

The legs are elongate *(Epilachnini, Coccinellini, Bulaeini, Tytthaspini)* or short *(Hyperaspini, Stethorini,* partially *Scymnini).* The femur and tibiotarsus are usually cylindrical, but sometimes the femora are wide and almost quadrate (some *Hyperaspini, Coelopterini*). A group of setae and a claw are situated at the apex of the tibiotarsus. In most larvae there is a narrow or wide tooth at the base of the claw.

The armature of both thorax and abdomen is rather variable and complex; it has a considerable taxonomical significance. In describing coccinellid larvae it is necessary to follow the strict definitions and precise terminology which has been established. If this is not used correctly, identifications at species, generic and sometimes higher taxonomical categories will be difficult.

We follow Gage's terminology for the different structures on the body of larvae: seta, chalaza, tubercule, struma, parascolus, sentus and scolus (Gage 1920). A seta is situated directly on the body surface. A chalaza is also a seta, but is mounted on a small cylindrical or round appendix, which is called a base. A tubercule is a round protuberance covered with setae. A struma is a round protuberance covered by chalazae with low bases. A parascolus is an elongate protuberance covered with chalazae; if it is low then the bases of at least some chalazae are lengthened. A sentus is a horn-like appendix with short branches which bear setae. A scolus has numerous long branches with setae. Senti and scoli are similar in structure and the larvae of some species have structures which intermediate between senti and scoli.

In most species the pronotum has two or four sclerotized plates. Only in the *Lithophilinae* there are six sclerotized plates (Klausnitzer 1969). The notum of both the meso- and metathorax has two plates: they are oval and elongate; rarely round or quadrate. In the *Scymnini*, the plates on the thorax are only very slightly developed; in the *Platynaspini* and *Hyperaspini* they are absent.

The dorsum of the first eight abdominal segments bears a row of six structures (tubercules, strumae, parascoli, senti or scoli), which form dorsal (d), dorso-lateral (dl) and lateral (l) groups. By analogy, the tubercules on the ventral surface of the abdominal segments are called ventral (v), ventrolateral (vl) and paralateral (pl). Thus, for example dl II refers to the dorso-lateral scoli of the second abdominal segment.

Usually a definite type of armature structure is characteristic for each tribe. The larvae of *Hyperaspini* and *Platynaspini* are covered only with chalazae and setae, which also predominate in *Coelopterini.* The *Scymnini* and *Noviini* have tubercules, the *Psylloborini, Tytthaspini* and *Bulaeini* strumae, the *Chilocorini* senti and the *Epilachnini* scoli. Only the tribe *Coccinellini* is exceptional in that the larvae of its different genera have variable structures. They have tubercules *(Neomysia)*, strumae *(Propylaea, Coccinula,* parascoli *Coccinella, Adalia, Adonia, Hippodamia,* etc.), senti (*Anatis*) or scoli (*Harmonia*). Probably, the *Coccinellini* are a polyphyletic tribe.

Most of mature coccinellid larvae possess a well-developed body armature. If this armature is simplified then the larvae are either covered with a white wax-like exudation (*Hyperaspini, Scymnini*) or the sides of the abdominal segments have large gland foramina which produce a secretion (*Coelopterini*). In some tribes, there may be only one of the three above-mentioned types of integument. This suggests that all three types of integuments shown in the *Coccinellidae* possibly result from defensive adaptations, developed in the course of evolution.

There is consistency among coccinellid larvae in the morphological differences between instars, and changes occur in the coloration, proportions and armature of the body. Larvae of 3rd and 4th instars are brightly coloured and the head and legs are smaller in proportion to the body than in the monochrome 1st and 2nd instar larvae. The 3rd and 4th instar larvae also usually have well sclerotized plates on the thorax, and the abdominal segments have well-developed scoli, senti, parascoli, strumae or tubercules. In 1st and 2nd instar larvae these structures are weakly-developed. By contrast, the frontal suture in 1st and 2nd instar larvae is long and often reaches the bases of the antennae, while in the larvae of the 3rd and 4th instars it is shorter or absent. Antennae, mandibles, maxillae and labium are rather less subject to change with instar.

Because coccinellid larvae show such striking morphological changes with instar, all keys are as a rule suitable for identifying mature larvae.

## 2.2 Key to the subfamilies, tribes, genera and species of some larvae of the Palaearctic Coccinellidae

### 2.21 Key to the subfamilies, tribes and genera

1 (6) Frontal suture always developed, Y-shaped, with elongate epicranial suture. Front-clypeal suture complete. Mandibles with four to five large teeth at apex. Galea of maxilla elongate-oval with a truncate apex, bearing numerous sėtae. Antennae long, three segmented; the second segment longer than broad. Body covered with multi-branched scoli. (Fig. 1—9.) ............................ *Epilachninae*, one tribe *Epilachnini*

2 (3) Branches of scoli long, 5—7 times longer than wide. Thorns of scoli short, 2—3 times shorter than the branches or at most of the same length. (Fig. 2, 3, 5.) ........... ............................................................. *Henosepilachna* Li

3 (2) Branches of scoli short, as long or only slightly longer as wide. Thorns very long, thin, 8—10 times longer than the branches.

4 (5) Mandibles widened at base with two large and three smaller teeth at apex. d II scoli with two branches at apex. (Fig. 6, 7.) ........................................ ............................... *Subcoccinella* Huber [*S. vigintiquatuorpunctata* (L.)

5 (4) Mandibles narrow, elongate, only a little widened at base, with one large and three smaller short teeth at apex. d II scoli with four branches at apex. (Fig. 8, 9.) ............... ........................................ *Cynegetis* Redtb. [*C. impunctata* (L.)]

6 (1) Frontal suture absent or developed, V-shaped, epicranial suture absent (Fig. 50, 111) or present. If present, then the antennae short, with the second segment of about equal length and width. Mandibles at apex with one or two large teeth, below which one more small tooth or a row of smaller teeth are situated (Fig. 10, 15, 51, 109, 112). Galea of maxilla of different shape. Body with setae, chalazae, tubercules, strumae, parascoli or senti, if with scoli, they are two or three branched.

7 (82) Prothorax with two to four sclerotized plates or without any such plates. Apex of tibiotarsi with a more or less thick brush of setae.

8 (81) Terminal segment of maxillary palpi usually wide, robust, no more than thrice as long as wide (Fig. 18, 52). Foramina of secretory glands on abdominal tergites absent or only very slightly developed. Body bare or with long wax-like white exudations.

9 (14) Head of rounded triangular shape. Mandibles with six— seven teeth, the size of which gradually decreases towards the base. Submentum with several stick-like setae. Abdominal tergites with struma (Fig. 10—14). Fungivorous. ................ *Psylloborini*

10 (13) Prothorax with four sclerotized plates. d and dl strumae of different shape.

11 (12) d strumae oval in the horizontal plane; dl strumae round. Mandibles with six teeth. Tooth at base of claw rectangular. (Fig. 11.) ........................................ ....................................... *Thea* Muls. [*Th. vigintiduopunctata* (L.)]

12 (11) d strumae round, dl strumae oval in the longitudinal plane. Mandibles with seven teeth. Tooth at base of claw round. (Fig. 10, 12, 13.) ........................... ............................................................. *Halyzia* Muls.

13 (10) Prothorax with two sclerotized plates. d and dl strumae both round. Mandibles with

six teeth. Tooth at base of claw round. (Fig. 14.) ................................ *Vibidia* Mul. [*V. duodecimguttata* (Poda)]

14 (9) Head of different shape. Mandibles have no more than three teeth at apex. Submentum usually with thin, elongate, rarely short setae. Abdominal tergites with various structures.

15 (16) Frontal suture with acute angle. Mandibles with three teeth at apex. Abdominal tergites with quadrangular strumae. (Fig. 15, 16.) They are feeding on pollen of plants. ...... *Bulaeini* [*Bulaea lichatschovi* (Humm.)]

16 (15) Frontal suture rounded or absent. Mandibles with one or two teeth at apex.

17 (70) Mandibles with two teeth at apex; if one, then maxillary palps or antennae two segmented (in *Chilocorus* the mandibles have two apical teeth and the epicranial suture is present).

18 (19) Maxillary palpi and antennae two segmented. Body of elliptical shape. Abdominal tergites with tubercules. (Fig. 17—21.) ............... *Noviini* (*Rodolia* Muls.)

19 (18) Maxillary palpi three segmented. Antennae one or three segmented. Abdominal tergites with various structures.

20 (25) Antennae one or two segmented. Head usually rectangular, strongly and equally sclerotized. Abdominal tergites with senti. (Fig. 22, 49, 27, 31, 38.) ......... *Chilocorini*

21 (22) Epicranial suture well developed. Mandibles with two teeth at apex. Meso- and metathorax without sclerotized plates. Senti long, thin, 5—13 times as long as wide. (Fig. 23. 26, 27.) ........................................ *Chilocorus* Leach

22 (21) Epicranial suture absent. Mandibles with one tooth at apex. Meso- and metathorax with sclerotized plates. Senti short, only 3—4 times as long as wide.

23 (24) Prothorax with four sclerotized plates, if with two then the d senti short and triangular. like parascoli. (Fig. 28—33.) .............................. *Brumus* Muls.

24 (23) Prothorax with two sclerotized plates, d senti elongate. (Fig. 35, 39, 40.) ............ *Exochomus* Redtb.

25 (20) Antennae three segmented, the last segment sometimes only slightly developed. Head of rounded quadrangular shape.

26 (27) The last antennal segment cupola-like. Mandibles with thick short setae on inner margin. Maxillary palpi, galea and mentum all covered with numerous short setae. (Fig. 50—52.) ........................................ *Tytthaspini* (*Tytthaspis* Crotch.)

27 (26) The last antennal segment cupola-like or flat. Mandibles without setae on inner margin. Maxillary palpi, galea and mentum all covered by sparse long setae.

28 (31) The last antennal segment always cupola-like. IXth abdominal tergite elongate, posterior margin with a more or less well developed rounded excision. Abdominal tergites with strumae. (Fig. 54, 55.) .................................... *Coccidulini*

29 (30) Claws with a rounded quadrangular tooth at base. Setae on the body slightly knobbed (thickened at tip). (Fig. 56.) ............................ *Rhizobius* Steph.

30 (29) Claws without a distinct tooth, only widened at base. Setae on the body pointed. (Fig. 57.) ............................................ *Coccidula* Gyll.

31 (28) The last antennl segment cupola-like or flat. IXth abdominal tergite strongly transverse; if elongate, then posterior margin with a rounded protuberance or with an appendix. Abdominal tergites with scoli, senti, parascoli, strumae or tubercules. ............ *Coccinellini*

32 (37) Apex of IXth segment with a conical or triangular appendix.

33 (34) Apex of IXth segment with an elongate conical appendix. Inner margin of plates on prothorax straight. Plates of meso and metathorax rounded, as wide as long. Claw with a narrow tooth at base. (Fig. 58, 59.) ............................ *Propylaea* Muls. [*P. quatuordecimpunctata* (L.)]

34 (33) Apex of IXth segment with a shorter, triangular appendix (Fig. 60). Inner margin of plates on prothorax triangularly cut at base. Plates on meso- and metathorax of oval shape. Claw with a wide, robust tooth at base.

35 (36) Plates of thorax sclerotized, their external margins with four to five short finger-like senti and numerous chalazae. Abdominal tergites with senti. (Fig. 60—62.) .......... *Anisocalvia* Crotch [*A. quatuordecimguttata* (L.)]

36 (35) Plates of thorax slightly sclerotized, covered with numerous short setae, their external margins with only two to four chalazae. Abdominal tergites with parascoli. (Fig. 63. 64.) ............................................ *Calvia* Muls.

37 (32) Apex of IXth segment without an appendix.

38 (39) Abdominal tergites with a long cylindrical appendix, covered with numerous short chalazae and setae. Plates of thorax with numerous short setae, with chalazae at margins.

Larvae brightly orange with black spots. (Fig. 65.) ................................ ........................................ *Aiolocaria* Crotch [*A. mirabilis* (Motsch.)]

39 (38) Abdominal tergites covered with scoli, senti, parascoli, strumae and tubercules. Plates of thorax of different structure.

40 (41) Abdominal tergites with scoli. d and dl scoli two to three branched with a more or less long common base. (Fig. 66, 67.) ............................ *Harmonia* Muls.

41 (40) Abdominal tergites without scoli, covered with senti, parascoli, strumae or tubercules.

42 (45) Abdominal tergites with senti. Prothorax with two robust plates.

43 (44) d senti on abdomen 1.5 times as long as wide. Pronotum at hind corners and both meso- and metanotum on the outer margins with 3 senti and chalazae. Pale-coloured senti present in positions l I, l IV-VI, dl I ............................................ ............................................ *Sospita* Muls. [*S. vigintiguttata* (L.)]

44 (43) d senti on abdomen 3 times as long as wide. Hind margin of pronotum with 6 senti in a single row, meso- and metanotum with one sentus in each of the d, dl and l positions. Middle of pronotum with a red patch. Pale-coloured senti present at l I and l II (Fig. 68, 69.) .................................. *Anatis* Muls. [*A. ocellata* (L.)]

45 (42) Abdominal tergites with tubercules, parascoli or strumae. Prothorax with two or four plates without senti.

46 (51) Abdominal tergites with tubercules, covered with numerous slender setae or with sparse setae and one slightly developed chalaza.

47 (48) Tubercules of abdominal tergites large, with numerous slender setae. d I-II tubercules flat, d III-VIII conical. Plates of thorax also covered with numerous slender setae. Base of claw without tooth (Fig. 70—71.) ... *Neomysia* Casey [*N. oblongoguttata* (L.)]

48 (47) Tubercules of abdominal tergites small, only slightly developed, wider than high, with sparse setae and one slightly developed chalaza. Base of claw with a tooth. (Fig. 72).

49 (50) Tibia longer than dorsal edge of femur. d and dl. chalazae on the abdominal sclerites sparse and weak. ........................ *Myrrha* Muls. [*M. octodecimguttata* (L.)]

50 (49). Tibia as long as dorsal edge of femur. d and dl chalazae on the abdominal sclerites more dense and with broader bases. .................. *Aphidecta* Weise [*A. obliterata* (L.)]

51 (46) Abdominal tergites with parascoli or strumae.

52 (69) Inner margin of mandibles without fine teeth. Base of claw in most cases with a tooth.

53 (60) Antennae long, second segment significantly narrower and longer than the first. Third segment well-developed, cupola-like.

54 (55) Head elongated at front on account of the well developed clypeus. Prothorax with two quadrangular plates, truncate behind. Rear of IXth tergite covered with large upturned chalazae. Parascoli of abdomen strongly sclerotized with 10 large chalazae. Base of claw with a tooth. (Fig. 73, 74.) ........................................... .......................... *Spiladelpha* Sem. Dobr. (*S. barovskii kiritschenkoi* Bar.)

55 (54) Head not elongated at front, clypeus normal. IXth tergite without upturned chalazae.

56 (57) Base of claw with a well-developed, robust quadrangular tooth, surrounded with thick clavate setae upturned at apex. Prothorax with four sclerotized plates, if with two then each of them deeply excised in front or on one side. Plates of mesothorax oval, robust and with numerous chalazae. (Fig. 75, 80.) ................ *Semiadalia* Crotch.

57 (56) Claw only slightly widened at base and surrounded with setae which are straight or slightly thickened at the apex.

58 (59) Prothorax with four plates, lateral plates widely separated from medial ones. Bases of d parascoli oval with chalazae distributed uniformly or only along their internal margin. Larvae with orange spots. (Fig. 81, 82.) ........................... *Adonia* Muls.

59 (58) Prothorax with two or four robust plates, lying closely against each other. Bases of parascoli round with large chalazae situated at their apex. Larvae completely black or black with the IVth abdominal segment pale. (Fig. 83—85.) ...... *Hippodamia* Muls.

60 (53) Antennae short, second segment only slightly longer than the first. Third segment very small, indistinct.

61 (66) Prothorax with two ovoid-rectangular plates. If plates are excised on the front margin, then not deeply and the larvae are brightly coloured (orange-yellow with black).

62 (63) IXth abdominal tergite elongate-trapezoidal, its apex twice narrow as the base. Base of claw without a tooth. Larvae white with black d parascoli and white l parascoli. (Fig. 86, 87.) ................................ *Pseudoharmonia* (*P. montana* Sav.)

63 (62) IXth abdominal tergite wide and rounded. Base of claw with a tooth. Larvae coloured differently.

64 (65) Plates of mesothorax round to oval. Abdominal tergites with small strumae, bearing three to four large and several small chalazae. Larvae white or yellow with black spots.

(Fig. 88—90.) .......................................... *Coccinula* Dobzh.
65 (64) Plates of mesothorax slightly constricted at middle. Abdominal tergites with small parascoli. Larvae pink or red with white and black spots. (Fig. 91.) ................ .................................................... *Synharmonia* Ganglb.
66 (61) Prothorax with four or two plates, completely separated or connected in pairs with more or less wide connections. If two plates, then each is excised on both front and hind margins or very deeply on the front margin only and larvae are pale grey.
67 (68) Parascoli of abdominal segments high or low, but bearing numerous chalazae usually with elongated bases. If bases of chalazae are round, then parascoli have about 10 large chalazae or if less, then base of claw without a tooth. (Fig. 92, 93, 98, 99.) .......... ....................................................... *Coccinella* L.
68 (67) Parascoli of abdominal segments low bearing three to five large chalazae with distinctly rounded bases. Base of claw with a well-developed tooth. A yellow spot is situated between the d IV parascoli (Fig. 107, 108.) ........................ *Adalia* Muls.
69 (52) Inner margin of mandibles with fine and dense serrations. Base of claw without a tooth. Prothorax with four widely separated plates. Plates of mesothorax round. (Fig. 109, 110.) ............................ *Anisosticta* Dup. [*A. novemdecimpunctata* (L.)]
70 (17) Mandibles with one tooth, epicranial and in most cases also frontal suture absent. (Fig. 111, 112.)
71 (76) Head strongly transverse, twice as wide as long. Body oval or broadly oval, covered with chalazae and setae.
72 (73) Second antennal segment cylindrical. Body broadly oval, almost round, without a wax--like exudation. Chalazae situated only on pleurites, the tergites are bare. (Fig. 113, 114.) ............................... *Platynaspini* (*Platynaspis luteorubra* Goeze)
73 (72) Second antennal segment flat. Body oval, covered with a white wax-like exudation. Chalazae and setae situated both on tergites and pleurites. (Fig. 115, 116.) ........... ............................................. *Hyperaspini* (*Hyperaspis* Redtb.)
74 (75) Sides of prothorax with two sclerotized plates covered with numerous long setae. ..... ............................................... Subgen. *Hyperaspis s. str.*
75 (74) Prothorax without sclerotized plates. ................ Subgen. *Oxynychus* J. Lec.
76 (71) Head only a little wider than long, not transverse. Body elongate, with tubercules or strumae.
77 (78) Thorax with well-developed sclerotized plates. Tubercules on abdomen also strongly sclerotized. Dorsum without waxlike exudation. (Fig. 119.) ...................... ....................................... *Stethorini* (*Stethorus punctillum* Ws.)
78 (77) Sclerotized plates on thorax absent or only slightly developed. Tubercules or strumae on abdomen slightly sclerotized. Dorsum always covered with a white wax-like exudation. ............................................................ *Scymnini*
79 (80) Rear of IXth abdominal segment with two pairs of strong short pointed setae. Strumae of abdominal segments with one to two long setae. (Fig. 120.) ..................... ...................................... *Clitostethus* Ws. [*C. arcuatus* (Rossi)]
80 (79) Rear of IXth abdominal segment without strong pointed setae ........ *Scymnus* Kug.
81 (8) Terminal segment of maxillary palpi long, thin, five to seven times as long as wide. Foramina of secretory glands on abdominal tergites well developed. Body covered with a hardly noticeable wax exudation. (Fig. 129—131.) .............................. ............................................. *Coelopterini* (*Pharoscymnus* Bed.)
82 (7) Prothorax with six sclerotized plates. Apex of tibiotarsi with no thick brush of hairs, three simple setae are situated a little below the claw. (Fig. 136, 137.) (Klausnitzer 1969) .............................. *Lithophilinae* (*Lithophilus connatus* Pan.)

## 2.22 Key to the species

*Henosepilachna* Li

1 (2) Second antennal segment short, only a little longer than wide; the third segment only slightly developed. d scoli of prothorax with one branch at apex. (Fig. 1, 2.) .......... ....................................................... *H. argus* (Geoff.)
2 (1) Second antennal segment elongate, 1. 5—2 times as long as wide; the third segment cupola-like. d scoli of prothorax multi-branched at apex.
3 (4) d scoli of prothorax long, thin, only slightly thickened at base where there are few

branches which are shorter than those at the apex. (Fig. 3.) ........................................................................ *H. vigintioctopunctata* (F.)

4 (3) d scoli of prothorax shorter, significantly thickened at base, where there are numerous long branches, almost as long as those at the apex. (Fig. 4,5). ... *H. elaterii* (Rossi)

*Coccidula* Kug.

1 (2) Fore tibiotarsus almost to fully half (0.44—0.50) as long as the head is wide. Clypeo--labral suture somewhat obtusely concave. (van Emden 1949) ......... *C. rufa* (Herbst)

2 (1) Fore tibiotarsus stouter, slightly over two-fifths (0.42) as long as head is wide. Clypeo--labral suture straight. (van Emden 1949) .................. *C. scutellata* (Herbst)

*Rhizobius* Steph.

1 (2) Rear margin of pronotum with a row of chalazae in the middle; the bases of these chalazae only relatively low. Inner margin of meso- and metathoracic sclerites with a single chalaza ............................................ *Rh. chrysomeloides* (Herbst)

2 (1) Rear margin of pronotum with two rows of chalazae in the middle; the bases of these chalazae are relatively high. Inner margin of meso- and metathoracic sclerites with two chalazae ................................................ *Rh. litura* (Fabr.)

*Rodolia* Muls.

1 (4) Tubercules of abdominal segments covered with long and short setae.

2 (3) Tubercules with 4 to 5 long setae. l tubercules hemispherical. (Fig. 21) ............................................................................ *R. fausti* Ws.

3 (2) Tubercules with one long seta. (Fig. 19) .................... *R. cardinalis* Muls.

4 (1) Tubercules of abdominal segments covered with short. very thick setae. l tubercules elongate. at apex widened and narrowed at base. (Fig. 20) ....... *R. limbata* (Motsch.)

*Scymnus* Kug.

1 (2) Prothorax with two triangular sclerites which are connected together at the front margin. Meso- and metathorax each with two clear dorsal sclerites .................................................. Subgen. *Nephus* Muls. (*S. limonii* Donisth.)

2 (1) Prothorax with 2 medial sclerites which are separated. Meso- and metathorax without clear sclerites.

3 (14) Sclerotized ring of second segment of maxillary palp is about twice as long on the outer surface as on the inner. Width of the same segment less than twice the length (measured on the outer surface). Second antennal segment short and wide. (Fig. 121.) .......................................................... Subgen. *Scymnus s. str.*

4 (9) Prothorax extended forwards, sometimes in a trapezoidal shape. Prothorax without tubercules and covered with setae, which may form groups of 2—3.

5 (6) IXth segment transverse, straight behind and covered with setae only on margins. (Fig. 122.) ........................................ *S. rubromaculatus* (Goeze)

6 (5) IXth segment round, covered with numerous setae. Two d, dl setae, the internal one shorter than the outer.

7 (8) d, dl setae surrounded by a semicircle of numerous short thorns. (Fig. 123.) ........................................................................ *S. pusillus* Sav.

8 (7) Thorns around d, dl setae not numerous and irregularly placed (Fig. 124.) ........................................................................ *S. apetzi* Muls.

9 (4) Prothorax not extended forwards. Tubercules on prothorax more or less distinct.

10 (13) d, dl and l tubercules well developed with one long and several small setae. (Fig. 125.)

11 (12) Middle of pronotum with 2 setae. Tibia with setae set more sparsely. ............................................................................ *S. interruptus* Goeze

12 (11) Middle of pronotum with 6 setae. Tibia with setae set more densely. .... *S. frontalis* (F.)

13 (10) d tubercules only slightly developed, replaced by two setae. dl and l tubercules distinct; l tubercules with three long and several short setae. (Fig. 126.) ....... *S. nigrinus* Kug.

14 (3) Sclerotized ring of second segment of maxillary palp about 3—4 times as long on the outer surface as on the inner. Width of the same segment twice the length (measured on the outer surface). Second antennal segment elongate, usually cylindrical. (Fig. 127.) ............................................................ Subgen. *Pullus* Muls.

15 (16) Third antennal segment only slightly developed, transparent, indistinct. Legs. antennae and head-sclerites pale yellow (van Emden 1949.) ........... *P. suturalis* Thunbg.

16 (15) Third antennal segment well developed and distinctly noticeable.
17 (20) Second antennal segment longer than first.
18 (19) Legs, antennae and head-sclerites pale yellow (van Emden 1949.). . *P. auritus* Thunbg.
19 (18) Legs and ventrum green. Prothorax with 8 tubercules, each of them with one long and two small setae. d, dl and l tubercules with two setae. (Fig. 128.) ................................................................ *P. subvillosus* (Goeze)
20 (17) Second antennal segment not longer than first. Antennae brown, legs and head — sclerites brownish-red. (van Emden 1949.) ................ *P. haemorrhoidalis* Herbst.

Subgen. *Hyperaspis s. str.*

1 (2) Abdominal tergites covered with numerous long setae, short setae rare ................................................................ *H. desertorum* Ws.
2 (1) Abdominal tergites covered predominantly with short setae and with only a few setae of fair length ........................................ *H. reppensis* (Herbst)

Subgen. *Oxynychus* Lec.

1 (2) Sides of prothorax subdivided into three lobes. Abdominal tergites with long setae on the back margin. (Fig. 116, 117) ........................ *H. alexandrae* (Ws.)
2 (1) Sides of prothorax entire. Abdominal tergites with strong short setae on the back margin. (Fig. 118.) ................................................ *H. terrea* (Zasl.)

*Hippodamia* Muls.

1 (2) Prothorax with four closely adjacent plates. Larvae black with pale segment IV. (Fig. 83, 84.) .............................................. *H. tredecimpunctata* (L.)
2 (1) Prothorax with two plates, with a deep and narrow excision posteriorly. Larvae completely black. (Fig. 85.) ............................ *H. septemmaculata* (Deg.)

*Adonia* Muls.

1 (2) Medial plates of prothorax roughly triangular and separated from lateral plates by a narrow gap. d parascoli with six large chalazae of equal size with a triangular base. Larvae brown with small orange spots. (Fig. 81.) ...................... *A. amoena* (Fald.)
2 (1) Medial plates of prothorax roughly quadrangular and separated from lateral plates by a wider gap. d parascoli with five chalazae of different size, one of them noticeably the largest, this chalaza has a cylindrical base. Larvae grey, with yellow and orange spots. (Fig. 82.) ............................................ *A. variegata* (Goeze)

*Semiadalia* Crotch.

1 (6) Prothorax with four or two plates; if two, then each is deeply excised in front and the larvae are without orange patches.

---

Plate XIII

Fig. 1—53. 1 — antenna; 2 — d scolus of prothorax, *H. argus;* 3 —d scolus of prothorax, *H. vigintioctopunctata;* 4 — head, 5 — d scolus, *H. elaterii* 6 — mandible; 7 — d scolus, *S. vigintiquatuorpunctata;* 8 — mandible; 9 — d scolus, *C. impunctata* (Kapur, 1950); 10 — mandible, *H. tschitscherini;* 11 — d, dl strumae, *T. vigintiduopunctata;* 12 — d, dl strumae, *H. tschitscherini;* 13 — d struma, *H. sedecimpunctata;* 14 — d, dl strumae, *V. duodecimpunctata;* 15 — mandible; 16 — d, dl strumae, *B. lichatschovi;* 17 — antenna; 18 — maxilla; 19 — d tubercule, *R. cardinalis* (Kamiya, 1965); 20 — d, 1 tubercules, *R. limbata;* 21 — d, 1 tubercules, *R. fausti;* 22 — antenna, 23 — d sentus; 24 — tibiotarsus, *C. bipustulatus;* 25 — tibiotarsus, *C. geminus* (Zaslavskiï, 1962); 26 — sentus, *C. renipustulatus;* 27 — sentus, *C. rubidus;* 28 — prothoracic plate; 29 — d II, VIII senti, *B. octosignatus;* 30 — prothoracic plate; 31 — d II, VIII senti, *B. jacobsoni;* 32 — prothoracic plate; 33 — d II ,VIII senti, *B. mongolicus;* 34 — d, dl, 1 I senti, *E. quadripustulatus;* 35 — prothoracic plate; 36 — d, dl, 1 I senti, *E. mongol;* 37 — prothoracic plate; 38 — d, dl, 1 I senti, *E. kiritschenkoi;* 39 — mesothoracic plate; 40 — d VIII senti, *E. undulatus;* 41 — prothoracic plate; 42 — claw, *E. semenovi;* 43 — d I sentus; 44 — claw, *E. nigripennis;* 45 — prothoracic plate, 46 — d, dl, 1 I senti, *E. melanocephalus;* 47 — d, dl, 1 I senti, 48 — claw, *E. nigromaculatus;* 49 — antenna, *E. undulatus;* 50 — head, 51 — mandible, 52 — maxilla, 53 — d, dl, 1 strumae, *T. lineola.* (G. I. Savoïskaya del.)

1 2 3 4 5 6 7 8 9 10 11 12 13 14 15 16 17 18 19 20 21 22 23 24 25 26 27 28 29 30 31 32 33 34 35 36 37 38 39 40 41 42 43 44 45 46 47 48 49 50 51 52 53

2 (3) 4th instar larva with average lengths of 0.60 mm for front tibia and 0.79 mm for head-width. .................................................. *S. alpina* (Villa)

3 (2) 4th instar larva with average lengths of > 1 mm for front tibia and 0.90 mm for head-width.

4 (5) Prothorax with four plates. Parascoli high, elongate. d parascoli increase in height towards posterior body, especially this applies to the bases of the two large apical chalazae. Larvae orange. dl, 1 IV parascoli orange, the remainder brown. (Fig. 75, 76.) .................................................. *S. undecimnotata* (Schneid.)

5 (7) Prothorax with two plates, very deeply excised in front. Parascoli low, flat. d parascoli almost of the same structure on all segments. Larvae yellowish-brown. dl,1I and IV parascoli white or yellow, the remainder black. (Fig. 77.) ......... *S. notata* (Laich.)

6 (1) Prothorax with two plates, deeply excised at the front corners. Strumae high, elongate, especially parascoli d VIII, which is almost a sentus in structure. Prothorax red, the remaining segments orange. dl, 1 I and IV parascoli orange, the remainder black. (Fig. 79, 80.). .................................................. *S. przevalskii* Sav.

*Tytthaspis* Crotch.

1 (2) Larvae wholly pale except for blackish margins to some of the segments of the legs, setae dark brown. (van Emden 1949) .................... *T. sedecimpunctata* (L.)

2 (1) Some abdominal strumae pale in colour.

3 (4) d and dl strumae of I pale, all other abdominal strumae dark. Bases of the chalazae on the abdominal strumae are high and large .............. *T. trilineata* Weise

4 (3) Larvae light grey with small dark spots situated around the chalazae. Sides of head blackish. Prothorax with two closely adjacent plates, more or less uniformly covered with chalazae. d strumae oval, with 8 chalazae, 4 of which are larger than the others. (Fig. 53.) .................................................. *T. lineola* (Gebl.)

*Adalia* Muls.

1 (2) d and dl abdominal sclerites with areas of isolated setae which have no basal swellings .................................................. *A. conglomerata* (L.)

2 (1) d and dl abdominal sclerites with parascoli. Segment IV usually with a pale patch in the middle.

3 (4) Prothorax with two plates, deeply excised at the front corners. Mesothoracic plates also excised in front. Abdominal parascoli with five large and several small chalazae. Larvae creamy-white with black plates. dl I, 1 I—VIII strumae yellow, the remainder black. (Fig. 106, 107.) .................................. *A. decempunctata* (L.)

4 (3) Prothorax with four plates. Mesothoracic plates not excised in front. Abdominal parascoli with three large and some small chalazae. Larvae darker, abdominal tergites blackish. d IV, dl I parascoli orange, the other d and dl strumae black. 1 I, IV parascoli yellow, the other 1 parascoli white with grey or black apex. (Fig. 108.) ...... *A. bipunctata* (L.)

*Coccinella* L.

1 (2) d parascoli small with four chalazae, only two of which are large. Bases of chalazae rounded. Base of claw without or with only a very slightly developed rounded tooth. Larvae grey with yellow spots. (Fig. 92, 93.) .................................. .......................... Subgen. *Neococcinella* Sav. (*C. undecimpunctata* L.)

---

**Plate XIV**

Fig. 54—87. 54 — antenna, 55 — apex of abdomen, *Coccidula*, 56 — claw, *R. litura*, 57 — claw, *Coccidula* (van Emden, 1949); 58— IXth segment, 59— mesothoracic plate, *P. quatuordecimpunctata;* 60 — IXth segment, 61 — prothoracic plate, 62 — d, dl 1 II senti, *A. quatuordecimguttata;* 63 — prothoracic plate, 64 — d, dl parascoli, *C. punctata;* 65 — d, dl, 1 V, A. *mirabilis;* 66 — d, dl scoli, *H. axyridis;* 67 — d, dl scoli, *H. quadripunctata* (van Emden, 1949); 68 — prothoracic plate, 69 — d, dl senti, *A. ocellata;* 70 — d, I, VIII tubercules, 71 — claw, *N. oblongoguttata;* 72 — d, dl, 1 strumae, *M. octodecimguttata* (van Emden, 1949); 73 — IXth segment, 74 — d, dl parascoli, *S. barovskii kiritschenkoi;* 75 — prothoracic plate, 76 — d, dl parascoli, *S. undecimnotata;* 77 — d, dl parascoli, *S. notata;* 78 — prothoracic plate, 79 — d, dl parascoli, 80 — claw, *S. przevalskii;* 81 — d, dl parascoli, *A. amoena;* 82 — d, dl parascoli, *A. variegata;* 83 — prothoracic plates, 84 — d parascolus, *H. tredecimpunctata;* 85 — prothoracic plate, *H. septemmaculata;* 86 — IXth segment, 87 — claw, *P. montana.* (G. I. Savoïskaya del.)

54
55
56
57
58
59
60
61
62
63
64
65
66
67
68
69
70
71
72
73
74
75
76
77
78
79
80
81
82
83
84
85
86
87

2 (1) d parascoli larger, often very large, with no less and often more than 8 chalazae. Base of claw with a well developed tooth. Larvae brightly coloured with orange, red, yellow or white spots. .................................... Subgen. *Coccinella s. str.*

3 (14) Parascoli of abdominal segments high, finger-like, bases of chalazae elongate, cylindrical. Larvae large (7.0—12.8 mm).

4 (7) Medial prothoracic plates closely adjacent.

5 (6) Prothorax with two plates, not deeply excised in front and behind. Base of claw with a narrow, elongate tooth. Larvae velvet-black; prothoracic tergites (excepting the plates) and d, dl I—II parascoli yellow. (Fig. 94.) .................. *C. nivicola* Men.

6 (5) Prothorax with four plates, the medial and lateral plates connected by broad connections. Base of claw with wide tooth. Larvae black; small spots at the front and back angles of the prothorax and also d, dl I and IV parascoli orange. (Fig. 95, 96.) ... *C. reitteri* Ws.

7 (4) Medial prothoracic plates separated by a light strip, which is wider behind than in front.

8 (9) Medial prothoracic plates rather widened posteriorly and somewhat pear-shaped. d parascoli high, almost equal to abdominal segments by length. (Fig. 97, 98.) ....... .......................................... *C. distincta* Fald. (*divaricata* auct. nec Ol.)

9 (8) Medial prothoracic plates roughly octangular. d strumae smaller.

10 (11) Mesothoracic plates broadly oval, only slightly longer than wide. d parascoli orientated horizontally and very close to each other. Setae of parascoli strong, short, only a little longer than bases of chalazae. (Fig. 99.) ............ *C. transversoguttata* Fald.

11 (10) Mesothoracic plates oblong, significantly longer than wide. d parascoli orientated longitudinally. Setae of parascoli thin, long, 3 or more times as long as bases of chalazae.

12 (13) Base of claw with a narrow and very short tooth, slightly excised on the outer side. Larvae brick-red with black. (Fig. 100.) ................ *C. tianschanica* Dobz.

13 (12) Base of claw with a narrow but long tooth with a straight outer side. Larvae grey with yellow and orange spots. (Fig. 101, 102.) ................ *C. septempunctata* L.

14 (3) Parascoli of abdominal segments not high, sometimes flat. Bases of chalazae triangular, often with round corners; of equal height and width. Larvae smaller (4.5—7.0 mm).

15 (16) Prothorax with four plates, medial plates quadrangular. Setae of parascoli dark, thin, strong and very long, 5 times as long as bases of chalazae. Larvae black, dl I, IV, VI, VII and l I, IV parascoli orange. (Fig. 103.) ................ *C. quinquepunctata* L.

16 (15) Prothorax with two plates, each of them more or less deeply excised in front and behind. Setae of parascoli shorter, 3—3.5 times as long as bases of chalazae.

17 (18) Prothoracic plates slightly excised in front and behind. d parascoli not high, with 7 chalazae, the bases of which are elongate-triangular. Larvae black with small light spots. dl, l I, IV parascoli whitish. (Fig. 104.) .................... *C. hieroglyphica* L.

18 (17) Prothoracic plates broadly excised in front and behind. d parascoli flat, very low, with 10 chalazae having small, round bases. Larvae light-grey with large white spots and black plates and parascoli; though dl, l I, IV parascoli white. (Fig. 105.) .......... ........................................................ *C. trifasciata* L.

---

Plate XV

Fig. 88—137. 88 — prothoracic plate, 89 — d, dl, l II strumae, *C. elegantula;* 90 — mesothoracic plate, *C. principalis;* 91 — mesothoracic plate, *S. conglobata;* 92 — d, dl, l II parascoli, 93 — claw, *C. undecimpunctata;* 94 — claw, *C. nivicola;* 95 — prothoracic plate, 96 — claw, *C. reitteri;* 97 — medial prothoracic plate, 98 — d II parascolus, *C. divaricata;* 99 — d II parascolus, *C. transversoguttata;* 100 — claw, *C. tianshanica;* 101 — d II parascolus, 102 — claw, *C. septempunctata;* 103 — d parascolus, *C. quinquepunctata;* 104 — d parascolus, *C. hieroglyphica;* 105 — d parascolus, *C. trifasciata;* 106 — prothoracic plate, 107 — d II parascolus, *A. decempunctata;* 108 — d II parascolus, *A. bipunctata;* 109 — mandible, 110 — claw, *A. novemdecimpunctata;* 111 — head, 112 — mandible, *O. alexandrae;* 113 — antenna, 114 — abdominal segment, *P. luteorubra;* 115 — antenna, 116 — abdominal segment, 117 — prothorax, *O. alexandrae;* 118 — prothorax, *O. terrea;* 119 — d tubercule, *S. punctillum;* 120 — VIIIth and IXth abdominal segments. *C. arcuatus* (van Emden, 1949); 121 — antenna, *S. nigrinus;* 122 — IXth segment, *S. rubromaculatus;* 123 — d, dl, setae, *S. pusillus;* 124 — d, dl setae, *S. apetzi;* 125 — d, dl, l tubercules, *S. frontalis;* 126 — d, dl, l tubercules, *S. nigrinus;* 127 — antenna, 128 — prothorax, *P. subvillosus;* 129 — head, 130 — maxilla and labium, *P. balkhashensis;* 131 — foreleg, *P. tsharinensis;* 132 — metathoracic plate, 133 — abdominal segment, *P. pilosus;* 134 — abdominal segment, *P. heptapotamicus;* 135 — abdominal segment, *P. auricomus;* 136 — prothorax, 137 — tibiotarsus, *L. connatus* (Klausnitzer, 1969). (G. I. Savoïskaya del.)

88
89
90
91
92
93
94
95
96
97
98
99
100
101
102
103
104
105
106
107
108
109
110
111
112
113
114
115
116
117
118
119
120
121
122
123
124
125
126
127
128
129
130
131
132
133
134
135
136
137

*Coccinula* Dobz.

1 (2) Prothoracic plates closely adjacent, covered with numerous uniformly distributed chalazae and setae. Larvae grey, but the lst abdominal segment white, excluding the brown d strumae. Also small white spots are situated between the thoracic plates. (Fig. 88, 89.) ........................................ *C. elegantula* (Ws.)
2 (1) Plates of prothorax widely separated; chalazae and setae are not numerous here and mainly situated on the margins. Larvae are differently coloured.
3 (4) Larvae slender, coloured white with brown plates and strumae. Large white spots between thoracic plates. Two white strips along abdomen between d and dl strumae ........................................ *C. sinuatomarginata* (Fald.)
4 (3) Larvae more robust, yellow or orange with black plates and strumae.
5 (6) Prothoracic plates more or less deeply excised in front. Tibiotarsi slender, long, 9—10 times as long as wide. Larvae bright yellow, almost orange, with black plates and strumae. (Fig. 90.) ........................................ *C. principalis* (Ws.)
6 (5) Prothoracic plates with straight front margin. Tibiotarsi 5—6 times as long as wide. Larvae yellow with black plates and strumae. ......... *C. quatuordecimpustulata* (L.)

*Synharmonia* Gglb.

1 (2) Margins of prothoracic plates more or less uniformly covered by chalazae. Larvae reddish-brown with bright white and red spots. 1 parascoli red, white and black. ......... ........................................ *S. oncina* (Ol.)
2 (1) Chalazae on prothoracic plates principally situated on hind and hind outer margins. Between parascoli d IV there is a diffuse pale patch which stretches from the anterior to the posterior limit of segment IV.
3 (4) Segments I—VIII with a wide evenly defined pale mid-line. ........... *S. lyncea* (Ol.)
4 (3) Mid-line of segments I—VIII broken up segmentally into patches of varying size.
5 (6) Patch in mid-line of segment III large, extending forward past the front margin of the dorsal sclerites. Pale patch in mid-line of segment IV usually joined to the dl patches. ........................................ *S. conglobata* (L.)
6 (5) Patch in mid-line of segment III small, mainly restricted to the posterior margin of the segment. Pale patch in midline of segment IV usually not joined to the dl patches. ........................................ *S. impustulata* (L.)

*Harmonia* Muls.

1 (2) Parascoli of abdominal segments with elongate thin bases, d scoli have three branches, dl scoli two branches. Prothorax with six senti d, dl, 1 I; dl II—III: d, dl IV—V scoli all of orange colour, the remainder black. (Fig. 66.) ............ *H. axyridis* (Pall.)
2 (1) Parascoli of abdominal segments with short and wide bases, d and dl scoli have three branches. d IV, dl I—IV scoli yellow, the remainder dark. (Fig. 61.) ............... ........................................ *H. quadripunctata* (Pont.)

*Halyzia* Muls.

1 (2) d strumae with one large long chalaza. (Fig. 12.) ........ *H. tschitscherini* Sem.
2 (1) d strumae with several more or less large chalazae bearing thickening, short setae. (Fig. 13.) ........................................ *H. sedecimguttata* (L.)

---

Plate XVI

Fig. 138—160. 138—140: *Rhizobius chrysomeloides*, L4; 138 — Hind margin of pronotum; 139 — inner part of mesothoracic sclerite; 140 — inner part of metathoracic sclerite; 141—143: *Rhizobius litura*, L4; 141 — Hind margin of pronotum; 142 — inner part of mesothoracic sclerite; 143 — inner part of metathoracic sclerite; 144 — *Synharmonia impustulata*, L4, pigmentation of third and fourth abdominal segment; 145 — *Aphidecta obliterata*, L4, struma d 3 (by van Emden, 1949); 146—147: *Calvia decemguttata*, L4; 146 — ninth abdominal segment; 147 — struma dl 5; 148—153: *Sospita vigintiguttata*, L4, 148 — prothoracic sclerite; 149 — mesothoracic sclerite; 150 — metathoracic sclerite; 151 — sentus dl 6; 152 — sentus d 7; 153 — sentus l 3; 154—157: *Semiadalia alpina*, L4; 154 — mesothoracic sclerite; 155 — struma d 3; 156 — struma dl 3; 157 — struma l 3; 158 — *Adalia conglomerata*, L4, verruca d 2; 159—160: *Brumus oblongus*, L4; 159 — parascolus dl 7; 160 — parascolus l 3. (After B. Klausnitzer I. Kovář del.)

PLATE XVI

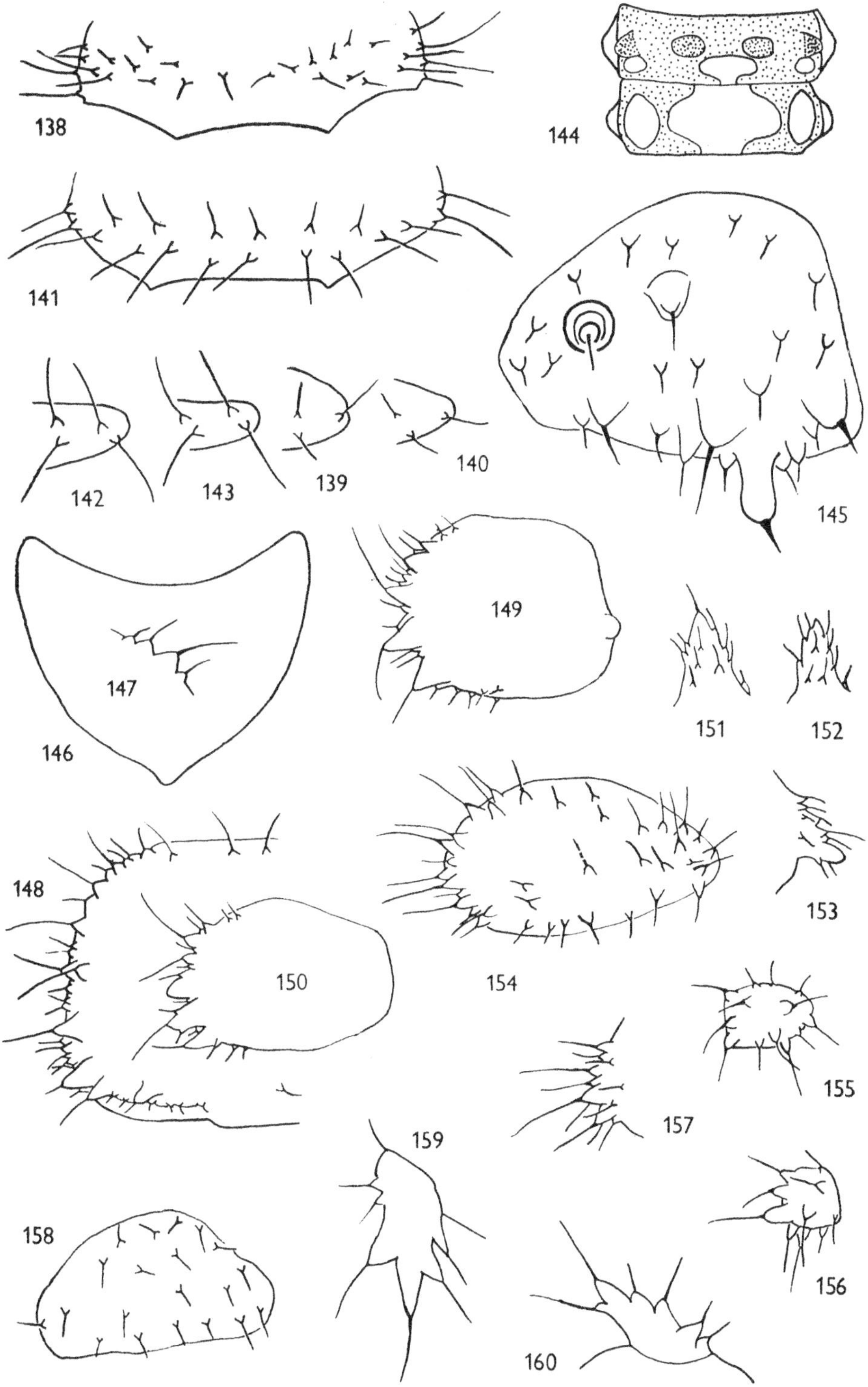
138
144
141
145
142
143
139
140
149
147
146
151
152
148
153
150
154
155
157
159
158
156
160

*Calvia* Muls.

1 (2) dl IV strumae dark, l strumae all pale. Larva bi-coloured and with general appearance resembling a *Psylloborini* larva. ........................ *C. decemguttata* (L.)
2 (1) Struma dl IV often more or less pale, but usually not white, at least strumae l II—III show darkening. Segments I—V with a white medial stripe. .................... ............................................... *C. quinquedecimguttata* (F.)

*Pharoscymnus* Bed.

1 (2) Head elongate. 3/4 as wide as long. Prothoracic plate covered with numerous setae. Mesothoracic plates oval. Larvae black. (Fig. 129, 130.) ...... *Ph. balkhashensis* Sav.
2 (1) Head rounded, as wide as long. Prothoracic plate with few setae.
3 (4) Frontal suture rounded posteriorly. Front tibiotarsi thin, elongate, 8 times as long as wide; femora quadrate, four times as wide as tibiotarsi. Larvae black. (Fig. 131.) ..... ............................................... *Ph. tsharinensis* Sav.
4 (3) Frontal suture pointed posteriorly. Front tibiotarsi robust, only 3 times as long as wide; femora thinner, only 1.5—2 times as wide as tibiotarsi.
5 (6) Metathoracic plates oval, only slightly narrowed at internal margin, straight behind, without a protuberance. Setae on abdomen on round sclerotized plates. Larvae black. (Fig. 132, 133.) ......................................... *Ph. pilosus* Sav.
6 (5) Metathoracic plates very narrowed at internal margin, with a round protuberance behind. Setae on abdomen without sclerotized plates at base.
7 (8) Front tibiotarsi almost uniformly thin, elongated, 5 times as long as wide. On abdominal segments 4 large setae are situated between the foramina of the secretory glands. Larvae brown with black head. (Fig. 134.) ............... *Ph. heptapotamicus* Dobz.
8 (7) Front tibiotarsi rounded and widened at base, only 3 times as long as wide. On abdominal segments 6 large setae are situated between the foramina of the secretory glands. Larvae black. (Fig. 135.) ..................................... *Ph. auricomus* Sav.

*Chilocorus* Leach.

1 (4) First abdominal segment white, senti not very long, only 5 times as long as wide, with about ten branches.
2 (3) Tibiotarsus 3.5 times as long as wide. Base of claw with a short wide tooth. (Fig. 23, 24.) ............................................... *Ch. bipustulatus* (L.)
3 (2) Tibiotarsus 5 times as long as wide. Base of claw with a long narrow tooth. (Fig. 25.) (Zaslavskiĭ 1962) ...................................... *Ch. geminus* Zasl.
4 (1) First abdominal segment black. Senti narrow, long and with more than 10 branches.
5 (6) Senti 7 times as long as wide. Branches of senti very short with long setae 10 times the length of the branches. (Fig. 26.) ................ *Ch. renipustulatus* (Scriba)
6 (5) Senti 11—13 times as long as wide. Branches of senti long, with short strong thorns of about the same length. (Fig. 27.) ......................... *Ch. rubidus* Hope

*Exochomus* Redtb.

1 (8) External margins of prothoracic plates with three large senti and numerous chalazae with elongate bases. Senti of abdominal segments large, bearing about 15 long branches ............................................... Subgen. *Exochomus s. str.*
2 (7) Plates of mesothorax of round oval shape with long multibranched senti on external margins. d VIII senti three to four times as long as wide.
3 (6) External margins of thoracic plates with senti and chalazae forming a continuous row. Branches of senti of abdominal segments cylindrical, elongate, with dark short strong setae which are shorter than the branches. Base of claw with a wide transverse tooth.
4 (5) Plates of prothorax with branched senti. d I—II senti of triangular shape, with one more or less large branch at apex. (Fig. 34.) ......... *E. quadripustulatus* (L.)
5 (4) Plates of prothorax with unbranched senti. d I—II senti elongate, with two distinct branches at apex. (Fig. 35, 36.) ............................ *E. mongol* Bar.
6 (3) External margins of thoracic plates bear senti separated by spaces. Branches of senti on abdominal segments of about equal length and width, slightly rounded at the apex, with long, light setae which are longer than the branches. Claw at base narrow, elongate. (Fig. 37, 38.) ......................................... *E. kiritschenkoi* Bar.
7 (2) Plates of mesothorax round, on external margins there are short senti having short branches. d VIII senti six times as long as wide. Base of claw with a robust, strongly

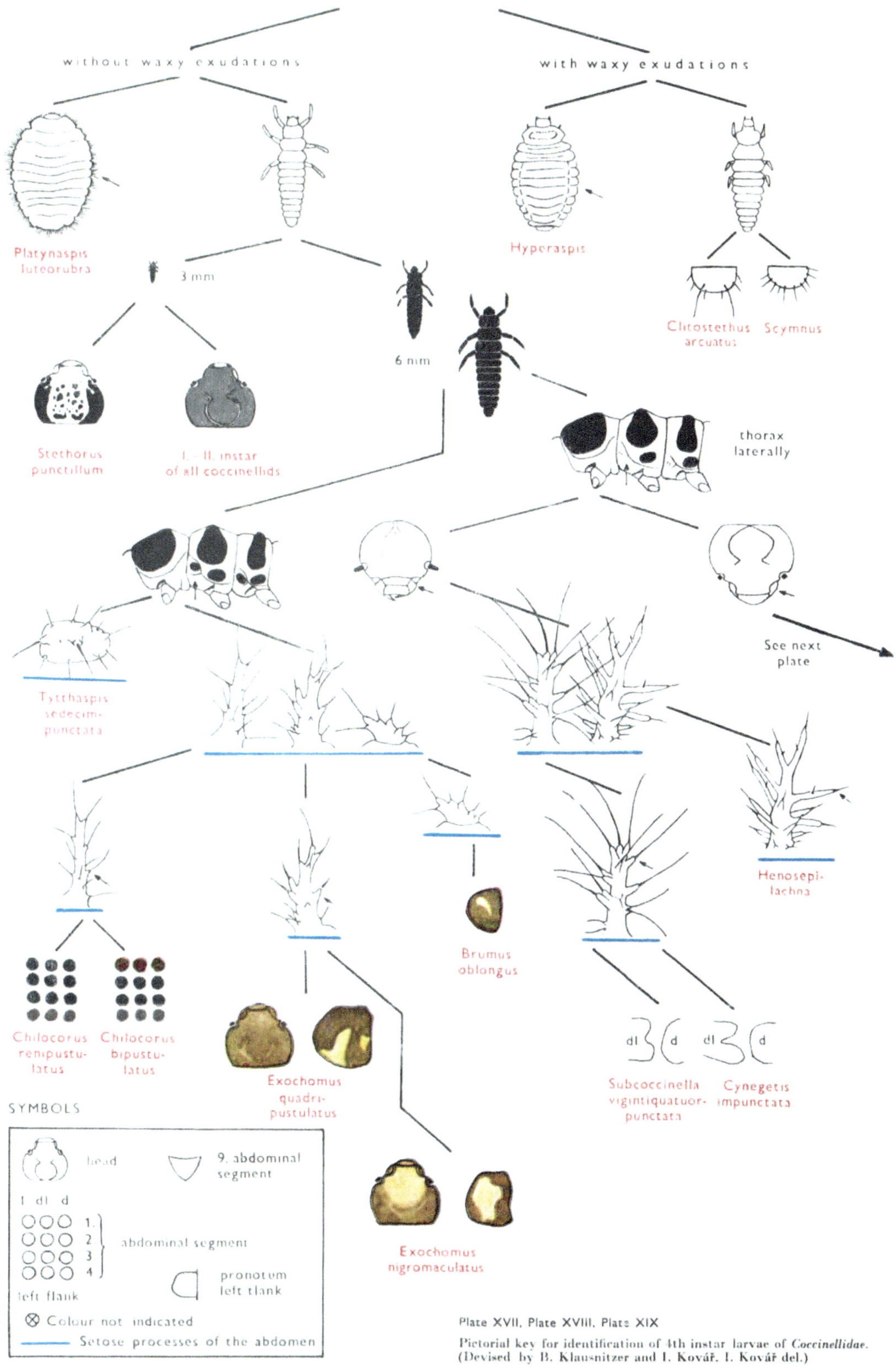


Plate XVII, Plate XVIII, Plate XIX

Pictorial key for identification of 4th instar larvae of *Coccinellidae*. (Devised by B. Klausnitzer and I. Kovář. I. Kovář del.)

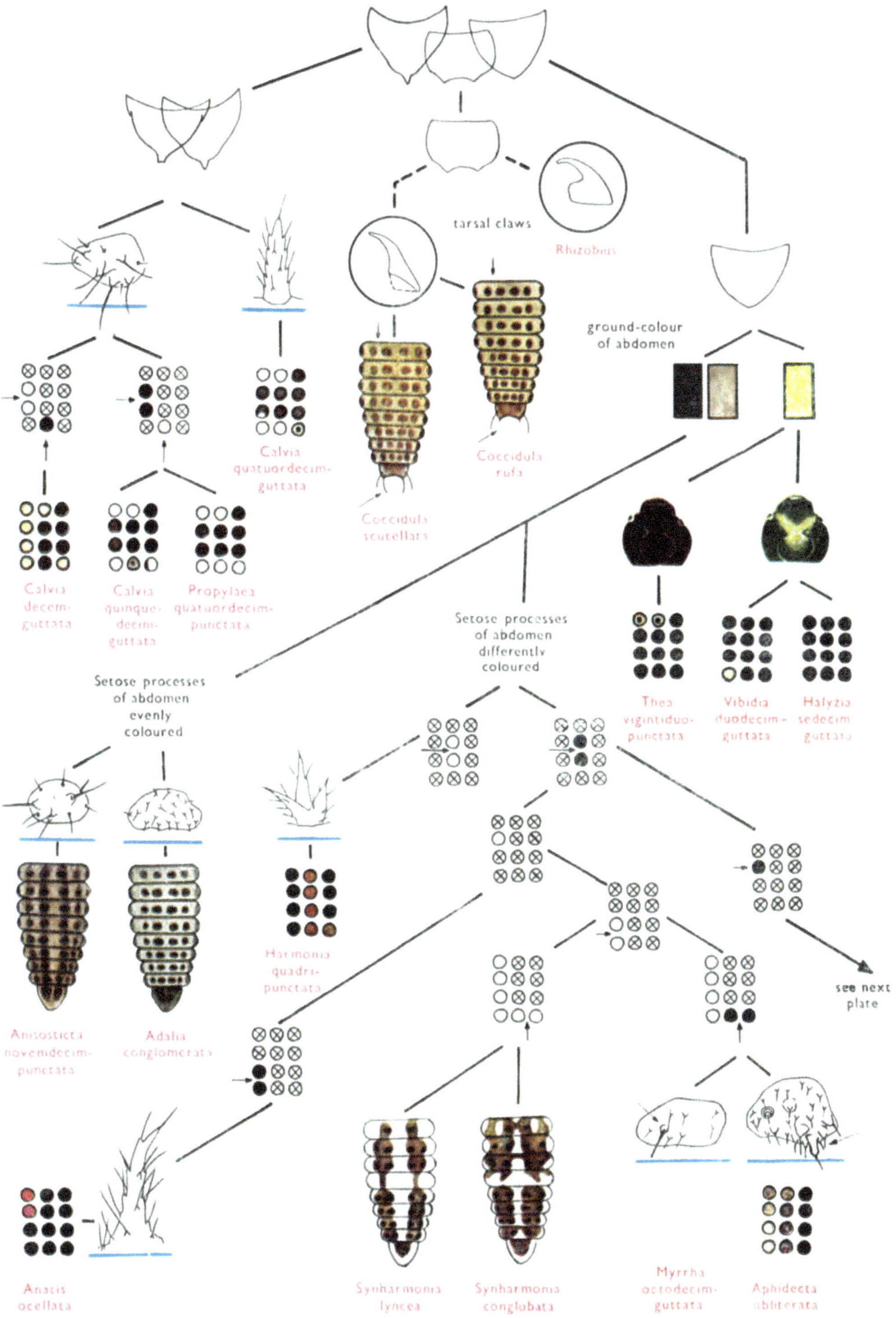
tarsal claws
Rhizobius
ground-colour
of abdomen
Calvia
quatuordecim-
guttata
Coccidula
rufa
Coccidula
scutellata
Calvia
decem-
guttata
Calvia
quinque-
decim-
guttata
Propylaea
quatuordecim-
punctata
Setose processes
of abdomen
differently
coloured
Thea
vigintiduo-
punctata
Vibidia
duodecim-
guttata
Halyzia
sedecim
guttata
Setose processes
of abdomen
evenly
coloured
Harmonia
quadri-
punctata
Anisosticta
novemdecim-
punctata
Adalia
conglomerata
see next
plate
Anatis
ocellata
Synharmonia
lyncea
Synharmonia
conglobata
Myrrha
octodecim-
guttata
Aphidecta
obliterata

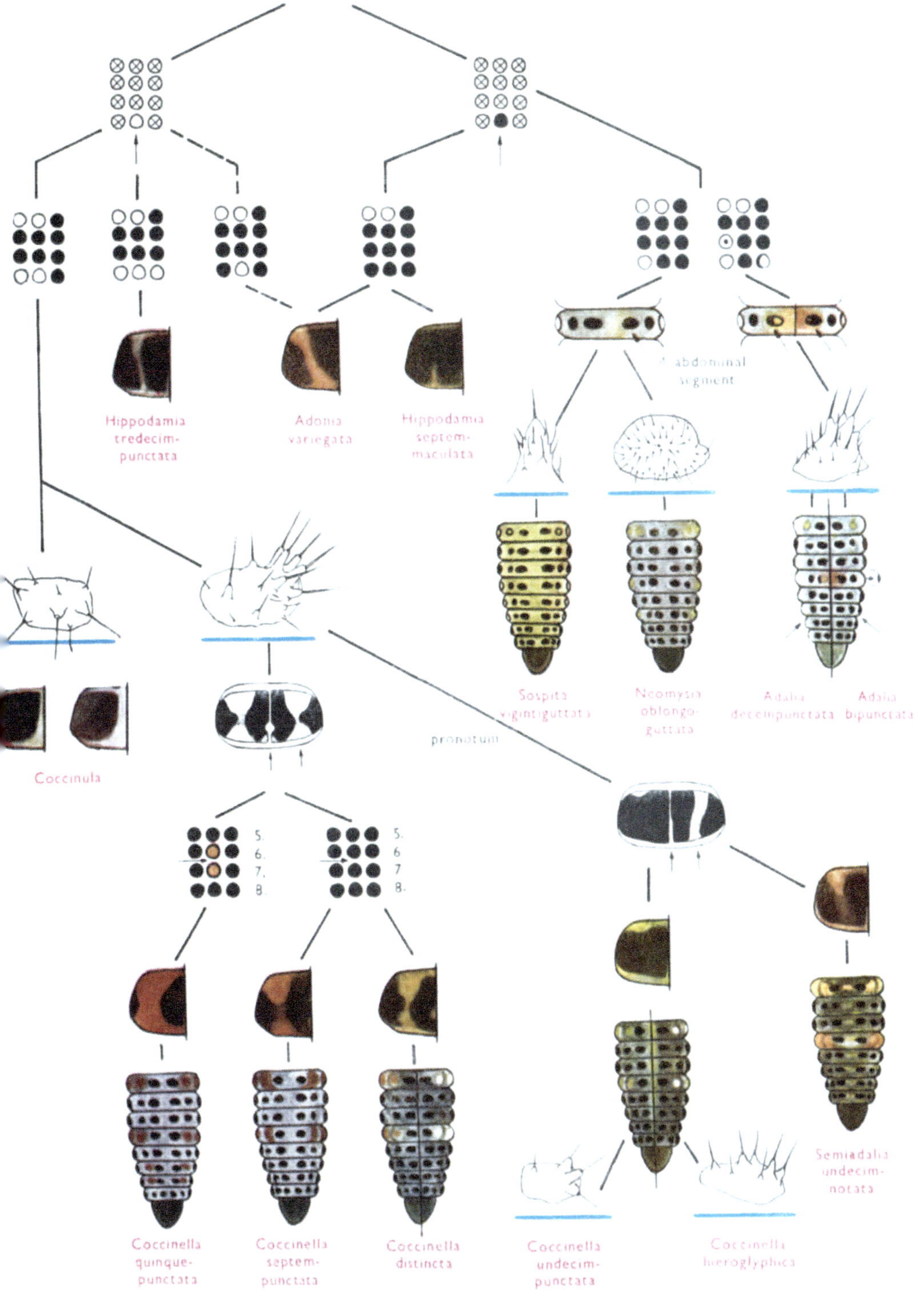
Hippodamia tredecim-punctata
Adonia variegata
Hippodamia septem-maculata
abdominal segment
Sospita vigintiguttata
Neomysia oblongo-guttata
Adalia decempunctata
Adalia bipunctata
Coccinula
pronotum
5.
6.
7.
8.
Coccinella quinque-punctata
Coccinella septem-punctata
Coccinella distincta
Coccinella undecim-punctata
Coccinella hieroglyphica
Semiadalia undecim-notata

transverse tooth. (Fig. 39, 40.) .............................. *E. undulatus* Ws.

8 (1) Plates of prothorax with impoverished armature, i. e. with only two low parascoli and several chalazae with short round bases on the external margins. Senti of abdominal segments not large, bearing about 10 short branches. ......... .................. ........................................... Subgen. *Parexochomus* (Bar.)

9 (14) Senti of abdominal segments with very short branches at apex of which long, thin setae are situated. Legs long, of tibiotarsus 6 to 8 times as long as wide.

10 (13) Parascoli of plates of prothorax with two straight-sided branches of equal size at apex.

11 (12) Base of claw widened with rounded sides, without a tooth. Larvae dark grey, almost black. (Fig. 41, 42.) ...................................... *E. semenovi* Ws.

12 (11) Base of claw with a well-developed, elongate narrow tooth. Larvae light. (Fig. 43, 44.) ................................................ *E. nigripennis* (Er.)

13 (10) Parascoli of plates of prothorax with one rounded branch at apex. Base of claw with a tooth. Larvae cream-coloured with black plates and senti. (Fig. 45, 46.) ........... ................................................ *E. melanocephalus* Zubk

14 (9) Senti of abdominal segments with distinct although small branches, at apex of which shorter strong setae are situated. Legs short, tibiotarsus only 3 times as long as wide. Base of claw with a wide transverse tooth. Larvae coloured light-cream with darker plates and senti. (Fig. 47, 48.) ...................... *E. nigromaculatus* (Goeze)

*Brumus* Muls.

1 (2) Pronotum with two sclerotized plates. .............. *B. oblongus* (Weidenbach)

2 (1) Pronotum with four sclerotized plates.

3 (4) Medial and lateral plates of prothorax completely separated. External margins of thoracic plates with senti. d senti slightly increase in length towards the back end of the body. Larvae orange in colour. (Fig. 28, 29.) ................. *B. octosignatus* (Gebl.)

4 (3) Medial and lateral plates of prothorax are connected in front or fused into one plate. External margins of thoracic plates with chalazae. d senti markedly increase in length towards the back end of the body. Larvae dark.

5 (6) Plates of mesothorax almost quadrate. d I—II senti caplike, d VII—VIII senti stick-like. All senti with short strong setae. (Fig. 30, 31.) .................. *B. jacobsoni* Bar.

6 (5) Plates of mesothorax oval, elongate. d I—II senti short, d VII—VIII senti elongate-triangular. All senti with long thin setae. (Fig. 32, 33.) ........ *B. mongolicus* Fleisch.

## 2.23 A simple key for field use

B. KLAUSNITZER and I. KOVÁŘ

As a supplement to the preceding key, a further key for 46 European species and genera has been drawn up which is based on the easiest characters, above all setae and colouring. A hand lens should provide adequate magnification. The key applies only to the 4th instar larvae. It aims to help those who are put off the identification of ladybird larvae by the complicated morphological examinations normally entailed. Users of the key who have no specialized knowledge of the systematics of coccinellid larvae will find that it gives them an outline knowledge of the subject. After some initial practice, the identification of larvae can be speeded up by the use of the separate dichotomous key for coloured symbols which also figures the most important morphological details which are frequently mentioned in the text.

1 Larvae with white wax secretions. ........................................ 2

— Larvae without wax secretions. ........................................ 4

2 Body elliptical. ........................................ *Hyperaspis*

— Body elongated. ........................................ 3

3 Tip of the 9th abdominal segment with a large seta on each side. ... *Clitostethus arcuatus*

— Tip of the 9th abdominal segment without large setae. ..................... *Scymnus*

4 Body broadly oval, with marginal setae. ..................... *Platynaspis luteorubra*

— Body elongate. ........................................ 5

5 Larvae small (up to 3 mm long). In doubtful cases they can be distinguished from I—II instars of all coccinellid larvae by the characteristic colouration of the head: see colour plate. ........................................ *Stethorus punctillum*

— Larvae usually substantially larger. .......... 6

6 Meso- and metathorax with two setose processes laterally. .......... 7

— Meso- and metathorax with only one setose process laterally. .......... 11

7 Upper side of abdomen with strumae. Body pale with lightbrown sclerites. .......... *Tytthaspis sedecimpunctata*

— Upper side of abdomen with senti or parascoli. .......... 8

8 Parascoli and pronotum: see colour plate. .......... *Brumus oblongus*

— Senti with distinct branches. .......... 9

— Senti without distinct branches. .......... 10

9 Upper side of head dark. L 1, dl 1 pale. Pronotum: see colour plate. .......... *Exochomus quadripustulatus*

— Upper side of head in front — pale. L 1 dark, dl usually dark. Pronotum: see colour plate. .......... *Exochomus nigromaculatus*

10 D, dl, l 1 pale. .......... *Chilocorus bipustulatus*

— D, dl, l 1 dark. .......... *Chilocorus renipustulatus*

11 Head with front-clypeal suture. Frontal suture Y-shaped. .......... 12

— Head without front-clypeal suture. Frontal suture Ʊ— shaped. .......... 14

12 Branches of scoli long, thorns of scoli short. .......... *Henosepilachna*

— Branches of scoli short, thorns of scoli very long. .......... 13

13 Base of d and dl abdominal scoli: see colour plate. .......... *Subcoccinella vigintiquatuorpunctata*

— Base of d and dl abdominal scoli: see colour plate. .......... *Cynegetis impunctata*

14 9th abdominal segment with caudal tip. .......... 15

— 9th abdominal segment rounded. .......... 20

— 9th abdominal segment ± truncate to weakly sinuous. .......... 18

15 Upper side of abdomen with senti. Dl, l 1; d, dl, l 4, l 5—7 white. .......... *Calvia quatuordecimguttata*

— Upper side of abdomen with strumae. .......... 16

16 D 2, l 3, pale, dl 4 dark. Colouration of abdomen: dl 1, d 4, and l 1—8 yellow. .......... *Calvia decemguttata*

— D 2, l 3, dark, dl 4 pale. .......... 17

17 1 with a small and 4 with a large yellow median spot. 2, 3, 5—8 with white median spots. Meso- and metathorax with yellow dorsal spots. Dl 1; d, dl 4; l 1, 4—8 white. .......... *Propylaea quatuordecimpunctata*

— Abdominal segment 1—5 with pale median stripe. Dl 1; l 4, white to yellowish, l 1, d 4 and l 5—8 white, dl 4 greyishbrown. .......... *Calvia quinquedecimguttata*

18 Larvae on deciduous and coniferous trees and in the herbaceous vegetation of dry habitats. Tarsal claw: see colour plate. .......... *Rhizobius*

— Larvae on riparian vegetation (banks of lakes, rivers, etc.) Tarsal claw: see colour plate. .......... 19

19 D, dl 1—8 dark. Setae on 9th abdominal segment: see colour plate. .......... *Coccidula rufa*

— D, dl 1—8 pale. Setae on 9th abdominal segment: see colour plate. .......... *Coccidula scutellata*

20 Ground colour of abdomen yellow. .......... 21

— Ground colour of abdomen light grey to black. .......... 23

21 Head dark brown, legs short, dl 1 yellow. .......... *Thea vigintiduopunctata*

— Head pale yellow with dark spots, legs long, dl 1 black. .......... 22

22 L 1—5 blackish, l 6—8 yellow. .......... *Halyzia sedecimguttata*

— L 1—3 black, l 4—8 yellow. .......... *Vibidia duodecimguttata*

23 All abdominal sclerites uniformly coloured. .......... 24

— Some abdominal sclerites distinctly paler or darker than the other ones. .......... 25

24 Body grey. D, dl 1—8 dark, l 1—8 pale. .......... *Adalia conglomerata*

— Body brown, thorax on each side with a dark brown spot. D 1—8 darker than dl, l 1—8. .......... *Anisosticta novemdecimpunctata*

25 Dl 2—3 pale. Parascoli trifurcate.
Dl 1—4, d 4 orange yellow. .......... *Harmonia quadripunctata*

— Dl 2—3 dark. .......... 26

26 L 2 pale. .......... 27

— L 2 dark. .......... 31

27 L 3—4 dark.
Upper side of abdomen with senti. L 1—2 white to red. Posterior margin of the pronotum with a median pale spot. .......... *Anatis ocellata*

— L 3—4 pale. .......... 28

28 D, dl 4 pale. Between the d 4 there is extensive pale spot. .......... 29

— D, dl 4 dark. .......... 30

29 Abdominal segments 1—8 with broad, pale median line. D 4; dl 1 and 4 pale. .......... *Synharmonia lyncea*

— Abdominal segments 1—8 with pale spots. Meso- and metathorax with white and vermilion median spot. D 4; dl 1 and 4; l 1—8 white. .......... *Synharmonia conglobata*

30 Ground colour brown to greyish brown. Chalazae of d, dl strumae weak and sparse. .......... *Myrrha octodecimguttata*

— Ground colour pale greyish brown. Chalazae of d, dl strumae with a rather large base and dense. .......... *Aphidecta obliterata*

31 Dl 4 dark. .......... 32

— Dl 4 pale. .......... 37

32 D, dl, l 4 dark. .......... 33

— Some of both d or l 4 completely or partly pale at least. .......... 34

33 Body greyish brown. Dl, l 1 yellow to orange. Pronotum: see colour plate. .......... *Adonia variegata*

— Body completely dark except only dl, l 1. Pronotum: see colour plate. .......... *Hippodamia septemmaculata*

34 Middle of 4th abdominal segment dark. D 4 completely dark.
Abdomen with senti or verrucae. .......... 35

— Middle of 4 th abdominal segment usually with pale spot. D 4 partly pale. Abdomen with parascoli. .......... 36

35 Abdomen with senti. Dl 1, l 1, 4—6 yellowish white. .......... *Sospita vigintiguttata*

— Abdomen with verrucae. Dl 1, l 1, 4, 6 reddish yellow. .......... *Neomysia oblongoguttata*

36 L 2, 3, 5—8 dark (sometimes l 4 is dark also); l 1, 4 yellowish to orange. Ground colour darker. Dl 1 and middle of the 4th abdominal segment yellowish organe. Colouration variable. .......... *Adalia bipunctata*

— L 1, 3—8 whitish; l 2 dark. Ground colour paler. Dl 1 yellow. Middle of 4th abdominal segment white to yellowish. .......... *Adalia decempunctata*

37 Dl 1, 4; l 1 pale. Pronotum: see colour plate. .......... *Adonia variegata*

— D 4; Dl 1, 4; l 1, 4 pale. Pronotum: see colour plate. .......... *Hippodamia tredecimpunctata*

— Dl 1, 4; l 1, 4 pale. .......... 38

38 Upper side of abdomen with strumae. Abdomen between the d and dl with a pale longitudinal stripe. Pronotum: see colour plate. .......... *Coccinula*

— Upper side of abdomen with parascoli. .......... 39

39 Prothorax with four plates: see colour plate. .......... 40

— Prothorax with two plates: see colour plate. .......... 42

40 Dl 6—7 red. Pronotum: see colour plate. .......... *Coccinella quinquepunctata*

— Dl 6—7 black. .......... 41

41 L of metathorax dark. Dl, l 1 and 4 orange red. Pronotum: see colour plate. .......... *Coccinella septempunctata*

— L of metathorax pale. Dl, l 1 and 4 whitish. Pronotum: see colour plate. .......... *Coccinella distincta*

42 Ground colour of abdomen orange — brown. Dl, l 1 and 4 dark orange. Pronotum: see colour plate. .......... *Semiadalia undecimnotata*

— Ground colour of abdomen grey or greyish-black. Dl, l 1 and 4 yellow or white. Pronotum: see colour plate. .......... 43

43 D parascoli: see colour plate. .......... *Coccinella undecimpunctata*

— D parascoli: see colour plate. .......... *Coccinella hieroglyphica*

# 3 VARIABILITY AND GENETIC STUDIES

A. HONĚK

## 3.1 Cytogenetic studies

The science of cytogenetics is concerned with the mechanisms of heredity and variability at the cellular level. Because of the crucial importance of chromosomal pattern, this pattern and its changes are of paramount interest.

### 3.11 Chromosomal complements and their variability

The chromosomal complements of many coccinellid species have been catalogued by Smith (1960). The number of 2n = 18 autosomes and the XX (♀) : XY (♂) sex determining mechanism, which are both typical in beetles are also retained in many coccinellids, though some changes (reduction or multiplication of autosomes, disappearance of the Y chromosome) in both these features have been found in several species.

Variability in chromosomal complements has been investigated by Smith (1956, 1959) in North American *Chilocorus* species. In this genus and some related coccinellids, all the standard chromosomes are metacentric; one arm is entirely euchromatic and the other entirely heterochromatic. This condition permits a reduction of chromosome numbers by "centric fusion" i. e. translocation of euchromatic arms between non-homologous chromosomes. In this way chromosomes arise, in which both arms are euchromatic, while the two heterochromatic arms are lost from the chromosomal complement. Thus the total number of chromosomes decreases by one, each time one centric fusion occurs (Smith 1965). *Chilocorus stigma* which inhabits the eastern part of North America, exhibits chromosomal polymorphism (Fig. 3.01). A population with 2n = 26 chromosomes inhabits Florida. As one moves northwest from Florida, three centric fusions appear to have been sequentially incorporated in the population; the first appears in Maine, the second in Ontario, the third in Manitoba. Thus within the Manitoba population, individuals are found with a chromosome number reduced to 2n = 20 as the result of all three fusions being homozygous. A complex of *Chilocorus* species is found in the western part of North America. *C. fraternus* and *C. orbus* (both in California) have 2n = 22, *C. tricyclus* (British Columbia and Washington) has 2n = 20 and *C. hexacyclus* (Alberta, Saskatchewan) has 2n = 14. All these species have a stable number of chromosomes. Crossing experiments have shown that they share a monophyletic origin. Each species underwent a characteristic number of centric fusions during its evolution. Smith (1959) suspects that the progressive increase in number of centric fusions which occurred both in *C. stigma* and the western *Chilocorus* complex confers some adaptive advantage during the recolonisation by these species of the northern regions after the glacial period.

Fig. 3.01 Proportion of centric fusions (dark) (maximum number per individual is six) in *Chilocorus stigma* populations (Smith 1956).

## 3.12 Interspecific crossing and sterility

Komai (1956) has sumarized the older data. Two field observations throw further light on this subject. In central Asia, near Tashkent, natural changes established contact between the distribution areas of *Chilocorus bipustulatus* and *C. geminus* (Zaslavskiĭ 1967). In the contact area (only 20—30 km wide) the $F_1$ hybrids are completely sterile. In both directions from this zone, only pure populations of one of the species occur. This sterility is the only isolating factor that isolates the populations, for the ecological conditions are similar on both sides of the contact zone and there is no geographical barrier. By contrast *Chilocorus tricyclus* has invaded the range of *C. hexacyclus* in Canada and succeeded in cross-breeding with the resident population (Smith 1966). The former species has the chromosome number 2n = 20, and the latter one has 2n = 14. Any hybrids therefore show chromosomal polymorphism and one part of population of hybrids is sterile for cytological reasons. Laboratory crossing is possible between all the North American species of *Chilocorus* but the percentage of viable eggs obtained is variable (Smith 1959).

## 3.2 Genetic studies

The only review is that of Komai (1956). It mainly stresses work with *Harmonia axyridis* from a number of Chinese and Japanese papers. The papers quoted by Komai are mostly omitted from our list of references.

### 3.21 Variability of colour patterns

The surface of most of coccinellids (particularly the elytra) has characteristic colour patterns, which show great variability within many species and have therefore been intensively studied.

Colour patterns are composed of a dark design on a light (brownish, yellowish, reddish or whitish) background. The dark pigment is melanin; the light pigments have been found to contain derivates of carotenoids (α—, and β— carotene and lycopene).

The basic unit of pattern is a 'spot'; the spots occur on exactly prescribed places on the elytra. The various aberrations are each characterized by the number of spots, by their position and by the number of their fusions. In this way, the variability of pattern is strongly discontinuous.

Variability can be classified in several alternative ways[1]):

1) According to the degree of elytral pigmentation, i. e. how much of the surface is covered by the dark design.

2) Forms may be placed into 'eunomic series' which are each defined by the particular sequential development of one character (e. g. increase of number of spots or number of fusions). Each series is distinct and forms a separate progression. Two individuals which differ from one another by one unit of a character change (e. g. by one spot or by one fusion) are placed next to each other in the 'eunomic series' (Zarapkin 1930).

3) The "Variationsbild" of Schilder combines the advantages of both methods 1) and 2) (Fig. 3.02). The beetles are classified according to the number of spots or number of fusions (degree of melanisation), into horizontal series (analogous to method 1). The way in which one form is thought to be derived from another is indicated by vertical lines (eunomic series as in 2). Schilder's technique thus arranges the variability in two dimensions (Schilder and Schilder 1951—2, Schilder 1952—3).

[1]) Both sexes ought to be treated separately, as sexual dimorphism is often expressed. Statistical differences in proportion of various aberrations between both sexes have been found. In *Aphidecta obliterata* variability is limited mostly to female sex, while males are almost uniform (Eichhorn and Graf 1971).

4) This approach bears an illusory resemblance to classical taxonomy. From the most 'primitive' (basic) pattern for a species other patterns may be derived by fusing or by eliminating spots (Fig. 3.03). Similarly basic patterns for a genus can be derived by comparing the basic patterns of the species within this genus (Johnson 1910).

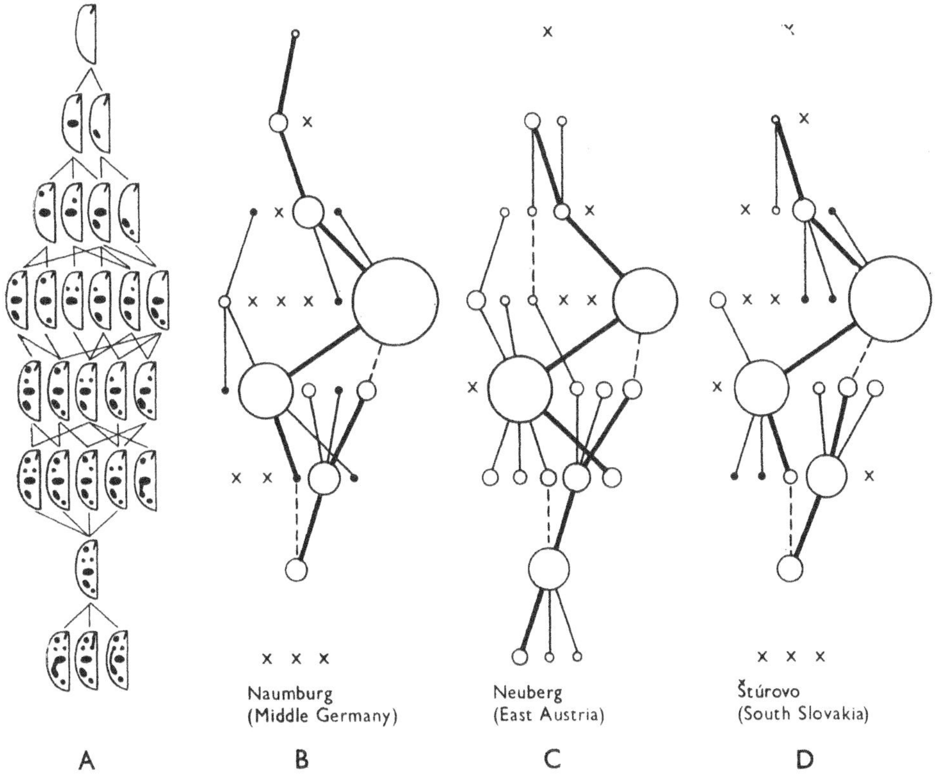


Fig. 3.02 Classification of variability in *Adonia variegata*. A — Schilder's „Variationsbild". B—D — Frequency of aberrations (indicated by diameters of circles) in three localities (Schilder 1951—2).

## 3.22 Ontogenetic manifestation of the genetic background

The genetic background of an organism is a kind of "blue-print" which materializes in the course of the ontogenesis of the animal, presumably by affecting the metabolism of the body structures. The end-point of the pathway of gene manifestation is a distinct morphological character. No more than a beginning has been made in elucidating the origin in colour patterns in coccinellids from this particular point of view.

Zarapkin (1938b) used statistics to study the colour patterns of coccinellid elytra. He found that the existing combinations of spot arrangement by no means exhausted the statistical possibilities. This indicates some biological restraint on randomness, a restraint which is probably genetical and physiological in origin. The action of the restraint can be well demonstrated in the ontogeny of colour pattern.

Fig. 3.03 Classification of variability (mode 4) in *Hippodamia convergens* (Johnson 1910).

It is well known that spots appear in a characteristic sequence in the course of several hours following the adult ecdysis. Zarapkin (1938a) described 7 different sequences in descendants of one pair of *Adalia decempunctata* (Fig. 3.04). In the population of descendants of the original pair, the various sequences were not equally represented

Fig. 3.04 Modes of ontogenetic development of colour patterns in *Adalia decempunctata* (Zarapkin 1938a).

in terms of numbers. Spots which formed earlier were larger than the later ones. The ontogenetic sequences of pigmentation development which Zarapkin established as most abundant in this laboratory-bred population were also the abundant eunomic sequences which can be found in the field.

### 3.23 Heredity of colour patterns

The heredity of colour patterns has been studied by the classical genetic methods, i. e. recording the ratios by which characters split after the first and the second generation etc. The most recent example of this type of work is that of Lusis (1971). Three forms of genus *Calvia* had been considered to be separate species according to their colour patterns (*C. punctata*, *C. obversepunctata*, *C. duplicipunctata*). However, Lusis ascertained that they were only forms of one polymorphic species, *C. punctata.* In the $F_1$ generation he obtained a typical Mendelian ratio, on the assumption that forms are genetically based on a series of three multiple alleles with the following sequence of dominancy:

$$c^d \text{ (punctata)} > c^o \text{ (obversepunctata)} > c^p \text{ (duplicipunctata)}$$

Individuals with all possible homo- and heterozygote combinations of alleles occure in the nature.

Similar polymorphism based on a series of multiple alleles has been found in *Adalia decempunctata*, *Coelophora inaequalis*, *Propylaea japonica*, *Harmonia axyridis* and *Hippodamia quinquesignata* (reviewed by Komai 1956).

Komai concludes that such series of multiple alleles determine the main colour types of elytra and ensure that the stable and distinct polymorphism within each species is maintained. The minor peculiarities of pattern, which undergo continual change are controlled by polygenes (modifiers). Ford (1964) considers it more likely that such a multiple allele effect as has been demonstrated in coccinellids is most likely caused by a sufficiently close juxtaposition of the loci of the genes so that crossing-over is most unlikely to separate them. Ford designates such a complex of loci as a 'supergene'.

Manifestation of genes may also be influenced by temperature, i. e. some characters may be produced either by hereditary or by somatic factors. In *Epilachna chrysomelina* the total area of spots decreases by 30—55 % (according to geographical origin of investigated population) when temperature increases from 20 to 35 °C. This variation in various geographic populations is due to differences in the rate of diminution of single spots; the shape of spots, however, remains constant (Timofeeff-Ressovsky 1941). The same increase in temperature also suppresses the melanisation of the ventral surface of the body (Zimmermann 1931). The critical stage for induction of colour changes by temperature is at the prepupa and early pupa. Other environmental agents (light, humidity) were without effect (Timofeeff - Ressovsky 1941). A similar temperature effect on gene manifestation has been found in *Harmonia axyridis.*

### 3.24 Geographic variability

The proportions of individuals with more or less pronounced pigmentation varies according to geographical position in a manner which is consistent among many coccinellid species (Dobzhansky 1933, Dobzhansky and Sivertzev-Dobzhansky 1927). The consistent variability of *Coccinella septempunctata*, *C. quinquepunctata*, *Adonia variegata*, *Synharmonia conglobata* is shown in figs. 3.05—3.08.

Another similar consistent trend in geographical variability occurs in the species *Anatis ocellata*, *Coccinella transversoguttata*, *Coccinula quatuordecimpustulata*, *Coccinella divaricata*, *Anisosticta novemdecimpunctata*, *Adalia bipunctata*.

Fig. 3.05 Geographical variability of the diameter of the macula discoldalis in *Coccinella septempunctata* (Dobzhansky and Sivertzev-Dobzhansky 1927).

For all the members of both 'groups' of species referred to, there exist centres with less pigmented populations and centres with more pigmented ones. Moreover, such centres for the different species roughly co-incide in geographical location.

The centre of light forms lies in Central Asia for the eastern hemisphere, and in California for the western hemisphere. A significant centre of dark forms is situated in the Old World in the Far East. However, the proportion of pigmented forms increases radially in all directions from each light centre, i. e. not only from Central Asia northeastwards to the dark centre in the Far East but also northwestwards to Europe and southwards to Africa as far as Assam in case of *C. septempunctata;* similarly in North America dark forms increase eastwards and northwards from California. Thus, in *Adonia variegata*, the most pigmented populations live simultaneously in Ethiopia and in the Far East. *Coccinella transversoguttata*, a Palaearctic and Nearctic species has similar duplicate geographical centres of light and dark forms in both regions.

The phenomenon of geographical variability parallel in several species (apparent also in other animal groups), is usually generalised by reference to various 'rules' (Allen's rule, Gloger's rule). Many workers have suggested that the individual biotypes are selected by different climatic factors (particularly temperature and humidity).

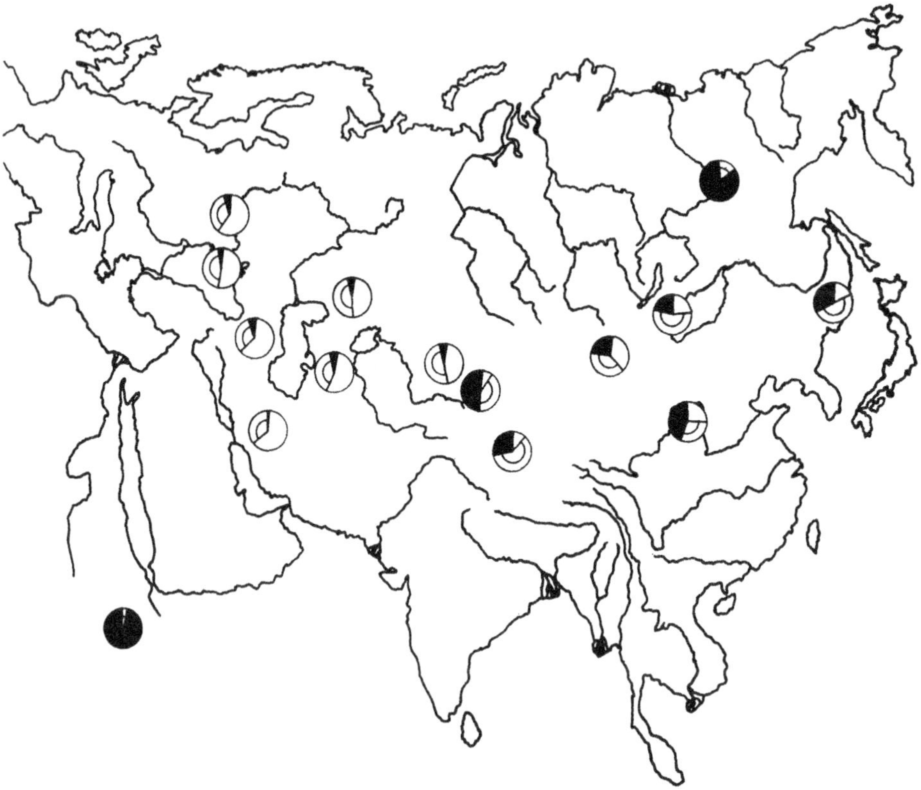

Fig. 3.06 Geographical variability by proportion of forms with 0 — 9 (white), 11 — 15 (white, with central stripe or 15 confluent (black) spots in *Adonia variegata* (Dobzhansky 1933).

The situation in coccinellids supports this suggestion only as far as humidity is concerned. The centres of light populations lie in arid areas, while pigmentation increases with the humidity of a region. The correlation between pigmentation and temperature of the respective regions is much less. For example, similar pigmentation is found in populations of *A. variegata* from Ethiopia (average temperature +30° C) and from Yakutsk (aver. temperature —10° C). Pigmentation is also similar in populations of *C. septempunctata* from Algeria or Italy (aver. temperature +15 to +20° C) and in those from the Far East (aver. temperature +5 to—5° C). Pigmentation itself need not have any selective value and may merely coincide with changes in the physiological processes of the organisms. On the other hand, it has been pointed out that the darker (i. e. more pigmented) species have an advantage in being able to make fuller use of insolation (Kryľcov 1956, see also *A. bipunctata*).

Differences in pigmentation of various forms have often been found to occur between closely adjacent local populations e. g. in *Adonia variegata* (Schilder and Schilder 1951—2) (Fig. 3.02). In principle, they can be explained by the above mentioned relationship between pigmentation and humidity. Kryľcov (1956) compared the pigmentation of coccinellids in two valleys in central Asia. These two valleys were relatively near to

Fig. 3.07 Geographical variability by proportion of forms with 0 — 3 (white, with central stripe), 5 (white), 7 — 10 (black) spots in *Coccinella quinquepunctata* (Dobzhansky 1933).

each other, but had a different climate, i. e. Chuiskaya valley had a dry continental climate and the Issyk-kul valley had a humid climate. Out of the 8 species studied 5 were darker in the more humid conditions (*Thea vigintiduopunctata, Bulaea lichatschovi, Propylaea quatuordecimpunctata, Adonia variegata, Hippodamia tredecimpunctata*), in 3 species (*Coccinella septempunctata, C. quatuordecimpunctata, C. sinuatomarginata*) no correlation between climate and pigmentation could be found and one (*Coccinula redimita*) showed the reverse tendency. Kryľcov claimed that the species which had a lighter background colouration were more likely to show local changes of melanisation.

Although the influence of humidity on pigmentation has been so often hypothetized, it has not yet been proved experimentally (see Chap. 3.23).

The species studied by Dobzhansky and mentioned earlier (see p. 62-64) provide information on the question of the origin of geographical races (Dobzhansky 1933). They can be ranked on the basis of some contemporary theories, to illustrate the steps by which they are thought to have originated. (1) First there are the species which, over the whole distribution area, largely only differ in the proportion by which individual forms are represented (*Adonia variegata*, *Anatis ocellata*). This is sympatric

Fig. 3.08 Geographical distribution of the typical form of *Synharmonia conglobata* (black) and its v. *buphthalmus* (white) (Dobzhansky 1933).

polymorphism, and the various forms have the taxonomic status of aberrations or varieties. (2) The next rank contains species which again show sympatric polymorphism but in which there are centres with quite different populations at the extremes of the area. However, the variability shows complete overlap and transition, so it is only possible to observe proportional changes in the forms between adjacent localities. (3) The extreme rank is the formation of geographical races, i. e. biotypes with allopatric distribution; these show either minimum areas of transition (*Hippodamia tredecimpunctata*) or completely isolated populations (*Synharmonia conglobata*). *Harmonia axyridis* is the special case; in the western part of the distribution area a monomorphic race occurs (consisting alone of the form *axyridis*), while the eastern part is inhabited by a mixture of forms (see also Komai 1956 for review of Japanese conditions).

## 3.25 The case of Adalia bipunctata

*Adalia bipunctata* is an example where the whole complex of problems has been shown in a single species. *A. bipunctata* is a polymorphous species. In principle, part of the population has a predominantly dark design (dark forms), while the other part has a design consisting only of individual spots (light forms). It is of interest that the North America populations show a very limited variability and dark forms are rare.

The first experiments with crossing the most abundant light aberration (*f. typica*)

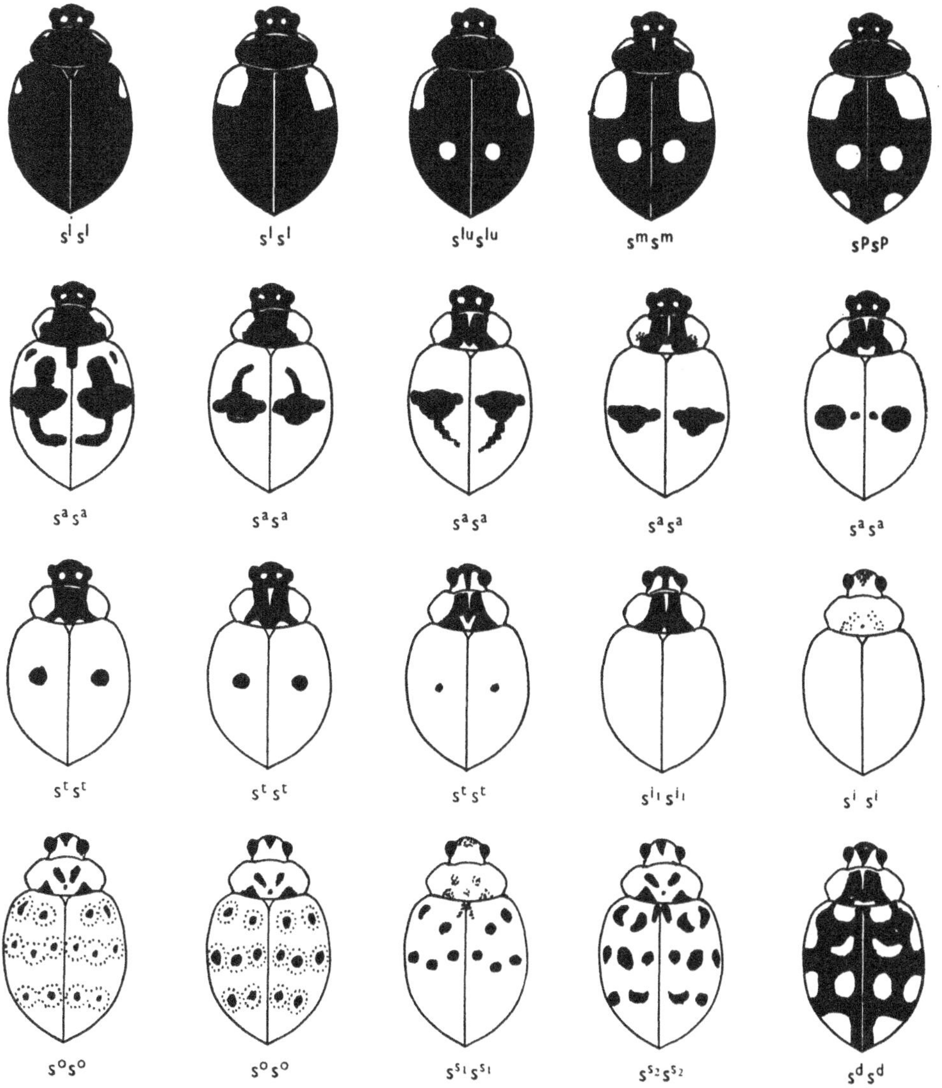


Fig. 3.09 Variability of *Adalia bipuncata*; homozygous forms (Lusis 1932)

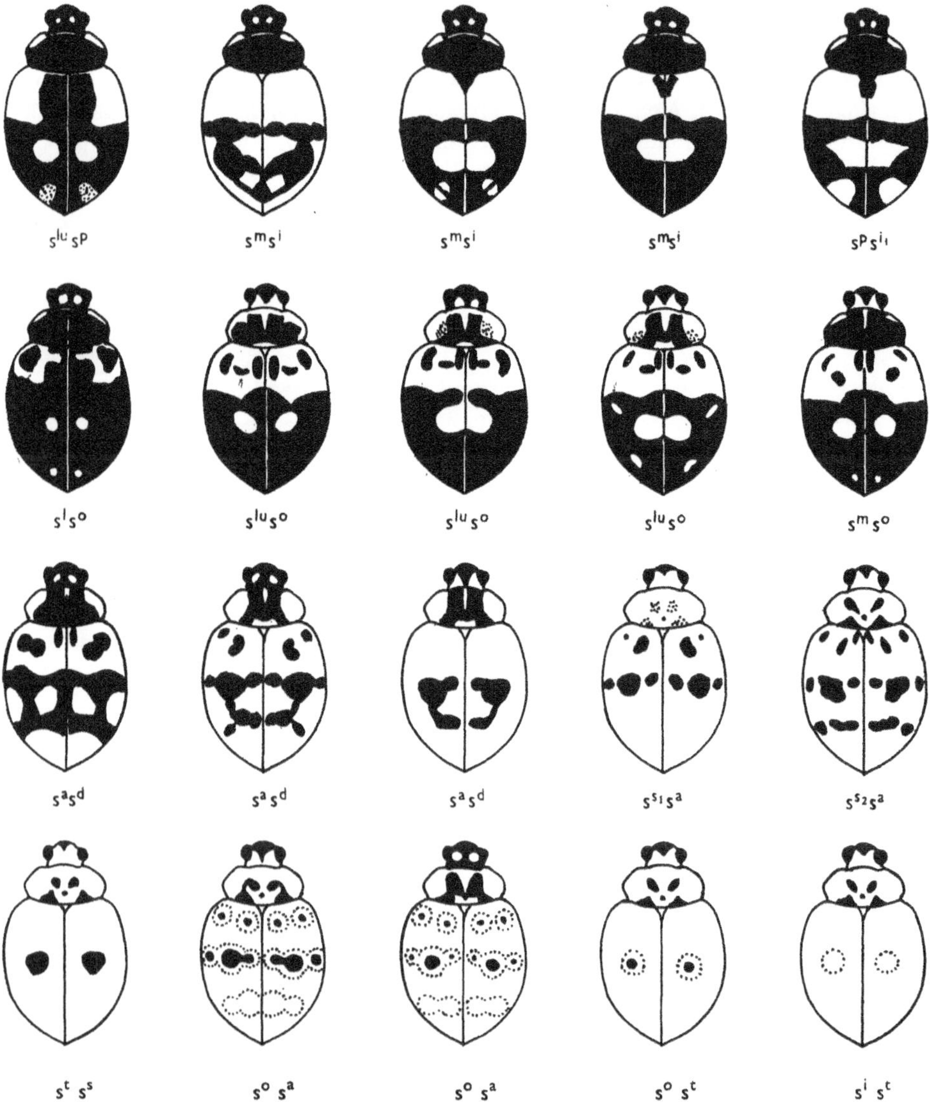


Fig. 3.09 Variability of *Adalia bipunctata;* heterozygous forms (Lusis 1932).

with two dark forms (*sexpustulata* and *quadrimaculata*) indicated that 'dark' is dominant over 'light'. This result however, cannot be generally applied to all forms. Lusis (1932) formed a series of twelve colour patterns on the basis of the homozygotous condition of twelve allelomorphic factors, (Fig. 3.09), and arranged allelomorphs according to dominancy:

$s^l$ $s^{lu}$ $s^m$ $s^p$ $s^i$ $s^{i1}$ $s^t$ $s^o$ $s^{s1}$ $s^{s2}$ $s^a$ $s^d$

This does not correspond to the sequence by which the total amount of melanistic pigment decreases:

(dark) $s^l$ $s^{lu}$ $s^m$ $s^p$ $s^d$ $s^a$ $s^{s1}$ $s^{s2}$ $s^o$ $s^t$ $s^{i1}$ $s^i$ (light)

Because of incomplete dominance, the colour pattern of heterozygotous combinations is intermediate (Fig. 3.09).

The heterozygote having alleles $s^l$ $s^{lu}$ $s^m$ $s^p$ combined with each other, or with $s^d$ $s^{s1}$ $s^{s2}$ $s^a$, tend to be dark. The combination of any allel with $s^i$ and $s^o$ considerably diminishes the quantity of deposited pigment. The homozygous combination *ocellata* ($s^o$ $s^o$) is lethal.

Both geographical and temporal variability in the proportions of the various biotypes have been shown to occur in natural populations. Populations from Central Asia, for example, include 27 % of form *impunctata*, 41 % of f. *typica* and only 18 % of dark forms, in contrast, European populations (Kiev-USSR, Germany) include 0 % of f. *impunctata*, 53—69 % of f. *typica*, and 28,5—44 % of dark forms (Dobzhansky 1924). This corresponds to general rules of geographical variability in coccinellids. Moreover the percentage of dark forms is very variable in European population. Lusis (1961) suggests that a larger percentage of dark forms live in humid coastal areas and also in industrial areas. The occurence of temporal variability has been verified by several observations. Dark forms prevail in summer and autumn; in spring the number of light forms increases markedly.

Several factors, which in summer and winter select for opposite tendencies play a role in maintaining polymorphism. A possible explanation (Lusis 1961) may be presented: It has been found that, in natural populations, the sexual activity of dark forms is greater. Thus, with 2—3 generations a year and frequent copulations, the number of dark individuals considerably increases towards the autumn. On the other hand, the dark forms have a much higher mortality during hibernation and thus the relative proportions get into balance again every spring. The advantage of dark forms lies in their better utilization of insolation for the increase of body temperature. This significant advantage is particularly apparent at foggy coasts and smoky urban localities with intermittent insolation. In consequence the rise in the percentage of the dark forms is most noticeable in both such areas.

*A. bipunctata* also has other interesting adaptations (Lusis 1947a, b). It has been discovered that the concentration of recessive alleles, which are lethal when homozygous, is high in all populations (inbreeding causes a very rapid decline in viability); practically every individual carries 1—3, and even more such lethal alleles. Therefore high heterozygosity appears necessary if the population is to be maintained. This heterozygosity is achieved partly by the behaviour of the species (frequent copulations) and partly by the fact that there occur in natural populations females whose progeny is exclusively female (male zygotes perish). Such females must then out-cross. Not only has this the advantage of reducing breeding between related individuals, but also the large production of females enables exceptionally rapid breeding in the spring. The percentage of females with only female progeny is constant within a given population and is evidently geographically variable. These females are more abundant in northern populations, which suffer a high mortality during hibernation. The small populations which emerge from hibernation are very vulnerable to inbreeding and also need to build-up rapidly. The heredity of this feature is very complicated. It is very likely that a complex of several dominant autosomal lethal genes, whose action is limited to the male sex, exists and that the effects of these genes are modified by maternal factors.

# 4 LIFE HISTORY AND BIOLOGICAL PROPERTIES

## 4.1 Developmental stages

### 4.11 Characteristics of individual stages

Coccinellids are holometabolous, i. e. they have a "complete metamorphosis", and pass through the following stages: egg, 4 larval instars, prepupa, pupa, adult.

**Egg.** — Eggs are usually oval or spindle-shaped, vary in colour from yellowish to reddish orange and are mostly laid in clusters with each egg attached to the substrate by the narrow end. These clusters are laid on the underside of a leaf or in bark-crevices. Some species, however, lay their eggs singly; for example, females of *Chilocorus rubidus* lay one egg at a time under the scale of a coccid larva (Pantyukhov 1968b). Females of *Exochomus flavipes* have occasionally been observed laying their eggs into empty pupae of their own species from which insect parasites had emerged (Geyer 1947). A few days before hatching the eggs become greyish.

**Larva.** — After hatching from the eggs, the larvae stay on the egg shells for up to one day, and often eat unfertilized nonviable eggs or larvae from later hatching eggs. Then they crawl about in search of their normal food. Usually 3 larval ecdyses separate the 4 larval instars. Species with 5 larval instars are rare, but this instar number has been observed in *Callicaria superba* (= *Pseudosynonycha japonica*) (Iwata 1932, 1965), and in the laboratory, in a certain proportion of larvae of *Coleomegilla maculata* (Warren and Tadić 1967) and *Chilocorus bipustulatus* (Yinon 1969). The larva stops feeding before ecdysis, uses its "anal organ" to fix itself with its head down to a surface and sheds its skin. Larvae of the genus *Chilocorus* aggregate for ecdysis (Fomenko 1970, Fomenko and Zaslavskiĭ 1970). Larvae of the Japanese species *Pseudoscymnus sylvaticus* live confined inside a closed aphid-gall. The gall only opens again when the newly emerging adults are ready to escape (Sasaji, in press). For several days before ecdysis to pupa, the 4th instar larva remains immobile and does not feed. Some authors separate this stage ("prepupa") from the other larval instars.

**Pupa.** — Different groups in the family *Coccinellidae* have different types of pupa. Coccinellids in the subfamilies *Coccinellinae* and *Sticholotinae* have an uncovered pupa— at the ecdysis to pupa the larval skin is sloughed from the pupa right up to the point where the cauda is attached to the substrate. The pupae in the tribes *Chilocorini* and *Noviini* are partly covered; they remain within the larval skin which splits lengthwise dorsally but is not shed. *Hyperaspini* and *Scymnini* have pupae completely covered by the larval skin. The pupa is not entirely immobile; if irritated, the head region is raised several times by upward jerks of the body. The colouration of pupae is much influenced by the environmental conditions. For example, *Coccinella septempunctata* may have light orange pupae under conditions of high temperature and low humidity (35 °C, 55%), whereas under the reverse conditions (15 °C, 95%) dark brown pupae are produced (Hodek 1958).

**Adult.** — The elytra of the emerged adult are at first soft, matt, light in colour and without pattern; the hind wings protrude from under them (Photo 5—7). The elytra acquire their normal appearance only gradually; most changes occur in a matter of hours, but the red colour remains a lighter shade for weeks or even months. It is thus easy to distinguish overwintered adults from the new generation for a long time.

## 4.12 Effect of temperature on the developmental rate

In common with all other insects the speed of development of coccinellids is dependent on the ambient temperature. Within the favourable range the developmental rate of all stages increases with temperature. As the upper threshold of thermal tolerance is approached, further temperature increases produce only a slight increase and above a certain thermal limit there is even a decrease in developmental rate.

Empirical values for the length of individual stages in coccinellids at different temperatures have mostly been presented only in the form of tables, and only values for embryonic development have been plotted as graphs.

A detailed survey of the efforts at a mathematical formulation of the dependence on temperature of developmental rate in poikilotherms can be found in an early review by Uvarov (1931) and in a series of physiological and ecological compendia, e. g. Wigglesworth (5th edition 1953), Allee et al. (1949), Andrewartha and Birch (1954), Schwerdtfeger (1963).

The simple classic formula suggested by Sanderson and Peairs (1913), Blunck (1914) and Peairs (1914) has been widely used by many workers, including Bodenheimer (1943) and Bodenheimer and Neumark (1955) for several coccinellids. The formula is based on the classic theory, originally produced by botanists, that the completion of any given stage in development requires the accumulation of a definite amount of heat energy. Because it is difficult to measure energy, the effective temperature (i. e. the temperature above the development threshold) is accumulated as day-degrees. The theoretical assumption, that the product of the effective temperature and time is always a constant, may be written as a formula: $D\,(T - a) = C$
where $D$ is the duration of the developmental stage at temperature $T$, $a$ is the development threshold, and $C$ is the thermal constant. This is the equation of an equilateral hyperbola, which — if calculated from two experimental results — is in agreement with the empirical data.

The reciprocal of development time, the speed of development, is expressed by the reciprocal of the hyperbola, i. e. by a straight line. This line crosses the temperature axis at the point "a" which is the theoretical lower development threshold.

The values of the development threshold and the thermal constant (for the periods of pre-imaginal development plus pre-oviposition period) for *Coccinella septempunctata* are 12.8 °C and 297 day-degrees (Bodenheimer 1943), for *Synharmonia conglobata* 12.4 °C and 323 day-degrees, for *Adalia decempunctata* 9.1 °C and 390 day-degrees and for *Chilocorus bipustulatus* 10.6°C and 697 day-degrees (Bodenheimer and Neumark 1955).

The linear relationship between temperature and the speed of development only holds for the middle range of temperatures; at either end of the favourable range the relationship is far from linear. The line expressing speed of development is in fact not straight but sigmoid. To get a close fit between the lines derived from the equations and empirical data over a greater proportion of the temperature range, other equations have been proposed (see the literature quoted above). Davidson (1942, 1944) found out

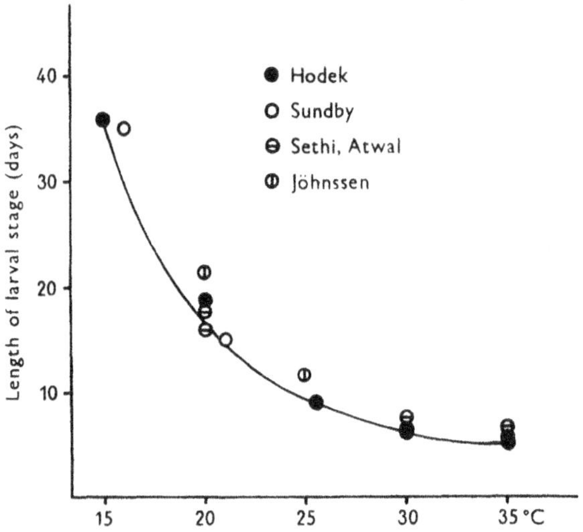


Fig. 4.01 Influence of temperature on the duration of larval development in *Coccinella septempunctata* (see also Tab. 4.01) (Hodek 1958).

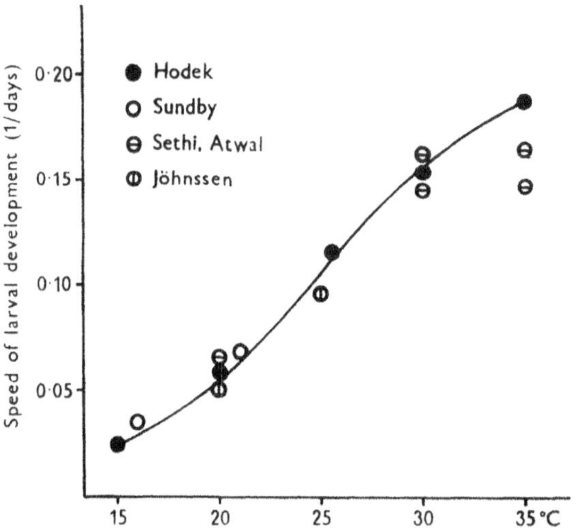


Fig. 4.02 Influence of temperature on the rate of larval development in *Coccinella septempunctata* (Hodek 1958).

that Verhulst's logistic curve si the most adequate description of the observed speed of development of insects, and the curve can be calculated directly from the empirical data.

The logistic equation may be written:

$$\frac{1}{y} = \frac{K}{1+e^{a-bx}}$$

where $y$ represents the duration of development at temperature $x$ (in °C), and $a$, $b$ and $K$ are constants. "$K$ defines the upper asymptote toward which the curve is trending; $b$ defines the slope of the curve and $a$ relative to $b$ fixes its position along the $x$ axis. This may be more easily comprehended by considering that the logistic curve is a bi-symmetrical sigmoid curve having a point of inflection whose coordinates are $1/y = K/2$ and $x = a/b$" (Andrewartha and Birch 1954).

Hodek (1958) found the logistic curve consistent with the empirical data for the preimaginal development of *Coccinella septempunctata* in the temperature range of 15—35 °C (Fig. 4.01 — 4.04).

One might expect that the relation to temperature varies, within one species, between populations separated very widely geographically. However, at least as far as the effect of the ambient temperature on the rate of development is concerned, this is not so even in the ecologically extremely plastic species *C. septempunctata*. The data for areas as far apart as Norway, central Europe and Punjab (India) are almost constant (Tab. 4.01).

This is even more striking when we take notice of the considerably different numbers found for central Europe at, for example, 25 °C by Jöhnssen (1930) and Hodek (1958). By contrast, only small differences between central Europe and Punjab have been reported by Hodek (1958) and Sethi and Atwal (1964).

Thus in central Europe (Hodek 1958) the rate of development was 1.8 days higher at 35 °C than at 30 °C (although the increase per degree was reduced in comparison with the central temperature range), whereas in the Punjab (Sethi and Atwal 1964) the total duration of pre-imaginal development was practically the same at both temperatures. As the upper limit of temperature tolerance is close to 35 °C, it is not surprising that some variability is evidenced.

The data of Ogloblin (1913) from Russia and Bodenheimer (1943) from Israel cannot be included in the comparison or the table, as the authors did not rear the coccinellids at constant temperatures.

Any differences between distant populations might be more likely to show themselves in survival near the upper temperature limit than in developmental rate. However, no data are available for such an interesting comparison.

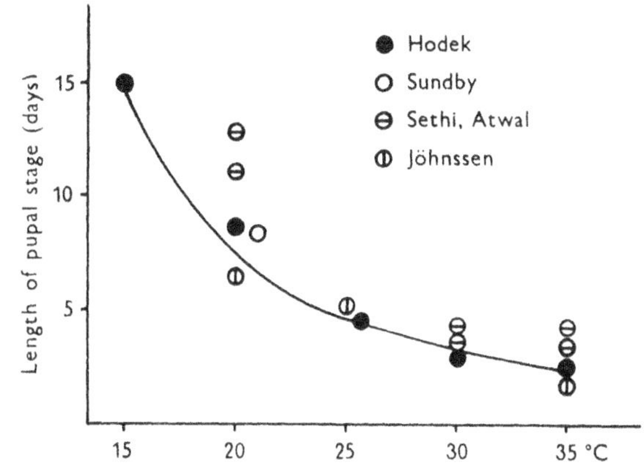


Fig. 4.03 Influence of temperature on the duration of pupal development in *Coccinella septempunctata* (see also Tab. 4.01) (Hodek 1958).

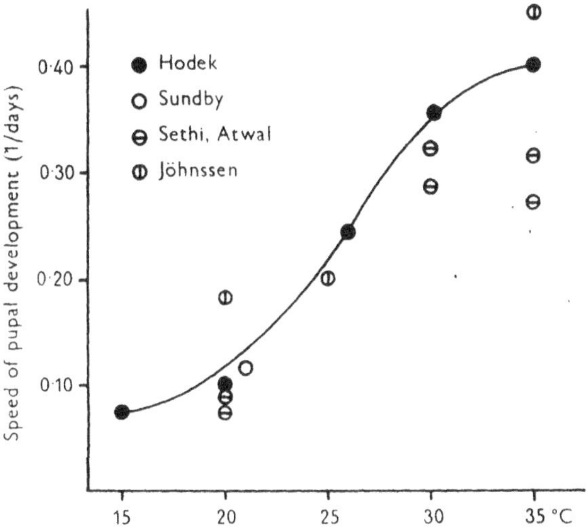


Fig. 4.04 Influence of temperature on the rate of pupal development of *Coccinella septempunctata* (Hodek 1958).

The east Asian race of *Coccinella septempunctata*, *C. septempunctata bruckii*, however, seems to be strikingly different from the nominate race in the dependence of developmental rate on temperature (Koide 1962). The curve for the dependence of the duration of larval plus pupal development on temperature is much more "recurved" than that of the nominate race (Fig. 4.05). The curves for the two races approach each other in the medium temperature range between 20 ° and 25 °C (and would identify between about 22 ° and 23 °C), but they separate from each other at both temperature extremes. The whole post-embryonal pre-imaginal development of *C. s. bruckii* is significantly slowed by 39 days when temperature decreases from 20 °C by 5° . At the other temperature extreme, development already slows from 30 °C so that at 33 ° it is 3 days longer than at 20 °. If the data can be repeated, this race would be strikingly more stenothermic than the nominate race.

The contrast between the ecophysiological constancy of the nominate race over such a large area of the world and the striking difference of the east Asian form has an analogy in the variability of colour patterns. Indian specimens from the Himalayas and Punjab

Tab. 4.01 Speed of pre-imaginal development of *Coccinella septempunctata* in relation to ambient temperature (Jöhnssen 1930, Hodek 1958—central Europe. Sethi and Atwal 1964 — northern India, Sundby 1966 — northern Europe)

| Author | Relative humidity [%] | Constant temperature [°C] | Duration of pre-imaginal stages (days) | | | | | N |
|---|---|---|---|---|---|---|---|---|
| | | | egg | larva | pupa | larva+pupa | all | |
| Hodek | (90) | 15 (14.5—16.3)[1] | 10.3 | 35.5 | 15.0 | 50.5 | 60.8 | 5[2] |
| Sundby | | 16 | | 35.0 | | | 60.4 | 30 |
| Johnssen | | 20 | 5.0 | 19—22 | 6 | | 30—33 | |
| Hodek | 80 | 20 (± 1.3) | 5.0 | 18.6 | 8.4 | 27.0 | 32.0 | 19[2] |
| Sethi, Atwal | 40 | 20 | 5.0 | 17.5 | 13.3 | 30.8 | 35.8 | 20 |
| | 70 | 20 | 4.8 | 17.2 | 12.9 | 30.1 | 34.9 | 20 |
| | 90 | 20 | 4.0 | 15.8 | 11.1 | 26.9 | 30.9 | 20 |
| Sundby | | 21 (± 1) | 5.0 | 15.0 | 8.5 | 23.5 | 28.5 | 67 |
| Jöhnssen | | 22—23 | 3.5—4 | 16—20 | 4—6 | | 25—29 | |
| Jöhnssen | | 25 | 3—3.5 | 10—11 | 4—5 | | 20—21 | |
| Hodek | 60—70 | 25.6 (±0.4) | (2.6) | 8.7 | 4.0 | 12.7 | 15.3 | 12[2] |
| Hodek | 60—90 | 30 (±2) | 1.9 | 6.7 | 2.9 | 9.6 | 11.5 | 18[2] |
| Sethi, Atwal | 40 | 30 | 2.0 | 6.5 | 3.0 | 9.5 | 11.5 | 20 |
| | 70 | 30 | 2.0 | 6.6 | 3.4 | 10.0 | 12.0 | 20 |
| | 90 | 30 | 1.7 | 7.3 | 3.7 | 11.0 | 12.7 | 20 |
| Jöhnssen | | 32—33 | 2 | 7—9 | 2.5—3 | | 12—14 | |
| Hodek | 60—90 | 35 (± 1.8) | 1.8 | 5.4 | 2.5 | 7.9 | 9.7 | 14[2] |
| Sethi, Atwal | 40 | 35 | 1.7 | 6.0 | 3.7 | 9.7 | 11.4 | 20 |
| | 70 | 35 | 2.0 | 7.1 | 3.1 | 10.2 | 12.2 | 20 |
| | 90 | 35 | 2.0 | 6.2 | 3.9 | 10.1 | 12.1 | 20 |

[1]) The temperature gradually increased during the two months of the experiment.
[2]) Number of adults hatched.

cannot be distinguished from the European specimens by the conspicuous enlargement of spots which is typical for the east Asian race (*C. s. bruckii*) from Japan and Korea (Günther 1958).

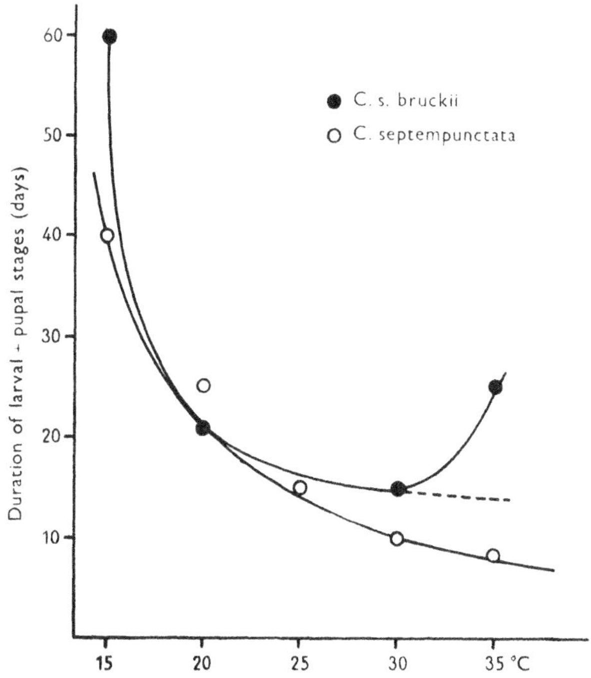


Fig. 4.05 Comparison of the effect of temperature on the postembryonal development of *Coccinella septempunctata* (Hodek 1958) and *C. s. bruckii* (Koide 1962).

## 4.2 Biological activities and properties of adults

**Copulation.** — Coccinellids copulate in the manner which is normal in beetles, i. e., the male uses its legs to hold itself onto the elytra of the female and is carried by her. Lim Sook Ming (1971) observed that newly emerged females (less than 1.5 — 2 days old) rejected the attempts of males to copulate: they moved rapidly away from the male, kicked the extended aedeagus with their legs and curved the last abdominal segments. The copulation posture may last as long as several days, as described by Iwata (1932) for *Aiolocaria mirabilis*.

The first copulation takes place a few days after emergence and is usually repeated several times during adult life, even though for most species (except for *Stethorus punctillum* which lacks a spermatheca (Moter 1959, Putman 1955), one copulation is generally reported as sufficient for permanent fertility of the female. Ellingsen (1969a) questions this assumption because a dying out of males was correlated with a decrease in the viability of eggs in *Adalia bipunctata*. In *A. bipunctata*, Sem'yanov (1970) proved that copulation had a stimulatory effect on the number of eggs laid. The maximum number of eggs was laid after the first copulation, oviposition then decreased but increased again following each copulation.

Diapausing coccinellids mate partly prior to diapause (when, apparently, some of the males die), but mainly at the end of hibernation before the aggregations disperse. The bringing together of the two sexes from different sites for hibernation probably has the adaptive advantages of preventing inbreeding in relatively common species and raising the probability that the sexes of rare species meet.

**Oviposition.** — Eggs are laid in the vicinity of prey, apparently because coccinellids oviposit where they have been feeding on essential prey. Other stimuli appear to be of minor value (Iperti 1965, Iperti in Hodek 1966, p. 121, Putman 1955). The close vicinity of the prey is not essential in the same way as it is for the oviposition of most syrphids. This feature causes a fairly high mortality of the first instar coccinellid larvae (see 6.31). Females bred on artificial diets do not lay eggs in Petri dishes, but only in cages containing parts of plants (Smirnoff 1958). There is considerable variability in the fecundity and size of egg batches of individual females.

**Longevity.** — The life span of coccinellids is also quite variable; in species with a long period of inactive life, longevity is about one year. There are, however, records of a second hibernation in *Calvia quinquedecimguttata* (Kanervo 1946), in *Propylaea quatuordecimpunctata* (El-Hariri 1966b), in *Stethorus punctillum* (Putman 1955) and in *Aiolocaria mirabilis* (Iwata 1932, Savoïskaya 1970b, d). Savoïskaya (1970b, d) ascertained that in *Harmonia axyridis* the longevity may reach three years in about 15—20 % of population. Females even oviposited in the third year, after the species was imported to the foot hills of Zailiĭskiĭ Alatau from the Far East.

# 5 DISTRIBUTION IN HABITATS

## 5.1 Terminology

Throughout this chapter, the word habitat is used frequently. Some information should therefore be given about this term, which is so often used but so poorly defined. Such explanation is necessary to make the chapter comprehensible in the context of the various different ecological terminologies.

In the German ecological literature a strict hierarchy for different types of "living space" (Lebensraum) has been established, the basic unit being the "Biotop" (as a living-place of a "Biocenosis", or "Zootop" for "Zoocenosis"), "Monotop", or "Demotop" (as a living-place of an animal or of a population). Smaller units than "Biotop" are "Stratum", a horizontal layer of a Biotop (e. g. in a forest: strata of tree crowns, of shrubs, of herbaceous plants, of surface of soil), or "Biochore" (point of concentration of individuals of a species) or "Merotop" (place where the conditions differ only slightly from the surroundings e. g. a slightly depressed or elevated place). A broader term is "Bioregion" (a living place for a "Biom", see Schwerdtfeger 1963).

Only the last two terms are used in the English ecological literature with the same meaning as in the German, although sometimes the words "biotic zone" are used instead of "bioregion" for the living-place of a "biome" (e. g. tundra, taiga, steppe etc.). The word "habitat" is used to include all German terms for lower categories, (i. e. all words with the ending "top") expanded by the prefixes "micro" and "macro". This seems a very useful device, as a classification precise enough to express the extremely complex structure of a "living-place" can really never be devised. The word "habitat" will be used in this book following Odum's (1953) definition: "habitat of an organism is the place where it lives". This use ignores the suggestion by Clements and Shelford (1939) that the word should be restricted to refer to "the physical and chemical factors that operate upon the community", and also Andrewartha and Birch (1954) who suggested the word should be replaced by the phrase "a place in which to live", which is cumbersome. In many standard textbooks and reviews the word is used in the sense defined by Odum, and for our purpose it is very useful that plants are also included in a habitat as a living-place for animals. The reader, accustomed to the German usage of terms, should replace the word "habitat" by the terms "Monotop" or "Zönotop" (Schwerdtfeger 1963), which appear to have a very similar meaning.

It seems, however, that the word "Monotop" also includes some aspects of the English term "ecological niche". As this latter term is often misinterpreted by non-English ecologists, it is worth quoting Odum's (1953) definition: "The ecological niche, on the other hand, is the position or status of an organism within its community and ecosystem resulting from the organism's structural adaptations, physiological responses, and specific behaviour. . . . . it may be said that the habitat is the organism's "address", and the niche is its "profession", biologically speaking."

The word eurytopic will be used to denote species which occur in many types of habitat (euryoecious species), and the word stenotopic will be used for the stenoecious species. Species will also be divided very roughly into xerophilous (inhabiting dry places), hygrophilous (humid places) and mesophilous (intermediate), according to their humidity requirements.

## 5.2 Typical habitat

Coccinellids show a wide range from stenotopic to eurytopic species. Many species of coccinellids are known to be firmly restricted to a certain, more or less strictly defined habitat. Among the common species, for example, the hygrophilous *Anisosticta novemdecimpunctata* is bound to vegetation of paludinous habitats (as marshes, fens and moist meadows), *Adalia conglomerata* to coniferous forests with preference for *Picea*, *Synharmonia conglobata* to deciduous trees, and the xero- and thermophilous species *Coccinula sinuatomarginata* to steppes and dry meadows. Only exceptionally is stenotopy so extreme, that the species is limited to one or a few plants in an extremely strictly defined habitat, e. g. *Coccinella hieroglyphica* is limited to forest peat bogs with *Calluna vulgaris*. By contrast, some species are very eurytopic and nearly ubiquitous, e. g. *Coccinella septempunctata* or *Propylaea quatuordecimpunctata*, although even these (esp. the former) breed preferentially in the herbaceous stratum of the various habitats they utilize.

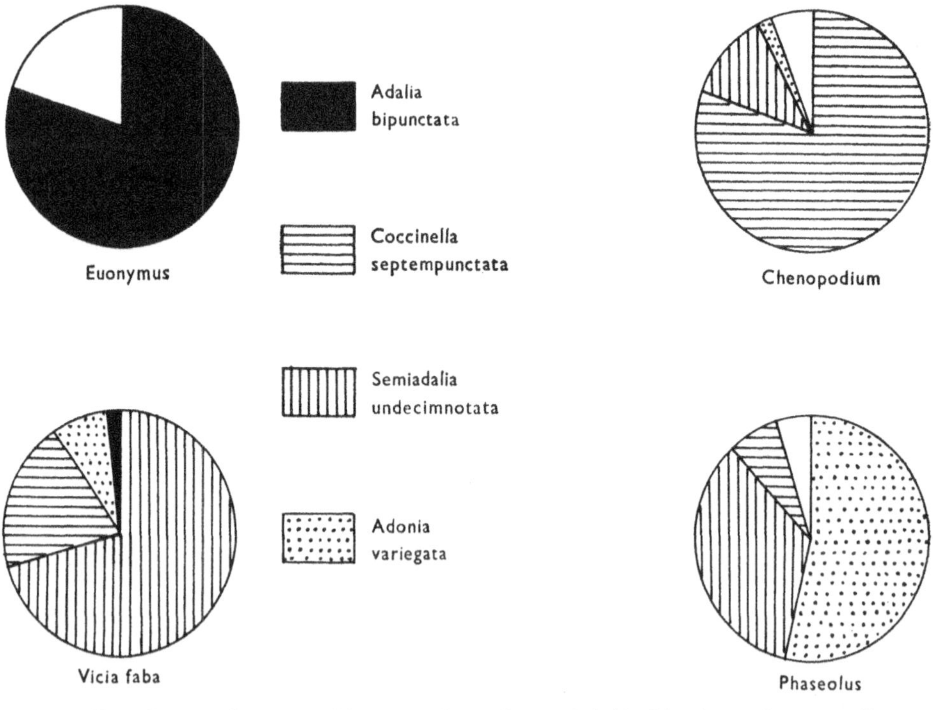


Fig. 5.01 Distribution of coccinellid larvae on host-plants of *Aphis fabae* in south-eastern France (Iperti, in Hodek 1966, p. 31).

The dependence of any species on a given habitat is mainly determined by the occurrence of essential prey and the physical conditions of the environment. In fact, for no coccinellid species is the interrelationship and differential importance of these two features of the habitat precisely known. There are some indications that for a number of species the microclimatic conditions may be of primary importance. Certain strata of vegetation are preferred by certain coccinellid species. Iperti (1965) has reported that in south-eastern France (1) *C. septempunctata* and *Semiadalia undecimnotata* prefer plants lower than 0.5 m; (2) *P. quatuordecimpunctata* and *Adonia variegata* prefer shrubs (0.5 — 2 m); (3) *Adalia bipunctata* and *Synharmonia conglobata* prefer trees (above 2 m). That this difference in distribution need not be a reflection of prey preference, is demonstrated by the different distribution of coccinellids on different plants infested with *Aphis fabae* (Iperti, in Hodek 1966, p. 31) (Fig. 5.01).

## 5.21 Criteria of the typical habitat

The usual criterion for considering a habitat typical for a particular species is that the latter will breed there (Tischler 1949). The adults of many coccinellids are not firmly restricted to a single habitat because of their great mobility and because they often visit non-typical sites while roving in search of food. The other criterion used for judging the relation of a species to a habitat is the abundance of adults present there. Often these two criteria will both apply in a given habitat.

In Karnevo's (1940) samples from *Alnus incana* (Hirvensalmi and Ruovesi, Finland), larvae were found whenever adults were abundant in four species: *Calvia quatuordecimguttata, C. quinquedecimguttata, Synharmonia conglobata* and *Propylaea quatuordecimpunctata.* The situation was different with two other species. In 1938 in Hirvensalmi, adults of *Coccinella distincta* were scarce, but were accompanied by larvae. On the other hand, *Halyzia sedecimguttata,* although the fourth commonest species in 1939 in Ruovesi, was present only as adults.

In North Dakota, *Hippodamia tredecimpunctata* and *H. convergens* were the most abundant species (76 and 18%) of all adult coccinellids (N = 617)[1)] on sweetcorn. All mature larvae found belonged to one of these two species (Havnvik and Frye 1969).

In the peach orchards of the Niagara peninsula (Ontario, Canada), Putman (1957, 1964) found ten species of "larger aphidophagous coccinellids". 46% of the adults were *Adalia bipunctata,* 22 % *Coccinella trifasciata perplexa,* 12 % *Coleomegilla maculata lengi* and 8 % *Coccinella transversoguttata richardsoni* (N = 888). 69 % of the larvae (N = 93) were *A. bipunctata,* other species present as larvae included *Coleomegilla maculata lengi,* and four others which only represented small fractions (1—4 %) among the adults found. Surprisingly no larvae of *C. t. perplexa,* the second most abundant species as adult, were found and no larvae of any other *Coccinella spp.* All fourth-instar larvae collected were *A. bipunctata;* this suggests that only this species completes its whole life-history in the peach orchards. The case of *C. t. perplexa* demonstrates that the abundance of adults is not a decisive criterion for evaluating the relationship of a species to a habitat.

[1)] Throughout this chaper, when percentage are quoted, the total number of coccinellids, of all species is given the form N = . . .

In 1949 and 1950 *Adonia variegata*, which became very abundant in the Ukraine, was found everywhere, including forests; the larvae, however, continued developing only in the fields. Of the numerous species found in the Ukrainian fields of cereals, only *C. septempunctata*, *A. variegata*, *C. divaricata* and less frequently *Coccinula quatuordecimpustulata* and *Hippodamia tredecimpunctata* produced larvae there (Dyadechko 1954).

The suitability of a habitat is also determined by reasons other than the presence of food or an adequate climate. Some plants may exert a mechanically deleterious effect on coccinellids. Putman (1955) found that the hooked trichomes on the foliage of scarlet runner bean, *Phaseolus coccineus*, tore the integument of *Stethorus punctillum;* the larvae were quickly killed and the longevity of adults was shortened by a laceration of the delicate membranes of the terminal abdominal segment. While studying the efficiency of *S. punctillum* as a predator of *Tetranychus cinnabarinus* in Israel, Plaut (1965) found that the coccinellid reproduced in several habitats (apple orchards, cotton, watermelon and sugar-beet fields), but on bush beans only adults without any larvae could be found. He supposed, similarly to Putman (1955), that the hooked trichomes on bush beans were deleterious for the larvae.

## 5.22 Adaptations to the typical habitat

In the stenotopic species, the continuing relationship with a particular habitat may be (and often is) reflected in specific morphological, physiological and ethological adaptations. These characteristics have seldom been studied experimentally and thus the causal relationship involved in the distributions of coccinellids remain almost completely hidden.

Climatic preference. — Ewert and Chiang, (in Hodek 1966, p. 195) tried to analyse the factors which might be responsible for the distribution of coccinellids among different microhabitats in corn fields. Apart from food relationship (see 5.41), distribution was governed by responses to physical factors. It was found that *Coleomegilla maculata* was brought to the lower strata of plants by a negative photoresponse and a positive response to high relative humidity, further increased after dehydration. The presence of *Hippodamia convergens*, which is tolerant to dessication, in the higher stratum of corn field (Fig. 5.02) was caused by its positive photoresponse and insensitivity to air moisture gradients. We may expect that strictly stenotopic species may show even more pronounced differences of this kind, e. g. a contrast between xerophilous and hygrophilous spp.

Changes in mobility. — In certain habitats flight can be an activity unfavourable for the survival of the species. The classic example is the fauna of small islands. Another example is given by *Spiladelpha barovskii kiritschenkoi* and *Coccinella reitteri* which live in exposed situations at high altitudes in the Zailiĭskiĭ Alatau (Kazakh SSR). Both species rarely fly, although they have normally developed wings (Savoĭskaya 1970a).

Adaptations to arboreal life. — A combination of morphological and ethological adaptations has evolved in arboreal species of coccinellids. Whereas the coccinellids which live on herbaceous plants usually fall from the plant very easily if disturbed, the arboreal species, particularly their larvae, stick to the surface. This is made possible both by morphological characters such as the form of the body and the pronounced adhesive organ on the 10th abdominal segment of the larvae (in *Adalia bipunctata* —

Putman 1964 or in *Anatis ocellata* — Kesten 1969), and by differences in behaviour. "The stout, short-legged larvae of *Coccinella spp.* tend to curl up and drop when disturbed. The more slender, active larvae of *Coleomegilla maculata lengi* and *Hippodamia spp.* ...... scramble about when disturbed and are easily dislodged. On the other hand, larvae of *A. bipunctata* cling tenaciously to leaves" (Putman 1964.)

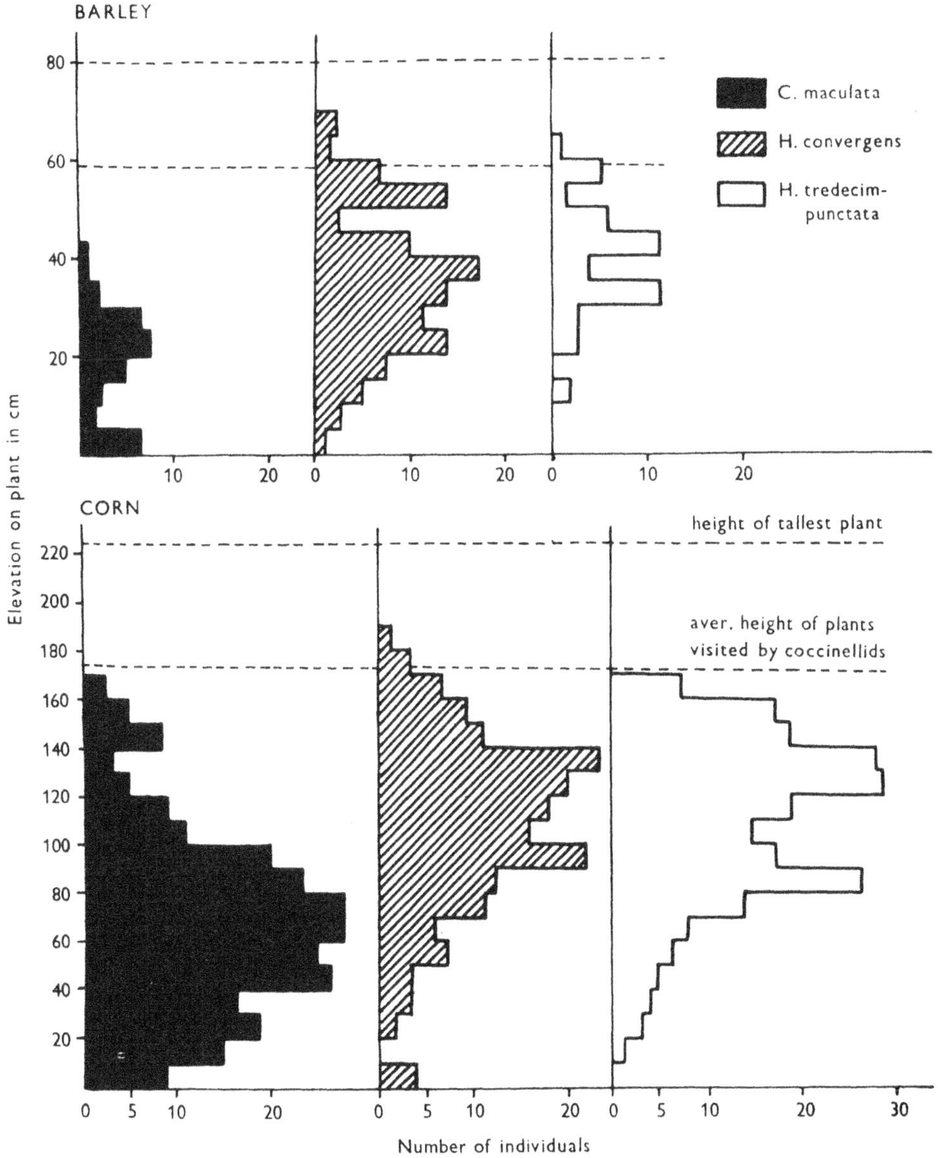


Fig. 5.02 Vertical distribution of three coccinellids on two crops (Ewert and Chiang 1966) (see also Fig. 5.04).

## 5.3 Changes of habitat with time

### 5.31 Sporadic migrations

It has long been well known that coccinellids sometimes occur in vast numbers. This phenomenon, which has often been a much discussed puzzle, has been partly explained by an understanding of the dormancy behaviour of *Coccinellidae* (see 7.3).

Some of the massive appearances of ladybirds, however, do not fit into a pattern of migration to or from dormancy sites or aggregation there. Such sporadic mass appearances have been convincingly explained by Hagen (1962). Favourable environmental conditions i. e. massive overpopulation of aphids and optimal weather may bring to realization the high potential fecundity of female coccinellids and lead to a large decrease in the mortality of larvae and pupae. When the young adults emerge (at an estimated density of e. g. about 54,000 individuals from one acre of alfalfa — Dickson et al. 1955) the aphids may be (and this has often been observed) disappearing from the habitat by the combined effect of their emigration as part of the normal seasonal life-cycle and the preying activity of the coccinellids. Due to a high competition for the remaining prey the mobility of the beetles increases tremendously upon emergence; a hot day may bring them into the air in enormous numbers and the air currents will blow them far away. This highly probable chain of events may indeed explain the nonperiodic appearances of coccinellids, whether migrating (Marriner 1939) or aggregating, both on mountains and in plains, or washed up on the seashore or edges of large lakes (see references in Hodek 1960a, Hagen 1962 and recent observations by Savoïskaya in Hodek 1966 p. 139). Generally only the aphidophagous coccinellids appear in such great numbers because of the ephemeral abundance of their prey in most habitats; with phyto-, myco- or coccidophagous species the same phenomenon is almost unknown.

### 5.32 Regular seasonal movements

The chain of events described above with reference to mass appearance of coccinellids also regularly occur on a lesser scale with many aphidophagous coccinellids living on herbaceous plants. These species, specialised to such an ephemeral prey as aphids, have a much higher tendency to dispersal than more polyphagous coccinellids. Ewert and Chiang in Hodek (1966, p. 195) ascertained in several ways that two aphidophagous *Hippodamia* species (*H. convergens* and *H. tredecimpunctata*) were much more mobile (15 times or 9 times, resp.) than *Coleomegilla maculata*, which feeds preferentially on pollen (see 6.111 and 6.12). Also the mobile aphidophagous species occurred in the higher stratum of corn fields, but *C. maculata* in the lower (see also 5.2).

When the prey becomes too scarce in one habitat, the coccinellids leave and search for food in neighbouring habitats. They may thus switch between several habitats with suitable prey during one vegetational season, in addition to their hibernation quarters. Sometimes, they also feed at the dormancy sites, or in their close vicinity.

In England, all the three abundant species (*Coccinella septempunctata*, *Adalia bipunctata*, *Propylaea quatuordecimpunctata*) produced their first brood on *Microlophium evansi* (= *Macrosiphum urticae*) infesting nettles, which are the most frequented breeding habitat of the coccinellids after emergence from hibernation quarters. Both the overwintering adults (zero generation) and their offspring (1st generation) then moved to bean plots infested with *Aphis fabae* and also bred there (Banks 1955).

In early spring *C. septempunctata* adults feed on aphids on bushes near their hibernation habitats in the large plains. Usually neither oviposition nor development of larvae occurs there. This contrasts strongly with *A. bipunctata*, which is found on the bushes together with *C. septempunctata* but is, of course, breeding there in its typical habitat. Gradually the overwintering adults of *C. septempunctata* move to the fields (mostly to forage crops at first, later to other crops) and oviposit. The larvae often manage to complete their development there. In central Europe, however, the emergence of new adults more or less coincides with a decline in the aphid population in the fields. Hence the young adults rove in search of food, mostly to weeds and frequently to more humid places (Hodek et al. 1962, 1966). Some authors even suggest (Telenga 1948) that *C. septempunctata* is attracted to the tree and shrub habitats at the forest edges, and thus return to near their hibernation quarters in its search for food.

Lusis (1961) has followed a similar gradual change of habitat in *Adalia bipunctata* in three regions (Riga, Leningrad, Moscow). This species first appears on *Padus* and fruit-trees in parks and orchards, then on ornamental shrubs and trees (*Rosa*, *Philadelphus*, *Caragana*, *Ulmus*) and still later on other trees as *Tilia*, *Populus* or *Salix*. After the disappearance of aphids from trees, *A bipunctata* moves to weeds (*Lappa*, *Artemisia*, *Chenopodium*, *Atriplex*).

In the foothills of south-eastern Kazakhstan (USSR), *A. bipunctata* shows a pronounced spring and early summer preference for orchard aphids, whereas in midsummer and autumn, when the aphids have disappeared from the fruit-trees, the coccinellids are quite evenly distributed over both herbaceous plants and trees (Savoĭskaya 1965). Right in the Zailiĭskiĭ Alatau-mountains (S. E. Kazakhstan), in midsummer, a part of the abundant early summer population of this species migrates from deciduous to coniferous forests (Savoĭskaya 1970a). In Georgia (USSR), *Adalia bipunctata* produces the first generation in peach orchards; afterwards the beetles move to *Prunus*, following the migration of *Myzus persicae* (Aleksidze 1970).

In early summer, *Propylaea quatuordecimpunctata* is the most abundant and economically most important coccinellid species in the stone-fruit orchards infested with *Hyalopterus pruni* and *Anuraphis persicae* in the mountains near Alma-Ata. In midsummer, *P. quatuordecimpunctata* moves from the fruit-trees to herbaceous plants and becomes very numerous on mountain meadows (Savoĭskaya 1965, 1970a).

A similar change of habitat, due to shortage of food, is undertaken by two aphidophagous coccinellids in a neighbouring region of central Asia (USSR, Karakalpakistan) (Yakhontov in Hodek 1966 p. 35). In spring they live in the orchards; the larvae of *Synharmonia conglobata* mainly feed on *Hyalopterus arundinis* and *Myzus persicae*, and the larvae of *Coccinella undecimpunctata* on *Aphis pomi*. In mid-July the aphids almost completely disappear from the fruit-trees, some of them migrating to their secondary host-plants, e. g. *H. arundinis* to *Phragmites*. The search for food leads to massive migrations by the second generation of *S. conglobata* to herbaceous plants or to other trees infested with aphids e. g. to poplars with *Chaitophorus jaxarti* and to *Salix* with *C. shaposhnikovi* or *C. saliceti*; *C. undecimpunctata* disperses to cotton and alfalfa fields (with *Aphis gossypii* or *Acyrthosiphon gossypii* and *Therioaphis ononidis* or *Acyrthosiphon pisum*).

Also in south-east France, many coccinellids can only utilize those habitats which have the most favourable microclimatic conditions during the spring period, when plenty of aphids can be found on all plants. In summer, scarcity of aphids forces them to search for food in non-typical habitats (Iperti 1965). Sometimes they can only find alternative food there. Thus, for example, *Adalia bipunctata*, which reproduces in the spring on trees (mainly fruit-trees: cherry, plum, peach, apple, but also shrubs of rose), can be found on *Zea mais* and *Phragmites communis*, but only as adults. This alternation of habitats is similarly undertaken by *Adalia decempunctata* and *Synharmonia conglobata.*

Thus in *A. bipunctata* (and in ecologically similar species) a similar alternation of habitats occurs not only in regions which have similar seasonal changes of climate (French Riviera, the foothill zone in central Asia), but also in northern regions of the European USSR with both sea and continental climate. This similar habitat alternation in spite of differing climates demonstrates convincingly that the alternation of habitat by aphidophagous coccinellids results from the seasonal changes of host-plants used by their prey — aphids. As the coccinellids do not find their essential prey in many of the habitats they visited, they frequently occur only as adults (see also 5.21 and 6.113).

Coccinellids with typical habitats other than the fruit orchards also alternate their sites of active life throughout the vegetational season. Sometimes both sites may be typical habitats. Both adults and preimaginal stages of *Coccinella divaricata* occur on coniferous trees infested with aphids in late summer, while normally the species lives on herbaceous plants near rivers in the Zailiĭskiĭ Alatau-mountains (Savoĭskaya 1970a).

*Coccinula quatuordecimpustulata* and *C. sinuatomarginata* disappear from the mountain steppes in mid-summer. One part of the populations migrates down to the agricultural zone to alfalfa fields while the other part flies up to the herbaceous habitats of the forest zone (Savoĭskaya 1970a).

In the subtropical and tropical regions, low humidity due to high temperature can result in distributional changes of coccinellids.

In the Nile delta regular habitat change by *Coccinella undecimpunctata* has been observed (Wiesmann 1955) between "bersem" (a kind of *Trifolium*), cotton and maize. After a period of inactivity in their hibernation sites (XII — I) the coccinellids reproduce on "bersem" until mid-May, when irrigation of this crop ceases. The drying of "bersem" induces the migration of all insects including *C. undecimpunctata* to cotton where, however, no reproduction was observed in 1951. From mid-May to late August the predators migrate to maize where they reproduce. The drying of maize causes the predators to move to shelter. The migrations in Egypt of *C. undecimpunctata aegyptiaca* from cotton to maize are reported also by Ibrahim (1955a).

During the hot period of July — September when the temperatures exceed 40 °C, the distribution of *Chilocorus bipustulatus* var. *iranensis* (introduced from Iran to Mauretania, W. Africa — Iperti et al. 1970, Laudého et al. 1970) becomes reduced to the relatively more humid places near permanent water.

Brown (1969) observed that in the Orange Free State coccinellids dispersed to other crops (such as sorghum, radish, maize, peaches, garden ornamentals) in order to survive periodic disruption of the habitat of wheat crop by harvest. The most abundant species *Lioadalia flavomaculata* was able to breed continuously for 8—10 generations per year because of its regular seasonal movements to its most important winter habitat (Japanese radish, *Raphanus sativus*) in late summer.

Bielawski (1961, p. 514) assumes that the "oversaturation" ("przesycenie") of the habitat by populations of different coccinellid species is the reason why they alternate habitats. In this connection he also refers to a "waiting-period" ("okres wyczekiwania") in the life-cycle of *Adalia bipunctata, A. decempunctata, Coccinula quatuordecimpustulata* and *Subcoccinella vigintiquatuorpunctata.* In both *Adalia spp.* a "waiting-period" of about one month has been observed, when the beetles remain on young pines until the time of the formation of leaves on oaks (*Quer-*

*cus*), usually in the course of April. They are thus "waiting" actively (feeding, copulating) in the habitat where they hibernated. The other two species "wait" before hibernation at the same dormancy habitat, sitting inactively among needles of young pines under which they will hibernate in the litter until the first frosts. The "waiting" period of *C. quatuordecimpustulata* and *S. vigintiquatuorpunctata* is, however, connected more with the onset of diapause than with any "oversaturation of the habitat" where they spend their active life.

## 5.33 Influence of adjacent habitats

The mobility of some coccinellid species is rather great, and is increased markedly by high ambient temperature, so that often very distant habitats may be reached not only in migrations connected with dormancy, but also in dispersal during search for food (see the previous chapter 5.31 and 5.32). The larvae are of course much less mobile,[1] and also the dispersion of adults may be delayed and slowed down, if they find food nearby, e. g. after emergence from hibernation quarters, particularly if the temperature is low. Hence the occurrence of coccinellids in a given habitat may be influenced to a great extent by the neighbouring habitats.

The proximity of the hibernation sites — whether these be forest edges (Fenjves 1945, Galecka 1966) or hedgerows (van Emden 1965a, b, c) — have a fundamental effect on the occurrence of coccinellids on agricultural areas. With low aphid infestations of potato fields (Galecka 1966) or brussels sprouts (van Emden 1965c), there was an increase in the population density of aphids and a decrease in coccinellid numbers toward the centre of the field away from the adjacent shelter. At high aphid populations, this correlation becomes obscured by the dependence of coccinellids on the population density of aphids.

In England, the population density of coccinellids on three bean plots infested with *Aphis fabae* has been shown to be dependent on the distance between the plots and both grass, which appeared to be the hibernation site of *Coccinella septempunctata* (Fig. 5.03, I, II) and patches of nettles (Fig. 5.03, a, b, c) where overwintered adults of *Adalia bipunctata* and *C. septempunctata* first concentrated. The shelter afforded by trees and buildings hindered the access of the beetles to the beans in plot A. Plot C always had a high population of coccinellids; this was attributed to its situation between the two nettle sites rich in coccinellids (Tab. 5.01, 5.02) (Banks 1955).

Galecka (1966) studied aphids and their predators on four potato fields in Poland for two seasons. In two fields, situated 0.5 km from each other in an unwooded region near Warsaw, there were both less species and specimens of coccinellids than in the two other fields situated in the forest area of Kampinos; one was at the forest edge and the other 1.5 km from the forest on the bank of the river Vistula. In the two fields of the unwooded region about 40 and 60 coccinellids were collected, belonging to three eurytopic species: *Coccinella septempunctata*, *C. quinquepunctata* and *Propylaea quatuordecimpunctata*. In the forest area, the field situated near the river yielded 90 specimens, and apart from the above three species, three more were found. The presence of the hygrophilous *Hippodamia tredecimpunctata* is explained by the vicinity of the river. The field at the forest edge had greatest numbers of coccinellids, as 120 were collected. As well as three eurytopic species, *Adalia bipunctata* was also present and reflected the neighbourhood of the tree-habitat. The two fields in the wooded region were inhabited also by two inhabitants of drier herbaceous habitats: *Coccinula quatuordecimpustulata* and *Adonia variegata*. This enrichment of the coccinellid fauna of potato fields by steppe species in the forest region is rather

[1]) A rather high mobility of coccinellid larvae has been demonstrated by the capture in pitfall traps of large numbers of larvae of *Anatis ocellata*, *Adalia conglomerata* and *Neomysia oblongoguttata* under an old spruce stand with no ground vegetation (Klausnitzer and Bellmann 1969; see also 5. 51).

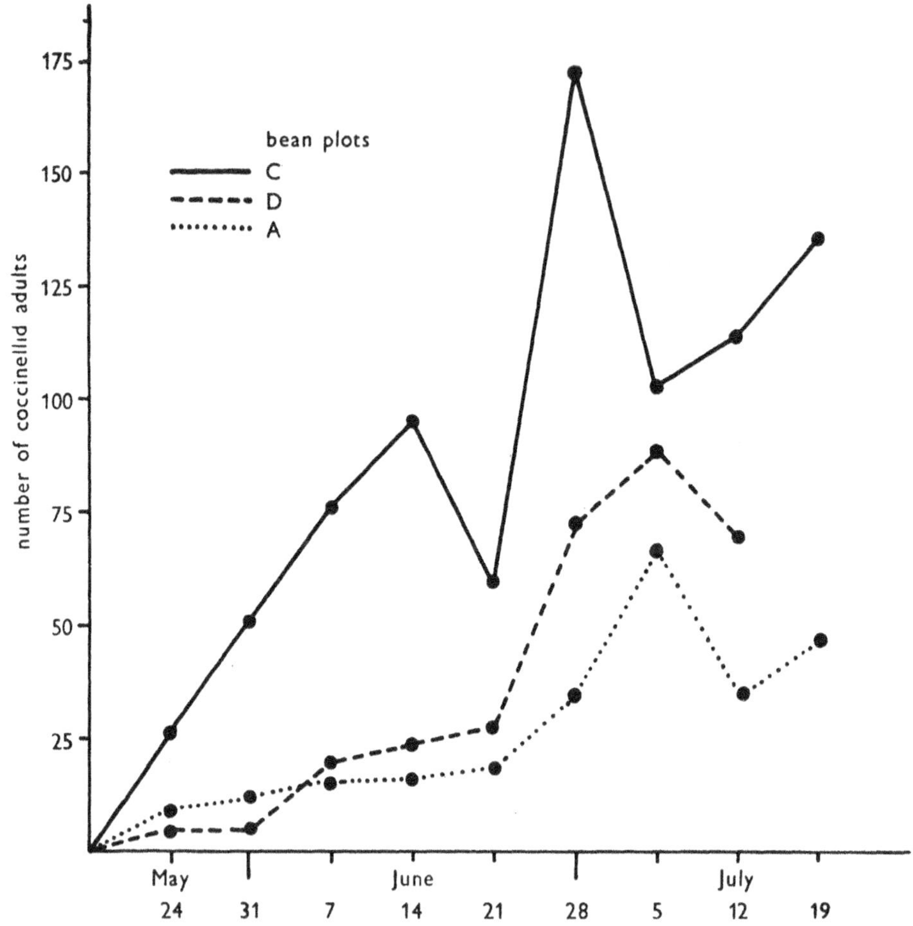


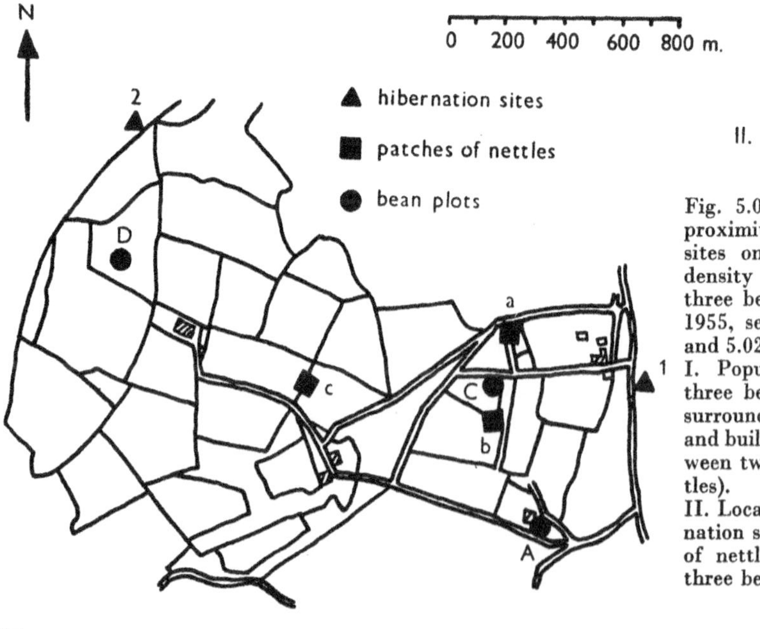


Fig. 5.03 Effect of the proximity of hibernation sites on the population density of coccinellids on three bean plots (Banks 1955, see also Tab. 5.01 and 5.02).
I. Population density on three bean plots (plot A surrounded by tall trees and buildings, plot C between two patches of nettles).
II. Location of the hibernation sites (▲), patches of nettles (■) and the three bean plots (●).

surprising. As there does not seem to be a difference in climate between the two regions, the presence of these two species may be assumed to be caused by provision of suitable quarter for hibernation in the nearby forest. Both species are known to hibernate in the litter of the forest edges in eastern Europe (Dyadechko 1954, Bielawski 1961).

Tab. 5.01 Maximum numbers of overwintered coccinellids at Rothamsted counted during April and May 1952 (Banks 1955)

| Locality | Habitat | No. | *C. septempunctata* | *A. bipunctata* | *P. quatuordecimp.* | *C. undecimpunctata* | Total |
|---|---|---|---|---|---|---|---|
| Harpenden Common West | grass | 1 | 111 | 0 | 0 | 12 | 123 |
| Barnfield | grass | 2 | 40 | 0 | 0 | 0 | 40 |
| Great Field III | nettles | a | 37 | 218 | 13 | 0 | 268 |
| Great Field II | nettles | b | 27 | 60 | 9 | 0 | 96 |
| Broadbalk | nettles | c | 108 | 137 | 47 | 0 | 292 |

Apart from providing hibernation shelter, adjacent uncultivated land may serve both as a source of prey, when the pest species have become scarce in the field, and as a source of re-inforcement of active natural enemies after the outbreak of the pest on crop. Uncultivated land may, therefore, enable beneficial insects to bridge failures in their synchronization to pest species (van Emden 1965a).

Adjacent parts of a crop may also have a similar function. Thus van den Bosch et al. (1959) report that adult coccinellids, moving from adjoining parts of the same crop, invaded those alfalfa plants where a quick build-up of aphids had occurred. This finding led to the idea of "strip cutting" of alfalfa (Schlinger and Dietrick 1960, see below and 9.225). By contrast, Skuhravý and Novák (in Hodek 1966 p. 167) could not show a similar movement of coccinellids when alfalfa was cut adjacent to aphid infested sugar-beet in central Bohemia. They failed to find an invasion of coccinellids from the mown alfalfa rich in coccinellid beetles, when a 300 m wide edge strip of the sugar beet field was sampled 3 and 8 days after the alfalfa had been mown. They believed that, even after the mowing of the alfalfa field, the coccinellids still found enough food there.

Uncultivated land is important in determining the effectivity of coccinellids and other natural enemies on crops. Van Emden (1965a) has raised the question whether pest problems in Great Britain are so much less than those in the United States for the very reason that in England crop fields are normally islands surrounded by large areas of less cultivated land. The study of these hitherto little known interrelations could lead to new methods of integrated control. The evaluation of the function of the adjoining crops (or the adjacent part of the same crop) as the source of mobile natural enemies — which include the vast majority of coccinellids — has already led to the introduction of strip-cutting, or strip-treating with pesticides; both methods seem very promising for integrated control (see 9.225).

Tab. 5.02 Total numbers of *Coccinellidae* of all species counted weekly on the bean plots A, C and D (Banks 1955)

| Plot | Week | | Adults | | | | | Eggs | Larvae | Pupae |
|---|---|---|---|---|---|---|---|---|---|---|
| | no. | 4th day | *A. bipuncta-ta* | *P. quatuor-decimp.* | *C. septem-punctata* | Others | Total | | | |
| A | 7 | 24.5. | 4 | 5 | 0 | 0 | 9 | 154 | 0 | 0 |
| | 8 | 31.5. | 2 | 10 | 0 | 0 | 12 | 327 | 13 | 0 |
| | 9 | 7.6. | 12 | 4 | 0 | 0 | 16 | 232 | 26 | 0 |
| | 10 | 14.6. | 13 | 3 | 0 | 0 | 16 | 571 | 51 | 0 |
| | 11 | 21.6. | 17 | 0 | 0 | 0 | 17 | 270 | 92 | 0 |
| | 12 | 28.6. | 25 | 4 | 3 | 0 | 32 | 1293 | 162 | 26 |
| | 13 | 5.7. | 47 | 10 | 3 | 5 | 65 | 979 | 276 | 70 |
| | 14 | 12.7. | 20 | 9 | 0 | 2 | 31 | 538 | 281 | 31 |
| | 15 | 19.7. | 29 | 4 | 10 | 1 | 44 | 21 | 146 | 32 |
| Totals | | | 169 | 49 | 16 | 8 | 242 | 4385 | 1047 | 159 |
| C | 7 | 24.5. | 13 | 13 | 1 | 0 | 27 | 456 | 0 | 0 |
| | 8 | 31.5. | 31 | 17 | 3 | 0 | 51 | 537 | 29 | 0 |
| | 9 | 7.6. | 61 | 12 | 3 | 0 | 76 | 1392 | 28 | 0 |
| | 10 | 14.6. | 76 | 6 | 11 | 0 | 93 | 2444 | 261 | 0 |
| | 11 | 21.6. | 49 | 2 | 4 | 4 | 59 | 1883 | 330 | 4 |
| | 12 | 28.6. | 115 | 23 | 34 | 1 | 173 | 3527 | 687 | 85 |
| | 13 | 5.7. | 53 | 31 | 15 | 3 | 102 | 1193 | 445 | 328 |
| | 14 | 12.7. | 83 | 28 | 6 | 2 | 119 | 215 | 402 | 130 |
| | 15 | 19.7. | 96 | 2 | 32 | 2 | 132 | 6 | 103 | 63 |
| Totals | | | 577 | 134 | 109 | 12 | 832 | 11653 | 2285 | 610 |
| D | 7 | 24.5. | 0 | 4 | 2 | 0 | 6 | 249 | 13 | 0 |
| | 8 | 31.5. | 4 | 1 | 0 | 0 | 5 | 180 | 15 | 0 |
| | 9 | 7.6. | 11 | 8 | 0 | 0 | 19 | 324 | 51 | 0 |
| | 10 | 14.6. | 20 | 3 | 0 | 1 | 24 | 616 | 39 | 0 |
| | 11 | 21.6. | 20 | 3 | 3 | 1 | 27 | 527 | 99 | 1 |
| | 12 | 28.6. | 46 | 16 | 7 | 2 | 71 | 2922 | 270 | 58 |
| | 13 | 5.7. | 61 | 10 | 12 | 5 | 88 | 1794 | 249 | 112 |
| | 14 | 12.7. | 38 | 27 | 1 | 2 | 68 | 323 | 348 | 105 |
| Totals | | | 200 | 72 | 25 | 11 | 308 | 6935 | 1084 | 276 |

## 5.34 Succession of species as a habitat ages

Changes in the condition of annual plants (and herbaceous plants generally) are so quick that it is virtually impossible to study their direct effect on the spectrum of coccinellid species, and moreover such effects would certainly be masked by changes in aphid infestation. The succession could be more easily followed on trees, but even so scarcely any such studies have been undertaken.

In central Europe, only the young stands of pines have usually been sampled for coccinellids (Bielawski 1961, Klausnitzer 1967), and the larvae of *Myrrha octodecimguttata* have never been found; Bielawski did not even find any adults during the vegetation season. Still this species, which hibernates in bark crevices at the base of large pines (Bielawski 1961, Pulliainen 1966) (see 7.324), has been supposed to be typical for stands of pines. Klausnitzer (1968) investigated samples of pine crowns of mature trees and found a much higher incidence of *M. octodecimguttata* adults (27 %) (N = 188) than on young pines (1 %). He believes that the species breeds in the crowns of older pines and considers that the stratum of crowns in old pines[1] is a typical habitat for *M. octodecimguttata*. *Scymnus nigrinus*, the second most abundant species in a young Pinetum, but which also occurs in other habitats such as deciduous forests, meadows and fields (see 5.5), is extremely scarce in crowns (Tab. 5.03). However, total number of individuals of both *Scymnus spp.* in young pine trees and crowns is remarkably similar. The near absence of *S. nigrinus* in crowns is compensated by a very high incidence of *S. suturalis*. The fauna of pine crowns is also characterized by a complete absence of the other eurytopic species, *Coccinella quinquepunctata*, *Chilocorus bipustulatus* and *Propylaea quatuordecimpunctata*. The first of these species is apparently a more or less incidental visitor of young pines, while the last has been reported as breeding (Bielawski 1961) in this habitat. The rather eurytopic *C. bipustulatus* seems to prefer the deciduous shrubs and trees (according to the survey of faunistical data from eastern Europe) (see 5.51). The hypothetical assumption that *C. bipustulatus* breeds in pine crowns (Bielawski 1961) has been disproved by its absence there (Klausnitzer 1968). The coccinellid fauna of pine crowns also differs from that of young pines by the lower occurrence of *Exochomus quadripustulatus*. Other differences may be incidental (N = 188).

Gumoś and Wiśniewski (1960) also compared young (about 10 years old) and old (about 40 years old) pine stands. They apparently did not sample Scymnini and included autumn samples. They thus obtained high numbers of *C. septempunctata* in the young stand (70 %). Their results are interesting, for they show large catches of *Anatis ocellata*, four times higher in the old forest (81 %), and the presence of *Halyzia sedecimguttata*.

We may conclude that in central Europe the fauna of young pines basically contains more species with a broader ecological range than does that of the crowns of old pines which are inhabited by less species, bound more strictly to the Pinetum and rather stenotopic.

The succession of coccinellid species (Tab. 5.04) in the course of ageing of red pine stands infested with *Schizolachnus piniradiatae* in central Ontario, Canada, is apparently brought about by a gradual decrease in prey density (Gagné and Martin 1968). *Cocci-*

[1])Placing the discussion of the effect of the tree age is somewhat arbitrary as it could equally well be discussed in 5.41 as a difference between strata. However stands of older pines do not usually have young pines growing among them — thus, strictly speaking, young pines form rather a habitat more distinct from old pines than applies with young and older trees in a mixed deciduous woodland.

**Tab.** 5.03 Coccinellids on pines [in %]

| Coccinellid species | Poland (15 km N. E. of Poznan) Gumos and Wisniewski 1960 | | Poland (Warszawa-Bielany) Bielawski 1961 | German Dem. Republic Klausnitzer 1967b, 1968a | |
|---|---|---|---|---|---|
| | 9—11 years old 3 m high | 37—41 years old 16 m high | cca 15 years old[1]) | young pines[1]) N = 1465 | crowns of old pines N = 188 |
| *Subcoccinella vigintiquatuorpunctata* | | | 1.6 | | |
| *Stethorus punctillum* | | | 1.0 | | |
| *Scymnus auritus* | | | 2.1 | | |
| *S. suturalis* | | | 23.7 | 33.5 | 60.2 |
| *S. nigrinus* | | | 22.6 | 29.0 | 0.5 |
| *S. frontalis* | | | 1.0 | | |
| *Aphidecta obliterata* | | | 1.6 | | |
| *Adalia decempunctata* | | | 2.1 | | |
| *A. bipunctata* | | | 1,3 | | |
| *Coccinella septempunctata* | 70 | 4 | 5.2 | 4.8 | 2.1 |
| *C. quinquepunctata* | 2 | | 1.6 | 5.3 | |
| *Coccinula quatuordecimpustulata* | 4 | | 1.6 | | |
| *Synharmonia conglobata* | | | 0.3 | | |
| *Harmonia quadripunctata* | | | 2.3 | 4.4 | 1.6 |
| *Myrrha octodecimguttata* | | | | 1.0 | 27.1 |
| *Propylaea quatuordecimpunctata* | | | 27.3 | 6.2 | |
| *Neomysia oblongoguttata* | | 2 | | | |
| *Anatis ocellata* | 20 | 81 | | 1.7 | 4.3 |
| *Halyzia sedecimguttata* | 2 | 11 | | | |
| *Chilocorus bipustulatus* | | | | 2.7 | |
| *Exochomus quadripustulatus* | | | 4.4 | 9.2 | 1.1 |
| Other species | 2 | 2 | | 2.2 | 3.1 |

[1]) Only the samples from May — July included, to exclude the coccinellids for which the habitat served as a dormancy site only.

Tab. 5.04 Occurrence of *Coccinellidae* in red pine stands of different age [1]) (Gagne and Martin 1968)

| Species | Age of the stand | | | | |
|---|---|---|---|---|---|
| | newly planted | 5 years | 15 years | 26 years | 36 year |
| *Hippodamia parenthesis* | —[2]) | +[2]) | — | — | — |
| *Coccinella novemnotata* | — | + | (+)[2]) | — | — |
| *C. transversoguttata* | — | + | + | — | — |
| *C. trifasciata* | — | + | + | (+) | (+) |
| *Scymnus lacustris* | — | + | + | (+) | (+) |
| *Anatis mali* | — | (+) | + | + | + |
| *Mulsantina picta* | — | (+) | + | + | + |
| *Cycloneda munda* | — | — | (+) | + | + |
| *Hippodamia convergens* | — | — | + | — | — |

[1]) Only species breeding at least at one age are included.
[2]) — absent, + present and breeding, (+) present.

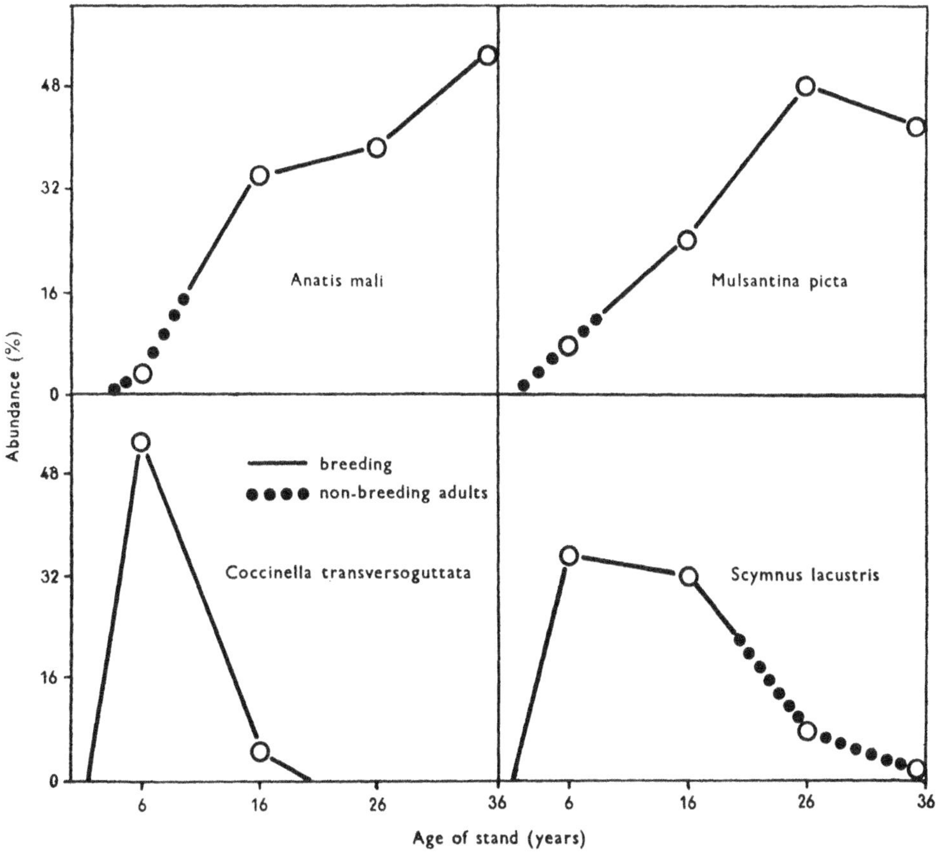


Fig. 5.04 Changes in relative abundance of coccinellids in red pine plantations of different age (Gagné and Martin 1968).

*nella transversoguttata* and *Scymnus lacustris* appear dominant in younger stands (Fig. 5.04), whereas *Mulsantina picta* and *Anatis mali* are the most common species in the older plantations. Survival of a coccinellid species in older stands with a lower prey density seems to require a greater searching capacity and mobility of the larvae. In fact, the larvae of *M. picta* and *A. mali* can be observed searching rapidly, while the larvae of *S. lacustris* are "slow moving and practically sedentary".

In Israel, the population level of *Chilocorus bipustulatus* appears considerably higher in mature citrus groves (23—30 years), than in young ones (7—9 years) throughout the year (Rosen and Gerson 1965). The same preference for older, more shaded groves has been found in date palm plantations (Kehat 1967a, 1968b). This feature, the depression of the population in summer, and the absence of *C. bipustulatus* in the warmer and drier regions of southern Israel (see 5.42), all are apparently related to the low resistance of the species to high temperature and/or low humidity.

## 5.4 Spatial differences

### 5.41 Differences between strata within a habitat

There is a great need for data on this subject. An isolated study has been undertaken by Ewert and Chiang (in Hodek 1966 p. 195). They found that the three dominant species of coccinellids were unevenly distributed at different heights in crops in Minnesota, USA. The difference in distribution was significant in corn and barley. Whereas *Hippodamia convergens* and *H. tredecimpunctata* occurred on higher portions in corn and barley, *Coleomegilla maculata* was distributed predominantly in the lower stratum of these crops (Tab. 5.05, Fig. 5.02).

The authors tried to reveal the causes underlying the differential distribution and

Tab. 5.05 Vertical distribution of adult *Coccinellidae* in four crops (Ewert and Chiang in Hodek, 1966)

| Crop | | Coccinellids | |
|---|---|---|---|
| Name | Average height of plants [1] [cm] | N | Average elevation on plants [cm] |
| | | *Hippodamia tredecimpunctata* | |
| corn | 172 | 242 | 112 |
| barley | 62 | 260 | 45 |
| sorghum | 87 | 148 | 64 |
| alfalfa | 42 | 105 | 30 |
| | | *Hippodamia convergens* | |
| corn | 172 | 230 | 114 |
| barley | 58 | 430 | 38 |
| sorghum | 86 | 175 | 57 |
| alfalfa | 44 | 66 | 34 |
| | | *Coleomegilla maculata* | |
| corn | 173 | 221 | 73[2] |
| barley | 59 | 83 | 26[2] |
| sorghum | 87 | 162 | 43 |
| alfalfa | 42 | 28 | 26 |

[1] Plants infested with coccinellids.
[2] Significantly different from *Hippodamia spp.* at the 0.01 level of probability.

they succeeded in finding several, at least for *H. convergens* and *C. maculata*. The aphidophagous *Hippodamia spp.* meet their prey in the higher stratum of the plants, while plant pollen, the preferred food of *C. maculata* (see 6.111 and 6.12) usually accumulates on the lower parts of plants. *H. convergens* and *C. maculata* also show opposite responses to physical factors (light, humidity) (see 5.22) and a contrasting resistance to desiccation. *H. convergens* is resistant, and *C. maculata* is sensitive to desiccation. This is in full accord with other factors and with their distribution in the crop strata. In *H. tredecimpunctata* the situation is more complicated. This species must search for its prey in the upper stratum and is directed in this respect by appropriate behavioural responses (negative geotaxis, positive phototaxis). The species is, however, not adapted to this microhabitat as it is sensitive to desiccation in the same way as *C. maculata* (Fig. 5.05). This is apparently related to its limited circumboreal distribution; the other two species occur in temperate and tropical zones.

As further evidence, it has recently been found that the coccinellids react differently to changes in the microclimate of corn plants as influenced by plant density (Smith 1971). *Coleomegilla maculata and Hippodamia tredecimpunctata* were most abundant at a density of 3.2 plants per $m^2$ (i. e. 9000 plants per acre) and two *Coccinella spp.* (*C. novemnotata* and *C. transversoguttata*) on even less dense plants (1.6 plants per $m^2$). The temperature was lower in the denser stand; the mean and range of differences were 2.1 (0.5—8.6) °C for the densities 1.6 vs. 11.4 plants per $m^2$.

Differences in the occurrence of coccinellids between the young trees and crowns of pines were discussed as age differences in chapter 5.34, as was the preference of *Chilocorus bipustulatus* for older groves of citrus trees and date palms in Israel. The same preference has been found for *C. b. iranensis*, introduced to Mauretania (W. Africa). Within the traditional stands of date palms with an undergrowth of younger palms, the coccinellid shows a clear preference for the upper crowns (Laudého et al. 1970).

Wylie (1958) reported that *Aphidecta obliterata* shows preference for a twig rather than a stem habitat on firs, which indicates its adaptation to *Adelges nüsslini* rather than *A. piceae*.

Hukusima (1949b, 1950, 1951a, b, 1955) systematically studied the effect of slight deviations in cultivation of potatoes on the temperature of the microhabitat and the consequent changes in the occurrence of, and the damage by the phytophagous *Epilachna vigintioctomaculata*. The temperature among the potato leaves increases and hence the feeding activity of the coccinellid decreases, if:

— the potatoes are planted more thinly, in furrows 75 cm apart (contrast 60 cm) (Hukusima 1949b);
— the plants are cultivated on higher ridges (20 or at least 15 cm, contrast 10 cm) (Hukusima 1950);
— the ridges run from east to west (Hukusima 1951a);
— intercropping or fencing cultivation of potatoes is used (Hukusima 1951b). These relationships are, naturally, only valid for the region studied. In different climatic conditions they would certainly be modified.

### 5.42 Differences between similar habitats in different biotic zones

The relation of coccinellid species to their habitats can vary in different regions of their distribution area. This may be due to the presence of different prey; but even if food conditions are the same or similar, the different temperature and/or humidity condi-

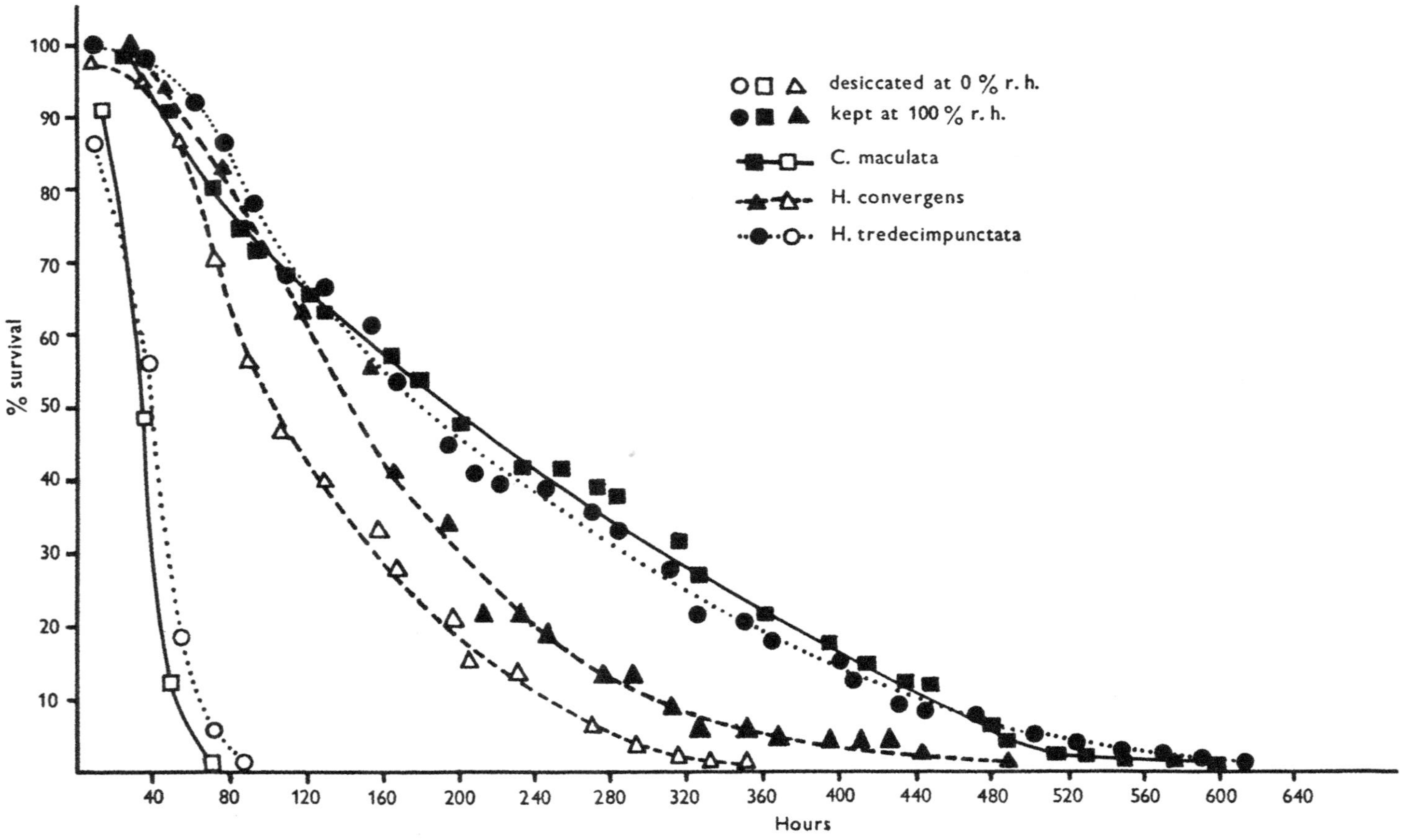


Fig. 5.05 Survival at 30 °C of three coccinellid species at different relative humidities (Ewert and Chiang 1966) (see also Fig. 5.02).

tions in another climatic zone may so drastically change the microclimate of a particular habitat that it becomes favourable for other coccinellid species.

Evidence that such differences exist is rather limited because only in rare cases has the fauna of coccinellids of the same habitat in different zones of one zoogeographical region been studied.

The material collected by Dyadechko (1954) in the Ukraine provides a possibility of seeking such a comparison. The area, except for a small part in the zone of mixed forests, belongs to the zone of the forest-steppe and the steppe zone.

Often the same typical coccinellid species occur in a certain type of habitat (see 5.5) regardless of the zone in which the habitat lies, but the zone may substantially influence relative proportions of the species. For example, in deciduous forests of the forest-steppe zone (Tab. 5.06), the most abundant species apart from *Stethorus punctillum* (33 %) were: *Synharmonia conglobata* (11 %), *Vibidia duodecimguttata* (10 %) and *Halyzia sedecimguttata* (10 %). In deciduous forests of the steppe zone, however, besides the similarly abundant *S. punctillum* (32 %), the most abundant species were: *Exochomus quadripustulatus* (16 %), *Propylaea quatuordecimpunctata* (16 %), *Scymnus (Pullus) subvillosus* (11 %) and *Hyperaspis reppensis* (9 %).

Tab. 5.06 Coccinellids in the mixed deciduous forests in the Ukraine (Dyadechko 1954)[1])

| Coccinellid species | Forest-steppe zone (Chernovcy, Kiev, Sumy) N = 1279 | Steppe zone (Nikolaev, Odesa Kharkov) N = 1205 |
|---|---|---|
| *Stethorus punctillum* | 32,8 | 32,4 |
| *Scymnus (Pullus) subvillosus* | 5,5 | 11,3 |
| *Hyperaspis campestris* | 3,7 | 6,0 |
| *Hyperaspis reppensis* | 1,1 | 9,4 |
| *Adalia bipunctata* | 2,1 | 2,6 |
| *Synharmonia conglobata* | 10,6 | 2,0 |
| *Calvia decemguttata* | 5,6 | 0,9 |
| *C. quatuordecimguttata* | 6,3 | 1,0 |
| *Propylaea quatuordecimpunctata* | 7,3 | 15,6 |
| *Halyzia sedecimguttata* | 9,9 | 0,9 |
| *Vibidia duodecimguttata* | 10,2 | 1,5 |
| *Exochomus quadripustulatus* | 5,0 | 16,4 |

[1]) In each zone at each of the three localities 400 sweepings were made (11. 6.—18. 7.).

Kehat (1967a) found a distinct difference in the occurrence of coccinellids in date palm plantations between the northern (Bet Shean, Jordan) and southern (Dead Sea, Elat — Arava) regions of Israel. Four species (*Pharoscymnus pharoides*, *Chilocorus bipustulatus*, *Rhizobius (Lindorus) lophantae* and *Scymnus (Nephus) bipunctatus* were absent in Arava, though they occurred in both northern regions. By contrast *Pharoscymnus setulosus* was evenly distributed. Two *Pharoscymnus* species had darker forms in the North and lighter ones in the South (Fig. 5.06) (see 3.24).

## 5.5 The coccinellid fauna of some diverse habitats

This chapter simply consists of an annotated list of coccinellid species in several habitats which have been chosen more or less arbitrarily, although emphasis has been put on habitats of economic importance. The chapter nevertheless includes most of the available data, particularly data suitable for comparison between habitats. Not many types of habitat have been studied sufficiently widely and most authors report the data of one single area. It is, of course, necessary to study the fauna of a certain habitat at several localities in the same region and over a long period to give a clear picture which excludes incidental effects such as adjacent habitats, climatic conditions in a particular season, etc.

The presenee of larvae and pupae of a species is a sign that the habitat is typical for that species (see chapter 5.21).

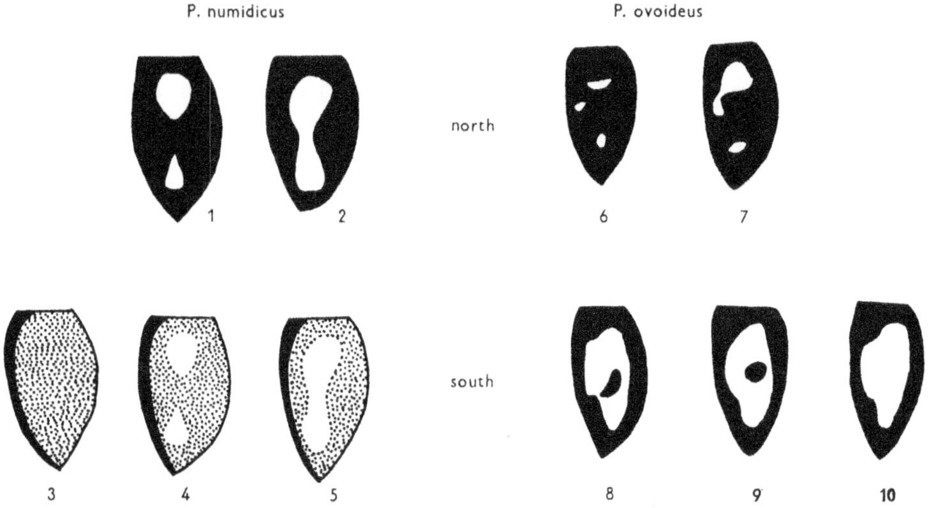


Fig. 5.06 Distribution of darker and lighter forms of two *Pharoscymnus spp.* in northern and southern Israel (Kehat 1967a).
1, 2, *P. numidicus* (nominate form); 3, *P. n. f. unicolor;* 4, 5, *P. n. f. junctus;* 6, *P. ovoideus f. guttatus;* 7, *P. o. f. hamifer;* 8—10 *P. ovoideus* (nominate form).

## 5.51 Forests

**Pine.** —This habitat is thoroughly discussed with respect to middle or eastern Europe and Canada in chapter 5.34.

**Spruce.** — A good picture of the coccinellid fauna of an old stand (110 years) in Eastern Germany (Tharandt near Dresden) was achieved surprisingly using pitfall traps (Barber traps) by Klausnitzer and Bellmann (1969). About one sixth of all *Coleoptera* larvae were the larvae of 8 coccinellid species (N = 1080). Most abundant was *Anatis ocellata* (55 %), which like *Neomysia oblongoguttata* (13 %) is not as typical of coniferous trees, as some of the other species which follow. *Adalia conglomerata*, which is the most typical inhabitant of spruce forests, accounted for 23 % of all larvae, and *Aphidecta obliterata*, which perhaps also prefers spruce, formed only 5.5 %. The rather eurytopic *Exochomus quadripustulatus* accounted for nearly 3 %. Three remaining species were apparently caught accidentally and together formed less than 1 %.

**Alder.** — (see 5.21).

## 5.52 Orchards, groves

**Apple.** — Palaearctic region. — Data on apple orchards are relatively abundant (Tab. 5.07), especially from western Germany: from near Hamburg (N = 739) (Speyer 1935), from near Köln (Redenz-Rüsch 1959), and from near Stuttgart (Asgari 1966). A striking difference between the data of Speyer and Redenz-Rüsch is the abundance of *Adonia variegata* (27.5 %) in the apple orchards near Hamburg. This typical inhabitant of herbaceous plants in drier places is absent in the samples of all other authors even in other countries. The only likely explanation is that during the sampling, which was in late season (VIII.—IX. 1934), the coccinellids temporarily invaded the orchard from adjacent fields, in which the crop had been harvested or was lacking aphids. Speyer's other aberrant result is a low incidence of *Propylaea quatuordecimpunctata* (1.7 %) which is usually quite abundant in orchards. The high incidences of *Adalia bipunctata* (38.5 %), *Coccinella septempunctata* (15.5.%) and *C. quinquepunctata* (14.5 %) are quite typical, although Redenz-Rüsch (1959) gives a much lower incidence of *C. quinquepunctata* (+) in comparison with *C. septempunctata* (+++). There are also variations in the incidence of *Adalia decempunctata*, rather abundant near Köln (++) and rare near Hamburg (1.3 %). *Stethorus punctillum* is absent in both places, Redenz-Rüsch (1959) reports rather high numbers of *Scymnus bipunctatus* (++). The rich occurrence of *Calvia quatuordecimguttata* near Köln is apparently caused by the proximity of forest (see below — Novák unpubl.).

Asgari (1966) listed natural enemies of *Aphidula pomi* in apple orchards around Stuttgart — Hohenheim without stating their abundance. His list contains seven likely species (*Coccinella septempunctata*, *Adalia bipunctata*, *A. decempunctata*, *Propylaea quatuordecimpunctata*, *Harmonia quadripunctata*, *Coccinula quatuordecimpustulata* and *Calvia quatuordecimguttata* but also seven non-existent species (*Coccinella 12-punctata*, *C. 12-maculata*, *C. quadrimaculata*, *C. 10-maculata*, *C. bimaculata*, *C. 6-punctata*, *C. 10-pustulata)* ! (The last three are presumably aberrations of *Adalia decempunctata*). Therefore even his statement that *Coccinella septempunctata* is the most abundant as well as the most voracious coccinellid preying on *A. pomi* must be questioned, especially as it is unsupported by any numerical data.

That *A. bipunctata* and *C. septempunctata* are present in apple orchards is confirmed by other workers in northern, southern and western Europe, but from their papers certain contrasts also emerge. Kanervo (1962) failed to find *P. quatuordecimpunctata* in Finland, and reported a relatively high occurrence of *Calvia quatuordecimguttata* (++) and *Coccinella quinquepunctata* (++). In orchards near Ferrara in Italy and in the Netherlands near Wageningen, *P. quatuordecimpunctata* was present (Principi et al. 1967, Evenhuis 1968). Both in Italy and in the Netherlands *C. quinquepunctata* was absent.

Novák (unpubl.) sampled the coccinellids in apple orchards at two localities in northern Bohemia, Czechoslovakia (N = 3345) for three years. At Holovousy, a mixed deciduous forest was adjacent to the orchards, whereas at Podsedice the orchard was situated in an open area.

The difference in the setting of the orchards is clearly reflected in the high incidence in the Holovousy orchards (11.5 %, contrast 0.1 % at Podsedice) of *Calvia quatuordecimguttata*, a typical inhabitant of deciduous forests. The less open surroundings caused the orchard at Holovousy to be understandably more visited by *Adalia bipunctata* (30.2 %, contrast 4.5 %) and *A. decempunctata* (4.6 %, contrast 0.4 %). The tenfold higher abundance of *Coccinella quinquepunctata* in Holovousy can be related to the neighbouring forest providing a dormancy site of this species, which often hibernates in the litter at forest edges. At Podsedice, by the far most numerous species was *Stet-*

Tab. 5.07 Coccinellids in apple orchards (and the average from five fruit-orchards in the Ukraine)

| Coccinellid species | Germany | | | Finland | Italy | Holand | Czechoslovakia | | Ukraine (USSR) |
|---|---|---|---|---|---|---|---|---|---|
| | Speyer 1935 N = 739 | Redenz-Rüsch 1959 | Asgari 1966 | Kanervo 1962 | Principi et al. 1967 | Even-huis 1968 | Novák Holo-vousy 1968—69 N=1637 | (unpubl.) Podse-dice 1969—70 N=1708 | Dyadech-ko 1954 N=1081 |
| *Subcoccinella vigintiquatuorpunctata* | | | | | | | | 2.3 | |
| *Stethorus punctillum* | | | | ++ | | | 5.4 | 61.2 | 17.2 |
| *Scymnus (Pullus) subvillosus* | | | | | | | | | 8.5 |
| *Scymnus (Nephus) bipunctatus* | | ++ | | | | | | | |
| *Hyperaspis campestris* | | | | | | | | | 5.1 |
| *H. reppensis* | | | | | | | | | 1.3 |
| *Hippodamia septemmaculata* | | + | | | | | | | |
| *Adonia variegata* | 27.5 | | | | | | | 0.1 | 1.9 |
| *Semiadalia undecimnotata* | | | | | | | | | 9.6 |
| *Adalia decempunctata* | 1.3 | ++ | + | | | + | 4.6 | 0.4 | 7.1 |
| *A. bipunctata* | 38.5 | +++ | + | ++ | + | + | 30.2 | 4.5 | 24.4 |
| *Coccinella septempunctata* | 15.5 | +++ | + | +++ | + | + | 8.1 | 3.9 | 0.6 |
| *C. quinquepunctata* | 14.5 | + | | ++ | | | 33.5 | 3.0 | |
| *Coccinula quatuordecimpustulata* | | + | + | | | | | | |
| *Synharmonia conglobata* | | | | | | + | 0.7 | 0.2 | |
| *Calvia quatuordecimguttata* | | ++ | + | ++ | | | 11.5 | 0.1 | |
| *Propylaea quatuordecimpunctata* | 1.7 | ++ | + | | + | + | 4.2 | 20.4 | 1.8 |
| *Anatis ocellata* | | | | | | | | 0.2 | |
| *Thea vigintiduopunctata* | | | | | | | | 1.3 | |
| *Chilocorus renipustulatus* | | | | | | | | | 5.4 |
| *C. bipustulatus* | | | | | | | | | 1.5 |
| *Exochomus quadripustulatus* | | + | | | | + | | | 13.9 |
| *E. flavipes* | | | | | | | | | 1.6 |
| Other species | 1.0 | | | | | | 1.8 | 2.2 | |

*horus punctillum* (61.2 %) which was ten times less abundant at Holovousy (5.4 %), although the occurrence of *Panonychus ulmi* was approximately the same at both localities. The second dominant species at Podsedice was *Propylaea quatuordecimpunctata* (20.4 %) and all other species were far less abundant.

The fivefold difference in the occurrence of *P. quatuordecimpunctata* at the two localities is apparently due to some hitherto unknown microclimatic preference of this species, which is commonly considered as eurytopic (Dyadechko 1954) or ubiquitous (Horion 1961). Its somewhat higher incidence in more humid localities, mentioned by Bielawski (1959), seems to receive support from Dyadechko's (1954) data. In his samplings from the Ukrainian SSR, *P. quatuordecimpunctata* was completely missing in field habitats, was rare in orchards (1.8 %, in 3 out of 5 orchards missing completely) but rather abundant in deciduous forests (occurring, however, more frequently in the steppe-zone — 16 %, than in the forest-steppe zone — 7 %). It cannot be excluded that in an area with both forests and orchards it may be more abundant in the forests because these are the more humid environment, whereas in an area with no forests at all, it may prefer orchards as more humid than fields. These considerations again stress how necessary it is to study the incidence of such mobile insects as coccinellids in neighbouring habitats also. As the food ecology of *P. quatuordecimpunctata* has not been sufficiently studied, there is the possibility that it is able to maintain itself on psyllids in the absence of aphids better than the other coccinellid species.

Dyadechko (1954) reported on the occurrence of coccinellids in Ukrainian orchards of all kinds. From a general picture produced by five samples from four localities (Kiev, Chernovcy, Odessa, Kherson) on different dates (5.6.—3.8.), three species appear prevalent: *Adalia bipunctata* (24 %), *Stethorus punctillum* (17 %) and *Exochomus quadripustulatus* (14 %). Dyadechko (1954) further lists those species he regards as typical for orchards: *Adalia decempunctata* (7 %), *Scymnus* (*Pullus*) *subvillosus* (8.5 %), *Chilocorus bipustulatus* (1.5 %), *Hyperaspis reppensis* (1 %) and *H. campestris* (5 %). Although he caught quite a number of *Semiadalia undecimnotata* (10 %), he si certainly quite right not to include this species into his list of typical inhabitants of orchards.

The data of Hukusima (1963a, b, 1966a, b) on the influence of various pesticides on the fauna of orchards give some idea about coccinellids on apple trees in Japan. He collected more than 1500 coccinellids, about 300 of them in unsprayed controls. The three dominant species were *Chilocorus kuwanae* (28 % and 26 %[1]), *Stethorus japonicus* (26 % and 11 %) and *Propylaea japonica* (21 % and 32 %). *Harmonia axyridis*, which accounted for only 11 % in the total catch, was relatively much more abundant in the unsprayed plots (21 %). *Scymnus hilaris* accounted for 10 and 9 %, while *Coccinella septempunctata bruckii* was very rare (2 and 1 %).

N e a r c t i c  r e g i o n. — In the apple orchards of Canada the Holarctic *Adalia bipunctata* is accompanied by several Nearctic species: *Coccinella transversoguttata quinquenotata*, *C. novemnotata*, *C. trifasciata perplexa* and *Hippodamia parenthesis* have been found in the Belleville area of Ontario (Smith 1958). Evenhuis (1960) mentioned only two species from Nova Scotia, *C. transversoguttata quinquenotata* and *A. bipunctata*.

**Peach.** — Putman (1964) reported ten species from the peach orchards of the Niagara Peninsula (Ontario, Canada). *Adalia bipunctata* was by far the most abundant (46 %), the next frequent were: *Coccinella trifasciata perplexa* (22 %), *Coleomegilla maculata lengi* (12 %), *Coccinella transversoguttata richardsoni* (8 %), *Hippodamia tredecimpunctata tibialis* and *Coccinella novemnotata* (3 %). Thus the coccinellid fauna of peach orchards

[1]) Two percentages are given, the one from the sprayed plots, followed by that of the unsprayed plots.

differs from apple orchards, as reported by Smith (1958), only by the presence of *C. m. lengi* and *H. t. tibialis*.

**Pear.** — Thirteen species of *Coccinellidae* were collected in southern Oregon in pear orchards infested with *Psylla pyricola* (Westigard et al. 1968). Six species accounted for the majority taken: *Coccinella transversoguttata richardsoni*, *Scymnus marginicoelis*, *Hippodamia convergens*, *Olla abdominalis*, *Cycloneda polita* and *Stethorus picipes*. Only the first species is in common with the coccinellid fauna of East Canadian orchards (apple, peach).

**Walnut.** — *Hippodamia convergens* (42 %) and *Olla abdominalis* (53 %) were the most abundant (N =905) coccinellid species in walnut orchards in northern California infested by *Chromaphis juglandicola* (Sluss 1967). Both breed in the orchards. *Adalia bipunctata* adults were much less abundant (4 %) and its larvae were never found on the trees, although some pupae were present. *Hippodamia quinquesignata punctulata* was very rare. *O. abdominalis* was reported from the same habitat and prey by Essig (1912).

**Citrus groves.** — Palaearctic region. — Both sets of data available from Japan come from the isle of Honshu. Nakao (1962, 1964) found nine predatory coccinellids in citrus groves near Fukuoka, where three important pests were present: *Aphis citricidus*, the coccid *Unaspis yanonensis* and the mite *Panonychus citri*. The most abundant predators were *Scymnus hareja*, *Telsimia nigra* and *Chilocorus kuwanae*. The prey-predator relations are depicted in Fig. 5.07. Other predatory coccinellids present but less abundant were: *Scymnus hilaris*, *S. dorcatomoides*, *Stethorus japonicus*, *Hyperaspis japonica*, *Rodolia cardinalis* and *Propylaea japonica*.

In the citrus groves of the Yamaguchi Prefecture, Nohara (1963) found 23 species of *Coccinellidae;* all were entomophagous with the exception of the apparently mycophagous *Thea cincta*. Eight species were important predators: *Scymnus hilaris* and *Hyperaspis japonica* reported as aphidophagous, *Propylaea japonica* both aphidophagous and coccidophagous, *Stethorus japonicus* both coccido- and acarophagous, and four other species coccidophagous (*Rodolia cardinalis*, *Scymnus hareja*, *Telsimia nigra*, *Chilocorus kuwanae*). The coccid pest *Unaspis yanonensis* was reported as being controlled by the coccinellids though apparently aphids and mites were also present. Although the prey-predator relationships differ, the lists of coccinellid species given by Nohara and Nakao are identical, with the exception that *S. dorcatomoides* is not mentioned by Nohara.

In Israel the most important predator of citrus scale insects is *Chilocorus bipustulatus* (Rosen and Gerson 1965, Ben Dov and Rosen 1969).

Nearctic region. — In Florida, USA, Muma (1953—1955) found a number of predaceous coccinellids in citrus groves. Their predator/prey relations are given in Fig. 5.08. Apart from the species included there, some others were collected which were less abundant. The introduced *Cryptolaemus montrouzieri*, feeding on mealybugs was not common on citrus. *Delphastus pallidus* a predator of *Aleyrodidae* could be found in relatively large numbers, though only for a short time. *Scymnillodes subtropicus* which also feeds on white flies was rare together with an incidental predator of scale insects and white flies, *Scymnillus aterrimus*. *Decadiomus bahamicus* which feeds on mealybugs was found only in one region, as was the aphidophagous *Leis dimitata quinquespilota*, introduced from California in 1925. The aphidophagous *Scymnus partitus*, once common, has apparently decreased in numbers and was found only in late summer and early autumn.

Neotropic region. — From southern Brazil, Wille (1926) reported *Curinus (Orcus) zonatus* as the most important coccidophagous species.

In **date palm** plantations in Israel (Kehat 1967a), four *Pharoscymnus spp.* (*numidicus, setulosus, ovoideus, pharoides*), *Chilocorus bipustulatus, Rhizobius* (*Lindorus*) *lophantae* and *Scymnus* (*Nephus*) *bipunctatus* were found reproducing. Three other coccinellids were found frequently as adults only (*Exochomus flavipes, Scymnus* (*Pullus*) *pallidivestis, Stethorus punctillum* and 15 occasionally.

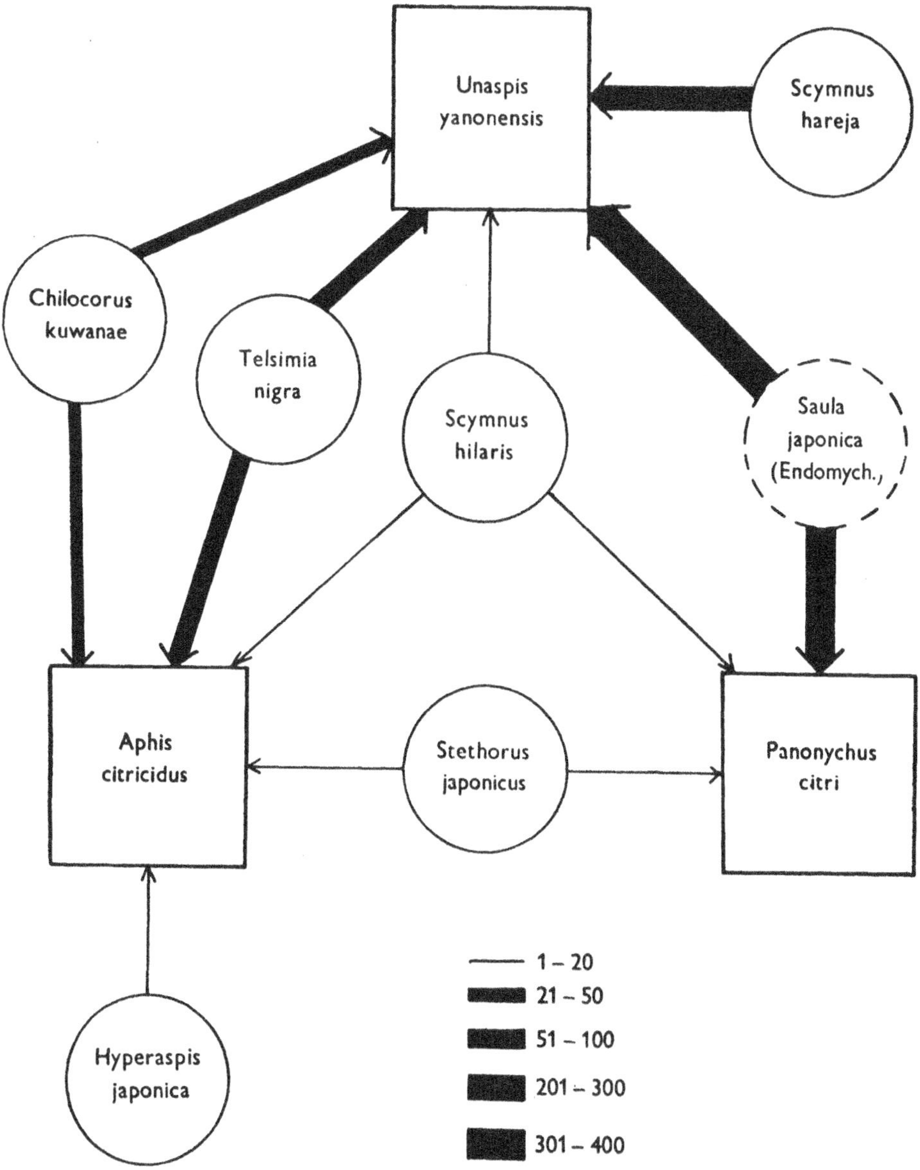


Fig. 5.07 Coccinellid predators (○) of the three most important pests (□) of citrus in Japan (Nakao 1964). The width of arrows shows the numbers of adult predators in Tachibana.

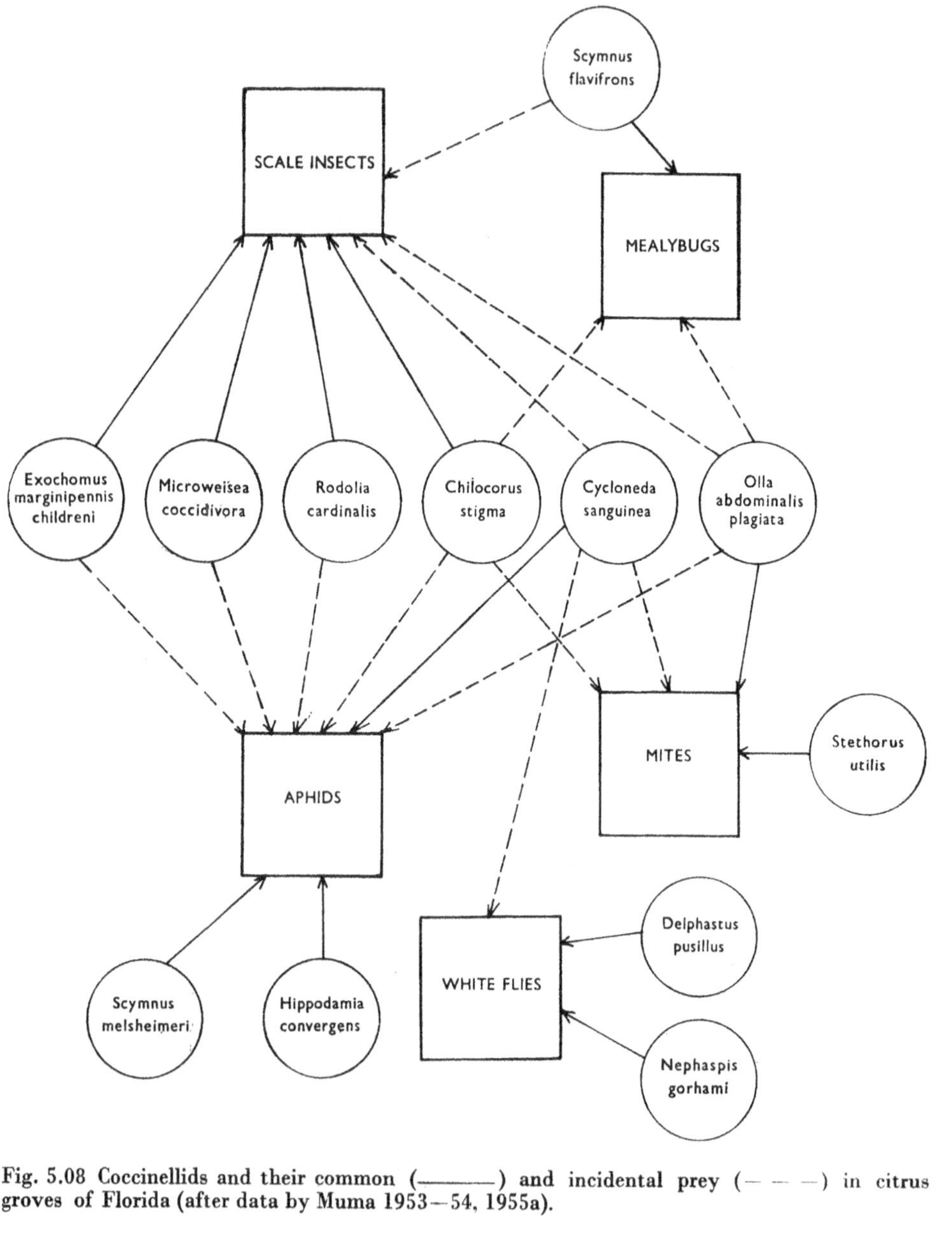


Fig. 5.08 Coccinellids and their common (———) and incidental prey (— — —) in citrus groves of Florida (after data by Muma 1953–54, 1955a).

## 5.53 Field crops

**Cereal crops.** — Palaearctic region. — In the Ukraine, *Adonia variegata* and *Coccinella septempunctata* are the typical species of cereal fields (Dyadechko 1954). In the steppe zone *C. quinquepunctata* seems to be absent in cereal crops and to be replaced by the south-palaearctic *Semiadalia undecimnotata* (Tab. 5.08).

**Nearctic region.** — In the USA, the most important species on corn and other cereals are *Hippodamia convergens*, *H. tredecimpunctata* and *Coleomegilla maculata* (Tab. 5.09). Their abundance varies significantly, as is shown by the data from corn in two neighbouring states (Minnesota, North Dakota). In the area of Belleville, Ontario, Canada, two *Coccinella spp.*, forming almost half of the specimens collected, were found in addition to the range of species observed in the USA, while *Hippodamia convergens* was only scarce (Smith 1971).

Tab. 5.08 Coccinellids on cereal crops in the Ukraine (USSR) (Dyadechko 1954)

| Coccinellid species | corn Odessa (July) N = 135[1]) [%] | | oats Sumy region (July) N = 166[1]) [%] | | winter wheat Kiev (June) N = 180[1]) [%] | |
|---|---|---|---|---|---|---|
| *Scymnus frontalis* | | | | | 10.6 | — |
| *S. apetzi* | 16.3 | + | | | | |
| *Hippodamia tredecimpunctata* | | | | | 6.1 | — |
| *Adonia variegata* | 30.4 | + | 21.1 | + | 55.0 | + |
| *Semiadalia undecimnotata* | 23.0 | + | | | | |
| *Coccinella septempunctata* | 10.4 | + | 40.4 | + | 11.7 | + |
| *C. quinquepunctata* | | | 21.7 | + | 9.4 | ? |
| *C. divaricata* | 11.9 | — | 16.9 | + | 2.8 | — |
| *Coccinula quatuordecimpustulata* | 8.1 | — | | | 4.4 | + |

1) + denotes presence, — absence of larvae.

Tab. 5.09 Coccinellids on cereals (Nearctic region)

| Coccinellid species | corn Smith 1958 Ontario | corn Schiefelbein & Chiang 1966 Minnesota [%] | sweet-corn Havnvik & Frye 1969 N. Dakota N = 617 [%] | corn Smith 1971 Ontario [%] | cereal crops Kieckhefer & Miller 1967 S. Dakota | oats Shade et al. 1970 North Indiana |
|---|---|---|---|---|---|---|
| *Coleomegilla maculata* | + | 31.0 | 0.2 | 17[5]) | | +++ |
| *Hippodamia convergens* | — | 62.0 | 18.0 | <1 | +[1]) | +++ |
| *H. glacialis* | | | | <1 | +[3]) | |
| *H. parenthesis* | — | | 0.8 | 2 | +[2]) | + |
| *H. tredecimpunctata* | + | 6.7 | 76.0 | 32[5]) | | + |
| *Cycloneda munda* | | | | | | |
| *Anatis mali* | | | | <1 | | |
| *Coccinella novemnotata* | + | | | 20[5]) | | + |
| *C. nivicola* | | | 2.5 | | | |
| *C. trifasciata* | + | | | 5 | | |
| *C. transversoguttata* | + | | | 24[5]) | | |
| *Adalia bipunctata* | — | | 2.5 | <1 | +[4]) | |

1) 2) 3) 4) — The numbers denote the approximate order of numerical importance.
5) — Reproducing on the corn field infested with *Rhopalosiphum maidis*.

**Alfalfa.** — Palaearctic region. — In eastern European fields of *Medicago*, which coccinellid species are dominant, is apparently related to air humidity (Tab. 5.10). Whereas *Propylaea quatuordecimpunctata* is by far the most abundant species in the zone of deciduous forests in Poland (Ruszkowski 1961), the species disappears completely in the alfalfa fields in both the forest-steppe and steppe zones of Ukraine (similarly as *C. quinquepunctata*) and the dominant position is taken over by the two south-palaearctic *Coccinula spp.*, which have a clear preference for dry environments (Dyadechko 1954).

Nearctic region. — The commonest coccinellid on alfalfa all over the USA is *Hippodamia convergens* (Tab. 5.11). Only in rather older records from Utah (Knowlton et al. 1938) are other three species similarly abundant.

**Potato.** — Palaearctic region. — Four common species form the bulk of the coccinellid fauna in potato fields in Czechoslovakia and Poland (Tab. 5.12): *Adonia variegata, Coccinella septempunctata, C. quinquepunctata* and *Propylaea quatuordecimpunctata.*

**Sugar-beet.** — Palaearctic region. — In Central Europe, the same four species are most abundant in sugar-beet (Tab. 5.13). In a dry and warm region (near Louny), however, *P. quatuordecimpunctata* is replaced by the south-palaearctic *Semiadalia undecimnotata.* In the Ukraine, there is apparently a similar climatic variation as occurs on alfalfa. In the moist oceanic climate of southern England *C. septempunctata* disappears and the dominant position is taken over by *Adalia bipunctata* which in continental Europe is a typical inhabitant of orchards, parks or edges of deciduous forests.

A general abundance of *A. bipunctata* in field crops has also been reported by Banks (1955) from beans in Hertfordshire and by Dunn (1960) from lettuce in Warwickshire. The distinct changes in habitat preferences between regions with different climate (e. g. typical continental climate of the Ukraine, the transition climate of central Europe, the oceanic climate of England) apparently reflect a distinct preference for particularly microclimatic conditions (most probably for different relative air humidity). As has already been stressed in chapter 5.22, there is a great lack of exactly that type of study which could elevate the subject of habitat preferences of individual species to a level of cause and effect.

Tab. 5.10 Coccinellids in the alfalfa fields (Palaearctic region)

| Coccinellid species | Ukraine (USSR) steppe (Kherson) N = 147 Dyadechko 1954 [%] | Ukraine (USSR) forest--steppe (Chernovcy) N = 150 Dyadechko 1954 [%] | Poland (Lublin Poznan) N = 128 Ruszkowski 1961 [%] |
|---|---|---|---|
| *Subcoccinella vigintiquatuorpunctata* | | 4.7 | |
| *Scymnus frontalis* | 6.1 | 22.7 | |
| *Hippodamia tredecimpunctata* | | | 4 |
| *Adonia variegata* | 18.4 | 19.3 | |
| *Adalia bipunctata* | | | 1 |
| *Coccinella septempunctata* | 15.0 | 8.7 | 21 |
| *Coccinella quinquepunctata* | | | 8 |
| *Coccinula quatuordecimpustulata* | 50.0 | 42.0 | 11 |
| *Coccinula sinuatomarginata* | 10.9 | 2.7 | |
| *Propylaea quatuordecimpunctata* | | | 55 |

Tab. 5.11 Coccinellids on alfalfa fields (Nearctic region)

| Coccinellid species | Utah 1937 Knowlton et al. 1938 [1] [%] | Utah Knowlton 1939 [1] | Utah Knowlton & Goodarzy 1956 | Utah Delta 1956 Goodarzy & Davis 1958 [%] | Kansas Manhattan 1958 Simpson & Burkhardt 1960 [%] | California Imperial Valley 1954—55 Dickson et al. 1955 [%] | South. California V. d. Bosch et al. 1959 | Ontario Belleville Smith 1958 | Massachussetts Yadava & Shaw 1968 |
|---|---|---|---|---|---|---|---|---|---|
| *Coleomegilla maculata lengi* | | | | | 11.2 | | | + | + |
| *Ceratomegilla vittigera* | 5.5 | + | | | | | | | |
| *Cycloneda munda* | 4.8 | + | | | 1.7 | | | | |
| *C. sanguinea* | | | | | | 14.9 | | + | |
| *Hippodamia convergens* | 13.0 | + | + | 54.9 | 82.0 | 85.1 | + | — | + |
| *H. glacialis glacialis* | | | | | 0.6 | | | | |
| *H. parenthesis* | 4.3 | + | + | 9.4 | 3.9 | | + | + | + |
| *H. sinuata disjuncta* | | | + | 17.4 | | | + | | |
| *H. americana* | 14.0 | + | | | | | | | |
| *H. apicalis* | 13.0 | + | + | | | | | | |
| *H. lecontei* | 7.2 | + | | | | | | | |
| *H. quinquesignata* | 7.1 | | | | | | + | | |
| *H. tredecimpunctata* | 3.3 | + | | | | | | + | |
| *H. pleuralis* | | | + | | | | | | |
| *Olla abdominalis* | | | + | | | | | | |
| *Coccinella novemnotata* | 3.3 | + | + | 3.8 | | | | + | + |
| *C. transversoguttata* | 15.5 | + | + | 6.1 | | | | + | + |
| *C. trifasciata perplexa* | | | | | | | | + | + |
| *Adalia bipunctata* | | + | | | | | | | |
| Other species | | | | 8.5 | | | | | |

[1]) Pea and alfalfa fields infested with *Macrosiphum pisi*.

Tab. 5.12 Coccinellids in potato fields

| Coccinellid species | England Dunn 1949 | Czechoslovakia western Bohemia Skuhravý & Novák 1957 N = 461 | western Slovakia Korbel unpubl. N = 260 | Poland Galecka 1966 unwooded area | near forests |
|---|---|---|---|---|---|
| *Hippodamia redecimpunctata* | | | 0.4 | | + |
| *Adonia variegata* | | 23.2 | 26.5 | | + |
| *Adalia bipunctata* | ± | 1.3 | | | + |
| *Coccinella septempunctata* | + | 5.4 | 54.2 | + | + |
| *Coccinella quinquepunctata* | | 12.8 | 0.4 | + | + |
| *Coccinella undecimpunctata* | | | 1.9 | | |
| *Coccinula quatuordecimpustulata* | | | 1.5 | | + |
| *Propylaea quatuordecimpunctata* | | 57.2 | 15.0 | + | + |

Tab. 5.13 Coccinellids in sugar-beet fields

| Coccinellid species | W. Germany (Göttingen) seed sugarbeet Sol 1961 N = 57 | Czechoslovakia central Bohemia Hodek et al. 1966 N = 126 [%] | northern Bohemia N = 69 [%] | Bohemia + Slovakia | Ukraine (USSR) (Rostov) Minoranskii 1966 1) | S. England Heathcote 1969 2) [%] |
|---|---|---|---|---|---|---|
| *Adonia variegata* | | 0.8 | 2.9 | + | + | |
| *Semiadalia undecimnotata* | | | 46.3 | + | | |
| *Adalia bipunctata* | + | | | – | | 50 |
| *A. decempunctata* | + | | | – | | 6 |
| *Coccinella septempunctata* | + | 46.8 | 30.4 | +++ | + | 1 |
| *C. quinquepunctata* | | 47.6 | 20.2 | ++ | | |
| *Propylaea quatuordecimpunctata* | + | 4.8 | | ++ | | 26 |

1) The aphid pest was not *Aphis fabae*, as in other cases, but *Pemphigus fuscicornis*.
2) Coccinellids were caught in traps 1.5 m above the ground (not on plants).

# 6 FOOD RELATIONS

## 6.1 Food range

The food of *Coccinellidae* has always been actively studied, largely because of the economic impact of most species. Unfortunately the prey-predator relations have mostly been studied by observing what the coccinellids happen to be eating, sometimes only by noting occurrence together of predator and prey on the same plant, etc. The reliability of the lists of data accumulated in this way (Schilder and Schilder 1928, Balduf 1935, Fulmek 1957, Börner and Heinze 1957, Kaddou 1960, Kurir 1964,) was questioned already by Thompson (1951): "The various species of ladybirds do not actually feed or at least feed habitually on all the various host insects with which they are associated in the records." Thompson warned further that: "The gradual accumulation of such records in the literature finally gives a picture which may be completely inaccurate in so far as the real behaviour and food habits of the species are concerned."

A short but critical survey of the food relations of central European *Coccinellidae* has been made by Klausnitzer (1966). An exact picture of the relationship of coccinellids to food can only be gained by a systematic study, preferably experimental, of individual species. Then it becomes gradually apparent that even within individual tribes, there are various specialisations in the feeding of the various members. Thus, for example, in the generally aphidophagous tribe *Coccinellini* there are also non-predaceous species, such as the phytophagous *Bulaea lichatschovi* (Capra 1947, Dyadechko 1954, Savoïskaya in Hodek 1966, p. 141; 1970a) and the mycophagous *Tytthaspis sedecimpunctata* (Dauguet 1949, Turian 1969). Within the same tribe there are further species which specialize on immature stages of *Coleoptera*, such as *Aiolocaria spp.* (Iwata 1932, 1965, Savoïskaya 1970b) and *Calvia quinquedecimguttata* (Kanervo 1940) which feed on pre-imaginal stages of *Chrysomelidae*, or *Neocalvia spp.* which prey on larvae of mycophagous coccinellids of the tribe *Psylloborini* (Camargo 1937). Thus, not only entomophagy, but also phytophagy and mycophagy are represented in the tribe *Coccinellini;* phytophagy is normal in the subfamily *Epilachninae*, and mycophagy in the tribe *Psylloborini*. The classification of coccinellids is still partially artificial and thus discussing the food specialization of a whole tribe cannot but have its limitations (Tab. 6.01).

The natural food range of a certain coccinellid species can be ascertained by several ways. The classic method is the microscopic detection and identification of prey remnants from the guts or excreta (Forbes 1883, Putman 1964) which can be compared with the whole specimens of insects from the same community. Three more refined methods are rather time consuming, but they have general application. In the precipitin test (Dempster 1960, Loughton at al. 1963, Rothschild 1966) the specific proteins of the prey are identified by their reaction with the serum of a sensitized mammal (usually a rabbit). Limitations of this method are discussed by Southwood (1966, pp. 250—1).

Tab. 6.01 Food of *Coccinellidae* (Compiled from Balduf 1935 and Sasaji 1971)[1]

| | |
|---|---|
| Subfamily *Sticholotinae* | |
| tribe *Sukunahikonini* | coccids, *Diaspinae* |
| tribe *Serangini* | aleyrodids |
| tribe *Sticholotini* | coccids, *Diaspinae* |
| *(Pharini)* | *Diaspinae* |
| *(Microweiseini)* | principally coccids, chiefly *Diaspinae (Aspidiotus Chionaspis)* |
| Subfamily *Scymninae* | |
| tribe *Stethorini* | phytophagous mites |
| tribe *Scymnini* | 62 % coccids. 23 % aphids |
| individual genera | extensive specialisation |
| *Clitostethus*, *Lioscymnus* | *Aleyrodidae* |
| *Diomus*, *Nephus*, *Sidis*, *Parasidis*, *Cryptolaemus*, *Pseudoscymnus* | *Pseudococcus* |
| *Platyorus*, *Scymnus (Pullus)*, *Scymnus (Scymnus)* | aphids |
| tribe *Aspidimerini* | aphids |
| tribe *Hyperaspini* | 75 % coccids — *Coccinae* *(Pseudococcus, Phenacoccus, Ripersia)* |
| tribe *Ortaliini* | *psyllids*, *Flatidae* |
| Subfamily *Chilocorinae* | |
| tribe *Telsimiini* | coccids, *Diaspinae* |
| tribe *Platynaspini* | ? aphids |
| tribe *Chilocorini* | 75 % coccids, aphids |
| Subfamily *Coccidulinae* | |
| tribe *Coccidulini* | all but one sp. coccids |
| *(Rhizobiini)* | 51 % *Diaspinae*, 35 % *Coccinae*, 14 % *Lecaniinae* |
| tribe *Exoplectrini* | *Icerya* and its nearest relatives |
| tribe *Noviini* | *Icerya* and its nearest relatives |
| Subfamily *Coccinellinae* | |
| tribe *Coccinellini* | 85 % aphids, also psyllids, *Chrysomelidae* |
| *(Hippodamiini)* | 76 % aphids |
| *(Synonychini)* | aphids |
| *Neda* | coccids |
| *Archaioneda* | coccids |
| *(Cheilomenini)* | 72 % aphids, coccids, *Aleyrodidae* |
| *(Veraniini)* | aphids, phytophagous |
| tribe *Psylloborini* | *Erisyphe* (mildew) |
| Subfamily *Epilachninae* | |
| tribe *Epilachnini* | phytophagous |
| *(Azyini)* | *Diaspinae* |
| *(Oeneini)* | *Aleyrodidae* |
| *(Scymnillini)* | *Aleyrodidae* |

[1]) Tribes in brackets are not considered in the modern system of *Coccinellidae;* the position of the last three tribes is not clear.

The precipitin test is unsuitable for quantitative studies and is, therefore, of limited use with voracious predators; for these the method of radio-isotope labelled prey could be of value (surveyed by Southwood 1966). An approximate measure of the number of prey eaten might be obtained from the level of radioactivity in the predator.

Putman (1965) determined the food of predators of phytophagous mites (other than coccinellids) in orchards by paper chromatography.

In nearly the whole of this section (6.1) we shall be dealing with predaceous *Coccinellidae*; these constitute the majority of the species and are of interest from the point of view of biological and integrated control. Chapter 6.15 will be devoted to myco- and phytophagous coccinellids.

## 6.11 Food specificity of predaceous Coccinellidae

Predaceous coccinellids have a wide range of accepted food. Apart from feeding on *Homoptera* and phytophagous mites, they often prey also on young instars of *Lepidoptera*, *Coleoptera* and *Hymenoptera*, small nematocerous *Diptera* and *Thysanoptera*. The larvae always prey on the same prey as the adults. Hence it is the adults which select a certain type of food for the larvae while laying their eggs (Way in Hodek 1966 p. 91), for the larvae are much less mobile and even appear to be unable to choose between suitable and toxic food (Blackman 1967b) (see 6.1123). Furthermore, the selection of food depends to a great extent on the adult preference for a certain habitat (see chapter 5).

When suitable food is in short supply the adult coccinellids are able to starve for quite a long time, to take only a limited amount of food, or, what is most important, they can switch to a different substitutive food, whether this be of insect (as mentioned above) or plant origin. Coccinellids have even occasionally been reported biting fairly strongly into human skin (Svihla 1952).

Some cases of phytophagy which have been reported are probably the result of an error of observation; in reality minute insect prey is being eaten (e. g. larvae of *Thysanoptera*) but this was overlooked by the observer. In other cases the reported phytophagy may represent a form of drinking; this is particularly true for the feeding on young leaves in spring by adults which have emerged from hibernation quarters and the feeding very probably augments the reduced water level in the body (e. g. the observation by Brassler 1930 on *Coccinella septempunctata*).

### 6.111 Substitution food of plant origin

Among food of plant origin, pollen and nectar both from flowers and from extrafloral nectaries form immensely important food for even explicitly carnivorous coccinellids. This plant food allows the coccinellids to survive with a reduced mortality when insect food is scarce for a while and then they can resume oviposition immediately the insect prey re-appears, as has been described, for example, for *Coccinella undecimpunctata aegyptiaca* by Ibrahim (1955b) (see also 6.23). Similar data have been recorded for other coccinellids (for *Hippodamia tredecimpunctata* and *Anisosticta novemdecimpunctata* by Goidanich 1943, for *Exochomus flavipes* by Geyer 1947, for *Chilocorus bipustulatus* by Yinon 1969a, for *Coccinella septempunctata* and *Coccinula quatuordecimpustulata* by Savoïskaya 1970a and for *Anatis ocellata* by Allen et al. 1970).

Alternatively, feeding on pollen enables reserves to be accumulated for a long-term cessation of feeding during dormancy (Hagen 1962). Pollens from different plants have, of course, different compositions (e. g. the pollen of *Pinus* contains much less protein than other pollens used) and therefore are not equally adequate as food for coccinellids (Smith 1960, 1961b). *Coleomegilla maculata* may even complete its full development on pollen of a number of plants (*Zea mays*, *Betula populifolia*, *Cannabis sativa*, *Carpinus caroliniana*) equally successfully as on aphids (Smith 1960); the species may actually prefer pollen to aphids (Ewert and Chiang 1966), and this euryphagy is a characteristic of this species (see 6.12). Other species studied (*Cycloneda sanguinea*, *Coccinella trifasciata*) cannot successfully complete their development if fed only on pollen (Smith 1961b).

Particularly the high-altitude Alpine coccinellid species, such as *Coccinella reitteri* or *Spiladelpha barovskii kritschenkoi* are also adapted to feeding on pollen, often on edelweiss (*Leontopodium alpinum*), because their habitats often lack aphids (Savoïskaya 1970a).

Extrafloral nectaries were reported to provide substitution food for coccinellids as early as 1933 by Watson and Thompson (1933) for *Leis conformis* (the plant was *Crotalaria striata*) and more recently, for example, by Putman (1955) for *Stethorus punctillum* (on young leaves of peach trees) and by Ibrahim (1955b) for *Coccinella undecimpunctata aegyptiaca* (the nectaries are at the midrib of cotton leaves).

### 6.112 Unsuitable insect food

The great variety of food consumed by predaceous coccinellids led to the assumption that food specificity in coccinellids exists only between major taxonomic groupings. Observed acceptability has been mistaken for real suitability of prey, even by experienced workers. Balduf (1935) concluded, on the basis of a survey of accepted food, that *Coccinellidae* of the tribes *Hippodamiini* and *Coccinellini* are mostly aphidophagous and that "no special groups of aphids are selected by them." In a special study of food of coccinellids, Kanervo (1940) states that he treated the six species of aphids he used to feed his coccinellids as a complex because "it appeared that the experimental animals (i. e. the coccinellids) did not make any great distinction between individual species (of aphids)". Such an assumption has long been accepted even though contradictory evidence has been reported.

#### 6.1121 Rejected prey

Since as early as 1907, observations were published that some aphids are not acceptable to certain coccinellids. The example of the aphid *Macrosiphum aconitum* feeding on *Aconitum*, which contains the poisonous compound aconitin (Hawkess 1920), is particularly relevant to the following section (6.1122). In other examples, unpalatability seems to have been due to an intensive colouration of aphids (Johnson 1907) or a waxy surface, as in *Brevicoryne brassicae*.

Telenga and Bogunova (1936) observed that the coccinellid *Harmonia axyridis* refuses *B. brassicae* in the field. This has also been shown experimentally; adults which were transferred from the aphid *Hyalopterus pruni* to *B. brassicae* reduced their food consumption by 90 % within 8 days, and the females ceased oviposition. George (1957) also noticed that *B. brassicae* was avoided by *Coccinellidae*.

*Hyalopterus pruni* itself is rejected immediately a larva of *Adalia decempunctata* pierces the body wall. In subsequent attacks this aphid is rejected as soon as the coccinellid touches it with its palps (Dixon 1958). *H. pruni* was already described as unsuitable for *A. bipunctata* by Hawkes (1920) who wrote: "This aphis has a grey-green mealy exudation which fills the stomata of the larvae and so kills them". By contrast, Hodek (1959) found this aphid to be the essential food for *C. septempunctata* in spite of the waxy covering, for the larvae successfully completed their development when feeding exclusively on this aphid.

*Anisosticta bitriangularis* and *Hyperaspis spp.* refuse *Schizolachnus piniradiatae* and starve when no other food source is available (Gagné and Martin 1968).

A similar study was made over 40 years ago on the famous *Rodolia cardinalis*. This coccidophagous ladybird did not prey on its normal host *Icerya purchasi* if the latter had fed on *Spartium* or *Genista*. It was hypothesized that leafless plants failed to shade the eggs of the beetle from the harmful insolation (Savastano 1911) or that the smell of the plants repelled the beetles (Balachowsky 1930). However, if the coccids which had fed on *Spartium* or *Genista* were offered to the ladybirds in isolation from the plants they were still rejected (Poutiers 1930). It may therefore be assumed that *Icerya* takes some substances from the plant which decrease its suitability as food for *Rodolia*. The unsuitable host-plants contain the yellow pigment genistein and the alkaloid spartein.

In contrast to some relative coccinellid species, *Chilocorus rubidus* refuses *Chionaspis salicis* and *Lepidosaphes ulmi* (Pantyukhov 1968b). *S. punctillum* fed not only on phytophagous tetranychid mites (*Metatetranychus ulmi*, *Tetranychus bimaculatus* and *Paratetranychus spp.*), but also on predaceous mites of the genus *Typhlodromus* whereas it does not feed on *Bryobia praetiosa* (Putman 1955).

### 6.1122 Accepted but inadequate prey

In addition to observations on the rejection of a particular prey, there is also evidence that food may be taken which, however, does not seem to be adequate. In such cases oviposition is prevented or reduced, and a varying percentage of larvae die without completing development.

This may occur when coccinellids, specialized to certain taxonomic groups of prey, are fed with food from other groups, e. g. other food than mites is fed to acarophagous species or only non-aphid food to aphidophagous species etc. Although both the adults and larvae of *Adalia bipunctata* were occasionally found feeding on tetranychid mites (Robinson 1951) and the gut of *A. bipunctata* and of three *Coccinella spp.* contained some remains of *Panonychus ulmi* (Putman 1964), these coccinellids could notdevtelop on this prey (Robinson 1951, Putman 1957). Conversely, the adults of the acar phagous *Stethorus punctillum* did not oviposit when fed on aphids, and the larvae cannoo complete their development (Putman 1955).

The aphido- and coccidophagous *Leis conformis*, in spite of its considerable polyphagy, has been reported not to be able to complete its development on some coccids and aphids (Moursi and Kamal 1946), while on *Aphis durantae* it not only oviposits but its larvae also develop.

Larvae of *Coccinella undecimpunctata aegyptiaca*, when provided with eggs or young larvae of *Prodenia litura*, or nymphs of the mealybugs *Pseudocococcus filamentosus* or *Phenacoccus hirsutus*, either refused to feed or fed for a very short time, but failed to complete their development (Ibrahim 1955a).

Not only the prey of other taxonomic groups is inadequate, but also other species of the same group.

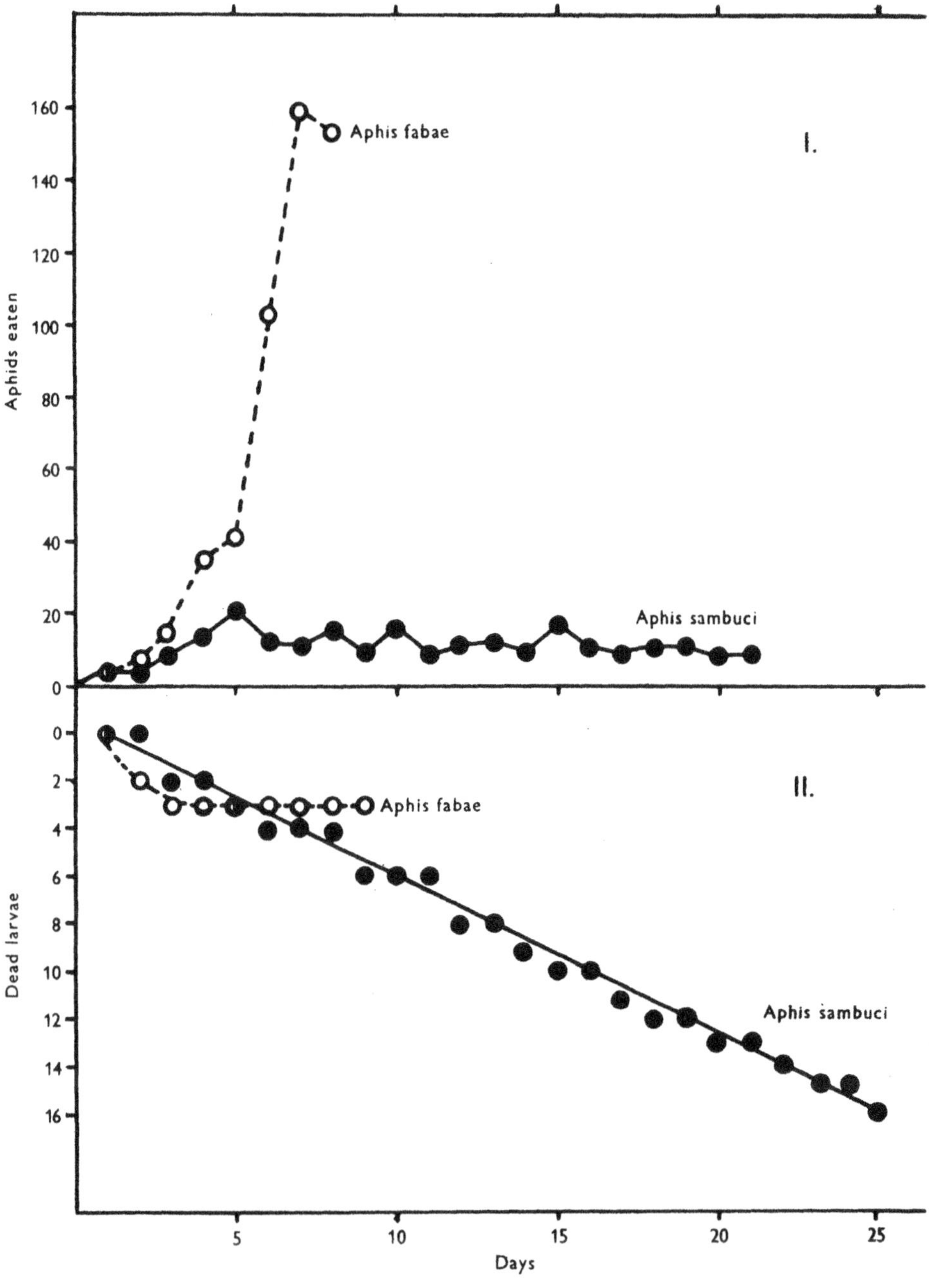


Fig. 6.01 Effect of *Aphis sambuci* on the larval development of *Coccinella septempunctata* (Hodek 1956). I. Daily feeding rate of larvae. II. Mortality of larvae.

Jöhnssen (1930) happened to observe a marked increase in feeding, when he changed the food of *Coccinella septempunctata* from *Aphis sambuci* to *A. hederae*. He did, however, not comment on this.

Dyadechko (1954) found *Aphis fabae* somewhat less suitable than *Toxoptera graminum* for *Coccinula quatuordecimpustulata*. The larvae fed less on *A. fabae* (the 4th-instar-larvae consumed 27 *A. fabae*, contrast 46 *T. graminum* per day) and the larval development was retarded (21 days on *A. fabae*, contrast 17 days on *T. graminum*).

Hodek (1956) found a rather low intake of *A. sambuci* (Fig. 6.01) by *C. septempunctata* and in a series of experiments showed that this aphid is inadequate food for both larvae and adults of *C. septempunctata*. The larvae could not complete their development and died within 25—26 days (Fig. 6.01). In the first experiment two thirds of larvae attained the 4th instar, whereas in the second experiment all but one died in the 3rd instar. This was consistent with the observed feeding rates in the two experiments. In the first experiment, the daily feeding rate after the 7th day exceeded 20 aphids, and later it ranged between 40—60 aphids. In contrast, in the second experiment the feeding rate remained consistently below 20 aphids.

This difference between the two experiments, carried out at naturally alternating temperatures, began to show on the 5th — 6th day. In the first experiment a large increase in temperature occurred at that time and the mean temperature then remained about 3 °C higher than in the second experiment. The effect may also have been caused by differences in the food. The chemical composition of aphids is probably different in early (first experiment, started 24. 6. 1953) and late summer (second experiment, started 22.7.) particularly as far as the presumed content of toxic substance is concerned (see later).

The adults, freshly emerged from the pupae, were sensitive to being fed on *A. sambuci* in the same way as the larvae. If, after emergence, they were fed only with this food, they died on average after 17.5 days. The control adults, fed on *Aphis fabae* or on *Uromelan aeneus* had at most a mortality of 16.6 %.

Overwintered adults fed on *A. sambuci* did not die more than those in the control culture, but did not lay eggs during the whole month of the experiment (Hodek 1957b). The ovaries, however, were found by dissection to be in the process of maturation. Not until suitable food had been fed for 9 days, did the adults begin to oviposit. In spring, adults of *C. septempunctata* are usually found on colonies of *A. sambuci* in the field, because the host-plant, *Sambucus nigra*, often grows at the hibernation quarters of this coccinellid.

As the feeding rate on *A. sambuci* is substantially lower than on the suitable food, *Aphis fabae* (Fig. 6.01), it might be assumed that the negative effect on the mortality of larvae is caused by insufficient intake of food which has little attraction. This possibility, however, can be discounted by following experiments which have shown (Hodek 1957a) that substantially reduced feeding rates on suitable food in the most voracious 4th instar did not cause any rise in mortality (Tab. 6.02). Of course, the reduction of the daily feeding rate to one-seventh prolonged the 4th instar by more than twice, but even so development was completed with about one-third of total food intake (in comparison with a surplus of aphids), and there was no increase in mortality.

If only the first three larval instars or only the fourth instar were fed with *A. sambuci* (Hodek 1957a), the majority of the larvae pupated; the mortality was respectively 32 and 24 %, (8 % in control) and larval development was prolonged to 9.2 and 11.2 days (in control 7.8 days).

The larvae of *Adalia bipunctata* were able to complete their development on *A. sambuci*, although at a somewhat slower rate than on *Aphis fabae* (Hodek 1957a). Perhaps an adaptation has evolved in this species, which breeds in central Europe in the same

Tab. 6.02 Larvae of *Coccinella septempunctata* fed on reduced daily rates of *Aphis fabae* (Hodek 1957a)

| Daily feeding rate in 4th instar (number of aphids) | Duration of development (in days) | | | | | Mortality [%] | N |
|---|---|---|---|---|---|---|---|
| | Larval instars | | Pre-pupa | Pupa | In total | | |
| | 1.+2.+3. | 4. | | | | | |
| surplus (aver. consumed 70) | 5.2 | 2.6 | 1.0 | 4.0 | 12.8 | 8 | 13 |
| 30 | 4.8 | 3.7 | 0.8 | 4.2 | 13.5 | 0 | 18 |
| 10 | 4.9 | 5.9 | 0.9 | 4.2 | 15.9 | 10 | 20 |

habitat as *A. sambuci.* In England *A. bipunctata* occurs abundantly on field crops (see 5.53).

On the basis of these observations and analogous concurrent findings on other coccinellids (Ibrahim 1955a, b, Putman 1955), criticism has been levied against the hitherto accepted way of evaluating the predator-prey relationship (Hodek 1959, 1962b). A warning was sounded that it is erroneous to judge this relationship by observing voracity, as this can confuse "acceptability" with "suitability". Such terms as stenophagy, polyphagy, etc., should not be improperly based on evidence of "acceptability" alone. Types of food can be divided into two main groups — essential and alternative. While essential foods ensure the completion of larval development and oviposition, alternative foods serve only as a source of energy and thus prolong survival in comparison with starvation. Naturally, there are transitions between both types; thus essential foods show varying degrees of favourability, etc.

In the last decade, a series of data have accumulated which confirm the validity of regarding food as "essential" or "alternative". *Semiadalia undecimnotata* can neither complete larval development nor oviposit when fed on *Aphis craccivora* (on *Vicia faba*) which is an essential food for *C. septempunctata* (Hodek 1960b). *Aulacorthum magnoliae* (on *Sambucus racemosa sieboldiana*) is detrimental to the larvae of *C. septempunctata bruckii* and *Harmonia axyridis*, while *A. craccivora* (on *Robinia pseudoacacia* as well as on *Vicia faba*) is only detrimental to the larvae of *H. axyridis* (Okamoto in Hodek 1966 p. 45). The same author (Okamoto 1961) found that *Brevicoryne brassicae* was less suitable for larvae of *C. septempunctata bruckii* than was *Rhopalosiphum prunifoliae.* He used pre-imaginal mortality, rate of development, resistance of resulting adults to starvation and their size as criteria of prey suitability.

Atwal and Sethi (1963) found that *Lipaphis erysimi* was more nutritious for larvae of *C. septempunctata* than were *Macrosiphum granarium* and *Aphis gossypii*, as both the larvae and resulting pupae were heavier and contained more dry matter.

Nine out of ten species of coccinellids (including the polyphagous *Coleomegilla maculata*) tested by Smith (1965a) failed to complete their larval development when fed on *Aphis fabae.* The tenth species, *Hippodamia tredecimpunctata*, reached the adult stage, but the beetles were smaller and their development time was significantly longer than when fed on *Acyrthosiphon pisum* or *Rhopalosiphum maidis.*

*Chilocorus bipustulatus* is unable to reproduce when feeding on *Saissetia oleae* (Huffaker and Doutt 1965.)

*Chilocorus stigma* has a markedly lower fecundity and survival and a prolonged pre-oviposition period if fed on *Lepidosaphes beckii* (Muma 1955b), whereas *Chrysomphalus aonidum* is a suitable prey.

If *Stethorus gilvifrons* preys on *Bryobia rubrioculus*, the adults do not oviposit and the larvae die in the 2nd and at the latest in the 3rd stage (Dosse 1967).

*Aiolocaria mirabilis*, a specialized predator on chrysomelids, has been reported as feeding both in the larval and adult stages on all pre-imaginal stages of several chrysomelids (Iwata 1932, 1965, Savoïskaya 1970b), but rejected *Agelastica coerulea* from *Alnus japonica* (Iwata 1965) and two central-Asiatic chrysomelids, mentioned only by their popular names (Savoïskaya 1970b). In Alma-Ata, its essential prey is *Melasoma populi*.

In a thorough study, Blackman (1965, 1967a) tested the suitability of several aphid foods for *Adalia bipunctata* and *C. septempunctata* (Table 6.03 and 6.04, Fig. 6.02 and 6.03). Although *Aphis fabae* and *A. sambuci* are natural prey for *A. bipunctata* in the field, they were found to be relatively unsuitable food for this coccinellid, as is shown by the longer larval development and the greater larval mortality. *A. bipunctata* had also a lower fecundity and fertility when fed on *A. fabae*. A preliminary observation suggested that *Brevicoryne brassicae* was even more unsuitable.

Tab. 6.03 Larval development and oviposition of *Adalia bipunctata* on different aphids (Blackman 1965, 1967a)[1])

| Aphid species | Larval development [days] | Larval mortality [%] | Weight of adult at emergence [mg] | Fecundity [total eggs laid] | Fertility [% of viable eggs] |
|---|---|---|---|---|---|
| *Myzus persicae* | 10.4 | 17.8 | 11.8 | 676.2 | 89.4 |
| *Aulacorthum circumflexum* | 9.5 | 16.7 | 11.9 | | |
| *Acyrthosiphon pisum* | 10.8 | 13.9 | 12.6 | | |
| *Microlophium evansi* | 10.6 | 9.1 | 12.4 | | |
| *Aphis fabae* | 13.0 | 27.6 | 7.9 | 249.6 | 55.9 |
| *Aphis sambuci* | 13.4 | 25.0 | 8.0 | | |
| *Brevicoryne brassicae*[2]) | 21; 23 | (66.7) | 5.1; 6.1 | | |

[1]) 20 °C const.. 16hr light per 24 hr.
[2]) Only 2 out of 6 larvae completed development.

Tab. 6.04 Development of *Coccinella septempunctata* larvae on different aphids (Blackman 1965, 1967a)[1])

| Aphid species | Development [days] | Mortality [%] | Weight of adult at emergence [mg] |
|---|---|---|---|
| *Myzus persicae* | 13.0 | 12.5 | 36.4 |
| *Aphis fabae* | 13.6 | 9.1 | 36.3 |
| *Acyrthosiphon pisum* | 13.3 | 18.6 | 37.2 |
| *Megoura viciae* | 14.8 | 13.4 | 33.5 |
| *Brevicoryne brassicae* | 16.1 | 26.1 | 30.9 |
| *Aphis sambuci*[2]) | 19.5 | (50.0) | 18.4 |

[1]) 20 °C const., 16 hr light per 24 hr.
[2]) Only 6 out of 12 larvae completed their development.

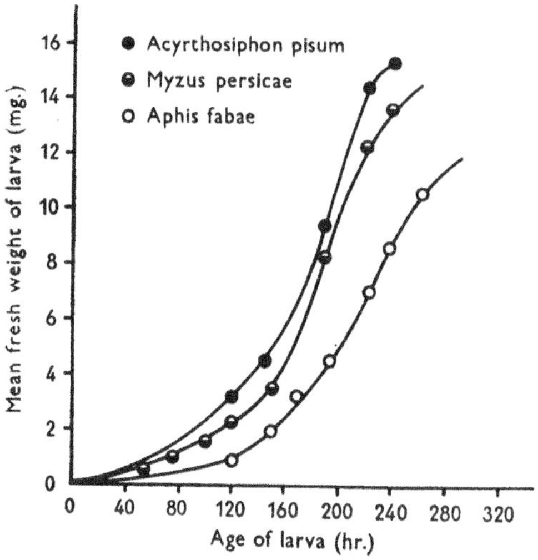


Fig. 6.02 Development of *Adalia bipunctata* larvae on different aphid foods (Blackman 1967a).

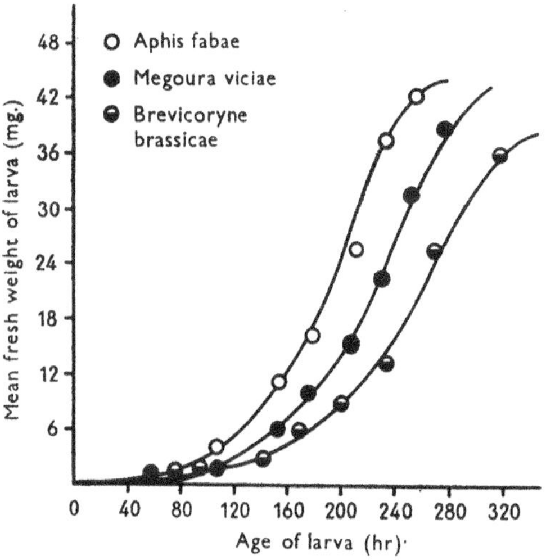


Fig. 6.03 Development of *Coccinella septempunctata* larvae on different aphid foods (Blackman 1967a).

In contrast to *A. bipunctata*, larvae of *C. septempunctata* were able to develop on *Aphis fabae* as successfully as on other favourable aphid foods. *Megoura viciae*, toxic for *A. bipunctata* (see 6.1123), was only slightly less suitable for *C. septempunctata* than other aphids. *A. sambuci*, although a very poor food, enabled half the specimens to develop to extremely small adults. Thus the effect of *A. sambuci* was less detrimental than in previous experiments with the same coccinellid, where mortality was complete (Hodek 1956).

*A. fabae* was shown to be a rather unsuitable food for *A. bipunctata* also by Hariri (1966a, b). When the larvae were fed on this aphid the emerged adults were lighter in weight, had less fat and glycogen content and their fecundity was halved (Tab. 6.05, Fig. 6.04). Iperti (1966) has also ascertained that *A. fabae* has an adverse effect on vitellogenesis in *A. bipunctata*, although it does not inhibit it entirely.

Hukusima and Kamei (1970) contributed to the understanding of food specificity in *Harmonia axyridis*. Three aphids out of ten (*Aphis pomi*, *Brevicoryne brassicae*, *Hyalopterus arundinis*) appeared slightly less suitable for the larvae, prolonging their development by approximately one-third compared with diets of the six more suitable species (Tab. 6.06). Larvae, however, succeeded in completing their development on these three relatively unsuitable aphid species. *Aphis craccivora* on most of its host- plants was an unsuitable food; larvae did not develop on this diet and died, and adults also died within 4—6 days. In contrast, *A. craccivora* on other host-plants (*Vicia sativa* or *Vigna catiang*) enabled larvae of *H. axyridis* to develop from the 2nd instar to adult. In contrast to *H. axyridis*, both larvae and adults of *Propylaea japonica* are resistant to the detrimental effect of *A. craccivora* from *Robinia pseudoacacia*.

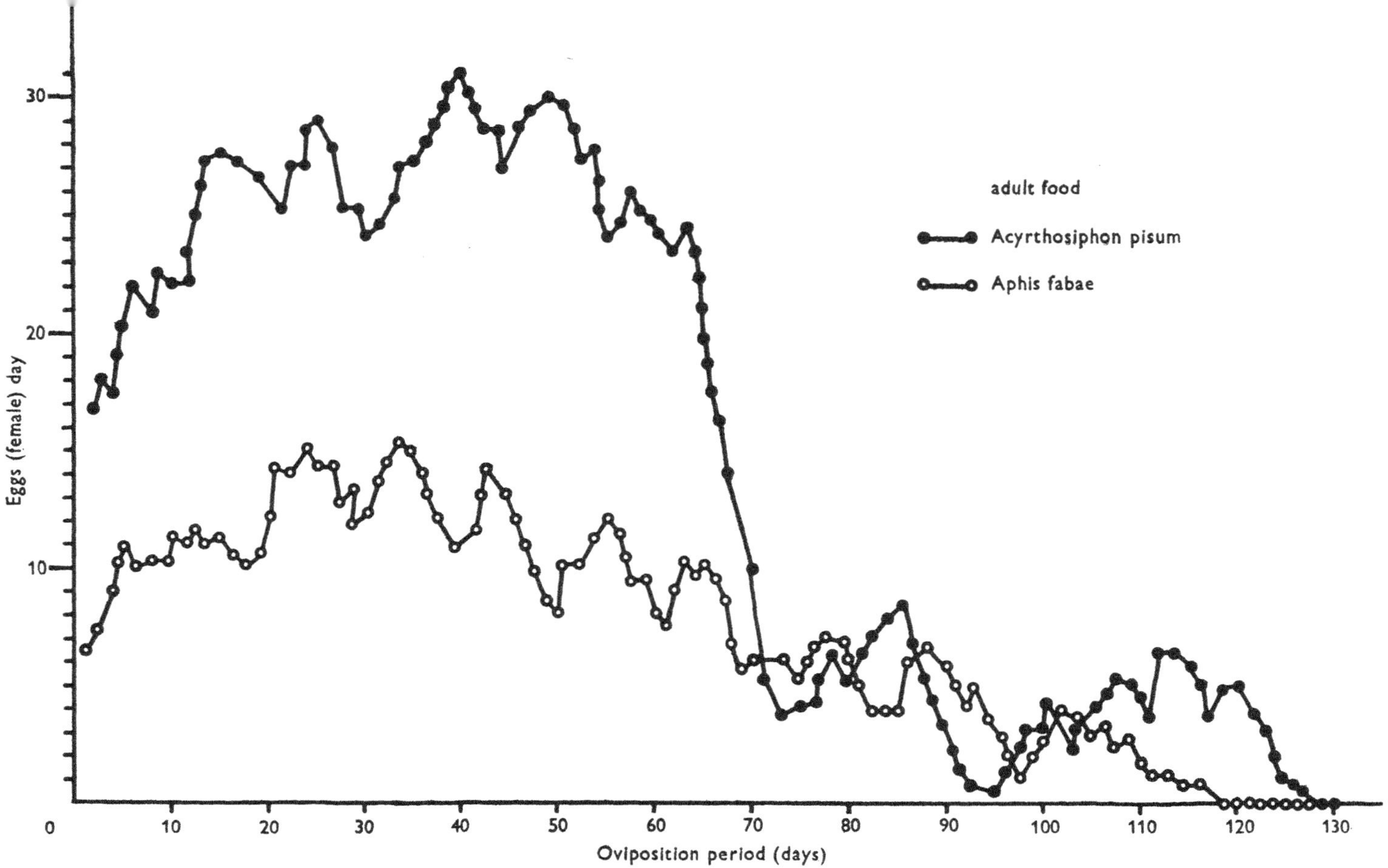


Fig. 6.04 Oviposition of *Adalia bipunctata* on different aphid foods (Hariri 1966a).

Tab. 6.05 Comparison of two aphid foods for *Adalia bipunctata* (Hariri 1966a, b)

| Diet for | | Newly emerged adult | | | Fecundity | | Longevity [days] | |
|---|---|---|---|---|---|---|---|---|
| larvae | adult | live weight [mg] | fat content [mg] | glycogen content [µg] | total [eggs] | per day [eggs] | ♂♂ | ♀♀ |
| *Acyrthosiphon pisum* | — | 14.2 | 0.7 | 26.1 | | | | |
| *Aphis fabae* | *Aphis fabae* | 9.2 | 0.2 | 11.0. | 738 | 9.3 | 81.7 | 76.9 |
| *Microlophium evansi* | *Acyrthosiphon pisum* | | | | 1535 | 20.4 | 77.1 | 79.4 |

Tab. 6.06 Various aphids as food for larvae of *Harmonia axyridis*[1]) (Hukusima and Kamei 1970)

| Aphid species | N | Total consumption [aphids] | N | Development [days] |
|---|---|---|---|---|
| *Acyrthosiphon pisum* | 17 | 89 | 17 | 7.7 |
| *Aphis pomi* | 20 | 369 | 18 | 12.2 |
| *Amphorophora oleraceae* | 17 | 101 | 17 | 7.2 |
| | 15 | 123[2]) | 15 | 15.1[2]) |
| | 19 | 140[2]) | 19 | 15.4[2]) |
| *Brevicoryne brassicae* | 14 | 276 | 14 | 13.3 |
| *Hyalopterus arundinis* | 12 | 297 | 12 | 13.2 |
| *Macrosiphum rosae ibarae* | 41 | 156 | 41 | 7.9 |
| *Megoura viciae japonica* | 20 | 160 | 20 | 7.9 |
| *Myzus persicae* | 24 | 189[2]) | 24 | 15.9[2]) |
| | 8 | 211[2]) | 8 | 14.1[2]) |
| | 23 | 218 | 23 | 7.5 |
| *Periphyllus californensis* | 21 | 214 | 33 | 9.5 |

[1]) At 30 °C.
[2]) At 25 °C.

#### 6.1123 Toxic prey

While a gradual poisoning is one of the possible explanations of the detrimental effect of some unsuitable aphids (as *Aphis sambuci* or *A. craccivora*) (see 6.1124), in some instances an acute toxic effect has been proved or at least strongly indicated.

Larvae of *Adalia decempunctata* attack and eat *Aphis fabae* and *Megoura viciae*, but after about two minutes the larvae frequently release the prey and regurgitate the gut contents. Six of twelve 4th-instar larvae provided with *M. viciae* died after a few days, although other aphids had then been offered as food (Dixon 1958).

The toxicity of *M. viciae* to *Adiaal bipunctata* has also been demonstrated (Blackman 1965, 1967 a). This aphid is lethal to all larval stages and to the adults. Neither

Tab. 6.07 Three aphids as food for adults of *Harmonia axyridis* (Hukusima and Kamei 1970)

| Temperature | 25 °C | | | | | | 30 °C | | | | | |
|---|---|---|---|---|---|---|---|---|---|---|---|---|
| Aphid species | Aver. daily consump-tion [aphids] | | Longevity [days] | | Eggs laid | | Aver. daily consump-tion [aphids] | | Longevity [days] | | Eggs laid | |
| | ♂ | ♀ | ♂ | ♀ | total number | per day | ♂ | ♀ | ♂ | ♀ | total number | per day |
| *Myzus persicae* | 22 | 65 | 92 | 98 | 2,310 | 27 | 30 | 64 | 57 | 50 | 778 | 22 |
| *Amphorophora oleraceae* | 15 | 45 | 152 | 170 | 3,819 | 25 | 18 | 40 | 101 | 86 | 945 | 14 |
| *Aphis craccivora* | | | | | | | | | | | | |
| on *Robinia pseudoacacia* | 36 | 30 | 6 | 6 | | | | | | | | |
| on *Astragalus sinicus* | 20 | 30 | 4 | 4 | | | | | | | | |
| on *Vicia hirsuta* | 24 | 27 | 5 | 4 | | | | | | | | |

larvae nor adults have the ability to avoid attacking this aphid when it is presented mixed with a suitable food. Even when fed on a mixture of *M. viciae* and *Acyrthosiphon pisum* in the ratio of 1 : 9, all of the larvae failed to reach the pupal stage. The lst-instar larvae perished more rapidly on this food than when starved. The 4th-instar larvae accepted *M. viciae* readily, but after about 4 min. of feeding they suddenly rejected their prey and vomited. When fed on this prey the adults all died within a week of emergence.

*Aphis nerii* from *Nerium oleander* is poisonous for most coccinellids tested (*Coccinella septempunctata*, *Semiadalia undecimnotata*, *Propylaea quatuordecimpunctata*, *Adalia bipunctata*) with the exception of *Adonia variegata*, which develops normally on this prey (Iperti 1966).

### 6.1124 Possible causes of unsuitability or harmfulness

The reasons why particular prey species are harmful for particular coccinellids are not yet clear. That death is caused simply by starvation, resulting from a very low feeding rate on unsuitable food, can be discounted since such reduced feeding rates repeated artificially with essential food do not cause a substantial rise in mortality (Hodek 1957a). There seem to be two other possibilities. Either the aphids contain some special substances (derived from the plants) poisonous for coccinellids, or the aphids are deficient in nutritive value.

For *Aphis sambuci* we can assume a passage of the glycoside sambunigrin[1]) (contained in the host plant *Sambucus nigra*) from the plant into the aphid. This glycoside may be split enzymatically into hydrocyanic acid in the body of the coccinellid, for enzymes which split glycosides have been identified in the body of *Coccinella spp.* (Kuznetsov 1948). The same explanation is possible for *Aulacorthum magnoliae*, where the host plant is *Sambucus racemosa* (Okamoto in Hodek 1966 p. 45).

The rapid response of *Adalia bipunctata* and *A. decempunctata* (obvious distress within two minutes) to *Megoura viciae* eliminated viral transmission or polypeptides as the factors responsible and makes it likely that a substance is obtained by the aphid from the host plant (Dixon 1958). Dixon et al. (1965) made a thorough analysis of the aphid, searching for physiologically active compounds. They could not, however, implicate any of the compounds found (aphin pigments, alkanes, fatty acid esters, amino-acids, peptides and two sugars D-glucose and D-ribose).

Similarly in *Macrosiphum aconitum*, which feeds on *Aconitum*, the presence of poisonous aconitin may be the reason why coccinellids reject this aphid, although Hawkes (1920) suspects the intensive colour. Also the rejection of *Icerya* from *Genista* or *Spartium* by *Rodolia cardinalis* may be explained by the yellow pigment genistein and the alkaloid spartein contained in the host plants (see 6.1121).

Hukusima and Kamei (1970) suggest the same explanation for the unsuitability of *Aphis craccivora* for *Harmonia axyridis* if the aphid was collected from one of seven host plants (*Robinia pseudoacacia*, *Astragalus sinicus*, *Vicia hirsuta*, *Capsella bursa-pastoris*, *Cardamine flexuosa*, *Rorippa palustris*, *Hemistepta carthamoides*). As *A. craccivora* was a favourable food when collected on *Vicia sativa* or *Vigna catiang* var. *sinensis*, the authors tried to explain the differential host plant effect by two-dimensional thin layer chromatography. They analysed the ninhydrine-positive compounds in *Aphis craccivora* collected from the different host plants. Aphids from *R. pseudoacacia* contained two extra aminoacids (glycine and arginine) and additionally three unidentified substances; aphids from *V. catiang* contained two additional unidentified substances.

[1]) The chemical structure of sambunigrin is D—cyanonitrileglukoside.

This influence of host plant on unsuitability of prey to some extent throws light on an earlier observation that the larvae of *Coccinella septempunctata bruckii* and *Propylaea japonica* die when fed on the aphid *Aphis craccivora* in the spring, but not in the summer (Takeda et al. 1964). Apparently different host plants were again involved, and the effect is not one of temperature or of other abiotic factors connected with the advanced season.

These findings provide a context for a brief note on some results from Central Asia. Ul'yanova (1956) reports unpublished observations of unnamed workers from the Tashkent Institute of Plant Protection concerning the deleterious effect of feeding "local coccinellids" on *Aphis craccivora* from *Robinia pseudoacacia*, whereas the same aphid from cotton or alfalfa was suitable. Ul'yanóva, however, doubted these results, for she obtained apparently different results, i. e. complete larval development of *C. septempunctata* and *C. undecimpunctata* on *A. craccivora* collected from *R. pseudoacacia*.

Some authors attempt to explain the unsuitability of certain foods by the second possibility mentioned earlier, i. e. by nutritive deficiency.

Atwal and Sethi (1963) suggest that the higher protein content of *Rhopalosiphum* (*Lipaphis*) *erysimi* makes this aphid more favourable than *Aphis gossypii* for *C. septempunctata*. Hariri (1966b) similarly supposes that *Acyrthosiphon pisum* is more nutritious for *A. bipunctata* than is *Aphis fabae*. However, he states that the causes of the adverse effect of *A. fabae* are "still unexplained". For the same relationship, i. e. *A. fabae* unsuitable for larvae of *A. bipunctata*, Blackman (1967a) assumes two factors: difficulty in ingesting the food and low nutritive value. The two may be related; nutritive value may be low because some essential nutrient remains behind in the uningested part of the prey.

Mechanical defences of the aphid are further phenomena which may be harmful to the predator. Large aphids of the genus *Macrosiphum* can smear the mouthparts of *Hippodamia convergens* with a glue excreted from the cornicles and thus cause starvation of the coccinellid (Palmer 1914). This defence by the aphid seems successful only at low temperatures; at higher temperatures the coccinellids can ingest the glue. A similar defence mechanism of aphids has been observed in the relation between *Hyalopterus pruni* and *Adalia bipunctata* (Hawkes 1920).

### 6.113 Criteria of suitability

In the preceding text enough evidence has been gathered together to show that coccinellids in spite of their considerable polyphagy as to accepted food, are very specific as far as essential food is concerned. Thus it is necessary to define the prey-predator relationship in these terms and to assess the possible impact of a coccinellid against any given pest, by testing the range of essential prey of the predator under consideration. Different criteria of suitability have been used.

The old — and as we already know — invalid criterion of food suitability was its acceptability. Kanervo (1940) evaluates the type of prey (principal, secondary, substitutive) according to the degree of readiness at the food intake in the laboratory and according to observations in the open. A somewhat better idea of the relationship between coccinellid and host aphid might appear to be deducible from preferences for certain prey; but Blackman's (1967b) experiments with *Megoura viciae* have shown that even this criterion is not safe.

The presence of developmental stages of the predator on a particular prey is frequently considered to be good evidence for evaluating food specificity in the field. Such evidence may, however, often be misleading because the predator usually lives in any given habitat together with several insect species, any of which may serve as its prey.

There are isolated cases where the relation of a predator to a prey may be safely identified by another indirect method: that of systematic and long term observation. Eastop and Pope (1966, 1969) found 99 % of specimens of *Pullus auritus* on oaks infested with *Phylloxera glabra*. In the course of five years (1963—1968), they also observed a close coincidence between the abundance of the prey and the predator.

In most cases, only experimental evidence is adequately unequivocal and precise. Under experimental conditions, responses of coccinellids to different species of prey in terms such as rate of larval development, larval mortality, weight of larvae or pupae or emerging adults, feeding rate, longevity of adults and fecundity may be studied. Thus one can determine the essential foods which are those that ensure the completion of larval development and oviposition, and one can determine their relative suitability.

### 6.114 Essential foods

Mention of some essential foods has been scattered throughout the previous sections of chapter 6.11 as contrasts to unsuitable foods. For a better orientation, these examples as well as other essential foods are brought together in Tab. 6.08.

Tab. 6.08 Essential food of *Coccinellidae*

| | | |
|---|---|---|
| *Adalia bipunctata* | | |
| | *Acyrthosiphon pisum* | Blackman 1965, 1967a, Hariri 1966a,b |
| | *A. pisum* (dry powdered) | Smith 1965a |
| | *Appelia tragopogonis* | Iperti 1965 |
| | *Aphis pomi* | Iperti 1965, Sem'yanov 1970 |
| | *Aulacorthum circumflexum* | Blackman 1965,1967a |
| | *Brachycaudus subterranea* | Iperti 1965 |
| | *Hyalopterus pruni* | Sem'yanov 1970 |
| | *Microlophium evansi* | Blackman 1965, 1967a, Hariri 1966a, b |
| | *Myzus cerasi* | Iperti 1965 |
| | *Myzus persicae* | Blackman 1965, 1967a |
| | *Rhopalosiphum maidis* (dry powdered) (slightly slower development) | Smith 1965a |
| | *Rhopalosiphum padi* (lower oviposition) | Sem'yanov 1970 |
| *Adalia decempunctata* | | |
| | *Aphis pomi* | Iperti 1965 |
| | *Brachycaudus subterranea* | Iperti 1965 |
| | *Cinara palestinensis* | Bodenheimer and Neumark 1955 |
| | *Matsucoccus josephi* | Bodenheimer and Neumark 1955 |
| | *Rhopalosiphum maidis* | Iperti 1965 |
| *Adonia variegata* | | |
| | *Aphis craccivora* | Iperti 1965 |
| | *Aphis nerii* | Iperti 1965 |
| | *Macrosiphoniella artemisiae* | Iperti 1965 |
| | *Myzus persicae* | Iperti 1965 |
| *Anatis mali* | | |
| | *Acyrthosiphon pisum* | Smith 1965a |
| | *Rhopalosiphum maidis* | Smith 1965a |

| | |
|---|---|
| *Anatis ocellata* | |
| *Myzus persicae* (less suitable) | Kesten 1969 |
| *Rhopalosiphum padi* (less suitable) | Kesten 1969 |
| *Schizolachnus pineti* | Kesten 1969 |
| *Anisolemnia (Caria) dilatata* | |
| *Oregma bambusicola* | Puttarudriah and Channabasavanna 1952 |
| *Aphidecta obliterata* | |
| *Adelges cooleyi* (lower fecundity) | Wylie 1958 |
| *Adelges nüsslini* | Wylie 1958 |
| *Calvia quinquedecimguttata* | |
| *Melasoma aenea* (pre-imag. stages) | Kanervo 1946 |
| *Chilocorus bipustulatus* | |
| *Chrysomphalus aonidum* | Yinon 1969a |
| *Chilocorus kuwanae* | |
| *Unaspis yanonensis* | Nohara 1962a |
| *Chilocorus rubidus* | |
| *Eulecanium caraganae* (eggs for larvae, eggs and larvae for adults) | Pantyukhov 1968b |
| *Chilocorus stigma (bivulnerus)* | |
| *Chrysomphalus aonidum* | Muma 1955b |
| *Coccinella septempunctata* | |
| *Acyrthosiphon pisum* | Blackman 1965, 1967a |
| *Aphis craccivora* (= *Pergandeida medicaginis*) | Hodek 1960b, Iperti 1965 |
| *Aphis fabae* | Hodek 1956, Blackman 1965, 1967a, Iperti 1965 |
| *Aphis gossypii* | Iperti 1965 |
| *Aphis urticae* | Iperti 1965 |
| *Hyalopterus pruni* | Hodek 1960b |
| *Lipaphis erysimi* | Atwal and Sethi 1963, Sethi and Atwal 1964 |
| *Longiunguis donacis* | Iperti 1965 |
| *Macrosiphoniella artemisiae* | Iperti 1965 |
| *Megoura viciae* (slightly less suitable) | Blackman 1965,1967a |
| *Myzus persicae* | Blackman 1965,1967a |
| *Uromelan aeneus* | Hodek 1960b |
| *Coccinella septempunctata bruckii* | |
| *C. s. bruckii* (eggs) (development prolonged) | Koide 1962 |
| *Macrosiphum granarium* | Hukusima and Sakurai 1964 |
| *Myzus malisuctus* | Hukusima and Sakurai 1964 |
| *Myzus persicae* | Hukusima and Sakurai 1964 |
| *Neophyllaphis podocarpi* | Maeta 1965 |
| *Rhopalosiphum prunifoliae* | Okamoto 1961, in Hodek 1966 |
| *Vesiculaphis caricis* | Takeda et al. 1964 |
| *Coccinella transversoguttata richardsoni* | |
| *Acyrthosiphon pisum* (dry powdered) | Smith 1965a |
| *Coccinella undecimpunctata* | |
| *Aphis pomi* | Harpaz 1958 |
| *Coccinella undecimpunctata aegyptiaca* | |
| *Aphis durantae* | Ibrahim 1955a |
| *Aphis gossypii* | Ibrahim 1955a |
| *Aphis laburni* | Ibrahim 1955a |
| *Aphis maidis* | Ibrahim 1955a |
| *Aphis nerii* | Ibrahim 1955a |
| *Aphis pseudobrassicae* | Ibrahim 1955a |
| *Hyalopterus arundinis* | Ibrahim 1955a |
| *Macrosiphum sanborni* | Ibrahim 1955a |
| *Myzus persicae* | Ibrahim 1955a |
| *Coleomegilla maculata* (ssp. *lengi* included) | |
| *Acyrthosiphon pisum* (also dry powdered) | Smith 1965c |

| | |
|---|---|
| *C. maculata* (eggs) | Warren and Tadic 1967 |
| *Hyphantria cunea* (eggs) | Warren and Tadic 1967 |
| *Rhopalosiphum maidis* (also dry powdered) (slightly less suitable) | Smith 1965c |
| ***Cycloneda limbifer*** | |
| *Aphis craccivora* | Zelený 1969 |
| ***Cycloneda munda*** | |
| *Acyrthosiphon pisum* | Smith 1965a |
| ***Exochomus flavipes*** | |
| *Dactylopius opuntiae* | Geyer 1947 |
| *Matsucoccus josephi* | Bodenheimer and Neumark 1955 |
| ***Halyzia hauseri*** | |
| *Podosphaera leucotricha* (apple powdery mildew) | Liu 1950 |
| ***Harmonia axyridis*** | |
| *Acyrthosiphon pisum* | Hukusima and Kamei 1970 |
| *Amphorophora oleraceae* | Hukusima and Kamei 1970 |
| *Aphis craccivora* | Mogi 1969 |
| *Aphis pomi* | Hukusima and Kamei 1970 |
| *Brevicoryne brassicae* (less suitable) | Hukusima and Kamei 1970 |
| *Hyalopterus arundinis* (less suitable) | Hukusima and Kamei 1970 |
| *Macrosiphum rosae ibarae* | Hukusima and Kamei 1970 |
| *Megoura viciae japonica* | Hukusima and Kamei 1970 |
| *Myzus persicae* | Hukusima and Kamei 1970 |
| *Neophyllaphis podocarpi* | Maeta 1965 |
| *Periphyllus californensis* | Hukusima and Kamei 1970 |
| *Rhopalosiphum prunifoliae* | Okamoto in Hodek 1966 |
| ***Hippodamia convergens*** | |
| *Therioaphis maculata* | Nielson and Curie 1960 |
| ***Hippodamia parenthesis*** | |
| *Acyrthosiphon pisum* (dry powdered) | Smith 1965a |
| ***Hippodamia quinquesignata*** | |
| *Macrosiphum pisi* | Kaddou 1960 |
| ***Hippodamia tredecimpunctata*** | |
| *Acyrthosiphon pisum* (dry powdered) | Smith 1965a |
| *Rhopalosiphum maidis* (dry powdered) (1/4 slower development) | Smith 1965a |
| ***Hyperaspis lateralis*** | |
| *Pseudococcus aurilanatus* | McKenzie 1932 |
| *Pseudococcus sequoiae* (eggs and young larvae of both species are preferred) | McKenzie 1932 |
| ***Leis conformis*** | |
| *Aphis durantae* | Moursi and Kamal 1946 |
| ***Pharoscymnus numidicus*** | |
| *Parlatoria blanchardi* | Kehat 1968a |
| *Prodenia litura (eggs)* | Kehat 1968a |
| ***Pullus auritus*** | |
| *Phylloxera glabra* | Eastop and Pope 1966, 1969 |
| ***Semiadalia undecimnotata*** | |
| *Aphis fabae* | Hodek 1960b, Iperti 1965 |
| ***Stethorus gilvifrons*** | |
| *Panonychus ulmi* | Dosse 1967 |
| *Tetranychus cinnabarinus* — complex (eggs) | Dosse 1967 |
| *Tetranychus urticae* (eggs) | Dosse 1967 |
| ***Stethorus punctillum*** | |
| *Metatetranychus ulmi* | Putman 1955 |
| *Tetranychus bimaculatus* | Putman 1955 |
| ***Synharmonia conglobata*** | |
| *Aphis pomi* | Iperti 1965 |
| *Brachycaudus subterranea* | Iperti 1965 |

| | |
|---|---|
| *Cinara palestinensis* | Bodenheimer and Neumark 1955 |
| *Galerucella lineola* (preimag. stages) | Kanervo 1946 |
| *Matsucoccus josephi* | Bodenheimer and Neumark 1955 |
| *Rhopalosiphum maidis* | Iperti 1965 |
| *Synonycha grandis* | |
| *Oregma bambusicola* | Puttarudriah and Channabasavanna 1952 |

## 6.12 Stenophagy and euryphagy

As has been stressed before, the well-known polyphagy of predaceous *Coccinellidae* only applies to their relation to accepted but inadequate foods; they in fact show a pronounced specificity with regard to their essential food, which is food that allows full development (both of larvae and ovaries). Profound studies evaluating the essential food of individual coccinellids are still so scarce that it is premature to try to compare the level of true stenophagy or polyphagy in individual species.

A very narrow range of prey nearing to monophagy may be presumed in stenotopic species, as e. g. *Coccinella hieroglyphica*, occurring on heather (*Calluna vulgaris*)(Table 6.11), but such cases are exceptional.

A good comparison of the polyphagy or stenophagy of several nearctic species has been obtained in the peach orchards of the Niagara Peninsula, Ontario (Putman 1957). Several of the common coccinellids were fed with either two tetranychid mites, crawlers of a coccid (Table 6.09) or eggs and larvae of the oriental fruit moth (*Grapholitha molesta*). A distinct polyphagy was revealed for *Coleomegilla maculata lengi*. The larval development of this species was not slowed down by feeding on *Tetranychus telarius* (compared with aphid food). All larvae completed their development, adult survival was very high and oviposition was not inhibited. *Metatetranychus ulmi* also appeared to be suitable food for larvae of *C. maculata*, and the adults fed much more on crawlers of *Pulvinaria vitis* than did the other coccinellids. In contrast to *C. maculata*, all three *Coccinella spp.* appeared to be rather stenophagous, and *Cycloneda* and *Adalia* were next in order in this respect. This experiment was, of course, designed to compare polyphagy among major taxonomic groups of prey.

It appears necessary that studies of this type should be taken further to assess polyphagy within individual groups of prey, i. e. the comparison of the relation of a series of coccinellids to several species of aphids or coccids etc. Some data in this direction have already been supplied for *Coccinella septempunctata* and *Adalia bipunctata* by Blackman (1967a, b) and for *Harmonia axyridis* and *C. s. bruckii* by some Japanese authors (Okamoto 1961, 1966, Hukusima and Sakurai 1964, Hukusima and Kamei 1970, Takeda et al. 1964).

Recently the value of several aphids has been compared in their effect on the fecundity of overwintered females of *Adalia bipunctata* (Tab. 6.10). *Rhopalosiphum padi* and *Acyrthosiphon caraganae* were found to be somewhat inferior to the other two species (Sem'yanov 1970). It was reported that, in spite of the different fecundities obtained, the rate of larval development was not significantly different when any of the first three species listed in the table were used as food.

The stenophagy of the *Coccinella spp.*, indicated by the findings of Putman (1957) (Tab. 6.09), seems confirmed by the failure of these coccinellids to complete their larval development on artificial diets (Smith 1965b) (see 6.14), and by the following results by this author involving natural foods.

Of three coccinellids tested on different plant pollens, *Coccinella trifasciata* gave the poorest results; only the 1st instar was completed (Smith 1961b). Two *Coccinella spp.*

Tab. 6.09 Effect of some prey on coccinellids of Ontario peach orchards (tabulated from data by

| Prey species | *Rhopalosiphum rufomaculatum* | | *Tetranychus telarius* (two-spotted spider mite) | | | | |
|---|---|---|---|---|---|---|---|
| Type of response / Coccinellid species | larval period [days] | initial number of beetles | larval period [days] | initial number of larvae | completion of larval development [%] | initial number of larvae | survival of adults at 20 days |
| *Coleomegilla maculata* | 11.7 | 10 | 12.9 | 9 | 100 | 17 | 85 |
| *Hippodamia convergens* | 12.0 | 8 | 21.8 | 8 | 61 | 23 | 33 |
| *H. parenthesis* | | | | | 80 | 15 | |
| *H. tredecimpunctata* | | | | | 60 | 15 | (50)[3] |
| *Adalia bipunctata* | 14.3 | 10 | 26 | 10 | 31 | 16 | (33) |
| *Cycloneda munda* [1] | | | | | 15 | 13 | 0 |
| *Coccinella transverso-guttata* | | | | | 0 | 16 | 0 |
| *C. novemnotata* | | | | | 0 | 16 | 0 |
| *C. trifasciata* | | | | | 0 | 16 | (17) |
| *Anatis quindecimpunctata* | | | | | 0 | 8 | |

[1]) In Putman 1957 *C. sanguinea:* in Putman 1964 corrected to *C. munda.*
[2]) Enough mites were available to feed the larvae for 6 days only; infestations of *M. ulmi* dense could not be found on foliage free of toxic spray residues.
[3]) Brackets indicate percentages which are considered unreliable because of the small number

Tab. 6.10 Fecundity of *Adalia bipunctata* on three aphids (Sem'yanov 1970)

| Aphid species | Generation | Total number of eggs laid/female | | | Number of pairs |
|---|---|---|---|---|---|
| | | max. | min. | average | |
| *Aphis pomi* | over-wintered (O). | 298 | 230 | 264 ± 20.1 | 5 |
| *Hyalopterus pruni* | | 262 | 215 | 241 ± 8.5 | 5 |
| *Rhopalosiphum padi* | | 186 | 147 | 161 ± 6.8 | 5 |
| *Hyalopterus pruni* | 1. | 98 | 86 | 91 ± 2.8 | 4 |
| *Acyrthosiphon caraganae* | | 42 | 36 | 39 ± 1.3 | 4 |

(*C. novemnotata, C. trifasciata*) could not even complete their larval development on dry powdered aphids. *C. transversoguttata* succeeded at least on one of the three aphids tested (on dry *Acyrthosiphon pisum*) (Smith 1965a).

*Coleomegilla maculata* is exactly the opposite and is perhaps the most euryphagous coccinellid known. Its polyphagy, however, has one peculiar feature. The microhabitat preference (lower strata of a corn field) already discussed in 5.41 is apparently related to the pollinivory of this species (Ewert and Chiang, in Hodek 1966 p. 195). Laboratory tests have proved that plant pollen serves as essential food for *C. maculata.* The pollens of *Betula populifolia* and *Carpinus caroliniana* were found to be superior to pollen of *Zea mays, Cannabis sativa, Juglans cinerea* and *Typha latifolia*, but all these pollens

Putman 1957)

| | | | *Metatetranychus ulmi* | | *Pulvinaria vitis* (crawlers) | | | |
|---|---|---|---|---|---|---|---|---|
| initial number of beetles | cessation of oviposition | initial number of beetles | larval development within 6 days[2]) | initial number of larvae | number of beetles | number eating crawlers | number of crawlers eaten max. | aver. |
| 13 | no cessation | 6 | normal | 10 | 11 | 9 | 123 | 24 |
| 12 | in 6 days | 5 | 50 % | 10 | | | | |
| | | | mortality | | 3 | 3 | 5 | 4 |
| 4 | in 2 days | 9 | | | | | | |
| 6 | in 6 days | 11 | 100 % | 10 | 15 | 8 | 26 | 4 |
| | | | mortality | | | | | |
| 10 | | | | | | | | |
| 6 | in 6 days | 14 | | | 10 | 3 | 13 | 7 |
| 3 | | | | | 5 | 2 | 5 | 1 |
| 6 | in 6 days | 7 | | | 10 | 6 | 36 | 11 |

enough to bring the larvae to maturity or to support the adults for an appreciable length of time of individuals involved.

enabled the completion of larvae development in *C. maculata* (Smith 1960, 1961b). By contrast, the pollens of coniferous trees, also those of *Quercus rubra* and *Ambrosia artemisiifolia* were unsuitable (Smith 1960, 1961b). When Atallah and Newsom (1966a) used stamens of cotton as a source of pollen, *C. maculata* larvae failed to develop and died within 4 days. Although *C. maculata* can develop satisfactorily also on different artificial diets (see 6.14), on mites (Putman 1957), live aphids (Smith 1965a), and also on lepidopterous eggs (Szumkowski 1955, Warren and Tadić 1967), it in fact shows a distinct specificity for powdered food. Only 30 % of the larvae completed development if fed with live *Acyrthosiphon pisum*, and the resulted adults weighed only 8.5 mg; the same prey in a dry powdered state, however, gave a much higher survival (90 %) and heavier adults (12.0 mg). Larval and pupal development were only slightly prolonged on living aphids (17.5 vs. 16.4 days) (Smith 1965a). In the comparison of dry aphids and pollen for suitability as food, dry aphids appeared marginally more suitable, but this difference was not consistent (Smith 1965a).

A limit to the polyphagy of *C. maculata* was shown in experiments where both adults and larvae refused nymphal leafhoppers of several species which were, however, eaten by larvae of *Coccinella novemnotata* and adults of *C. trifasciata* (Yadava and Shaw 1968). The suitability of this prey for the latter two species was, however, not determined.

*Synharmonia conglobata* shows tendency towards polyphagy: both aphids and pre--imaginal stages of chrysomelids were found to be essential prey for this species by Kanervo (1940, 1946). The same author studied *Calvia quinquedecimguttata* and claimed that this species was a specialized feeder on chrysomelids and that aphids were accepted as merely alternative prey (Tab. 6.11).

Tab. 6.11 Chrysomelids accepted as food by coccinellid predators of *Melasoma aenea* (Kanervo 1946) [1] [2]

| Coccinellid species | *Melasoma aenea* | | | *Phyllodecta atrovirens* | *Gastroidea polygoni* | | *Phaeodon cochleariae* | | *Plagiodera versicolora* | | | *Melasoma populi* | | | *Galerucella sagittariae* | | | *Galerucella lineola* | | *Galerucella tenella* | | *Agelastica alni* | | *Aphids psyllids* [3] |
|---|---|---|---|---|---|---|---|---|---|---|---|---|---|---|---|---|---|---|---|---|---|---|---|---|
| | E | L | P | L | E | L | E | L | E | L | P | E | L | P | E | L | P | E | L | E | L | E | L | |
| *Hippodamia septemmaculata* | (0) | (0) | (0) | — | — | — | 0 | — | — | — | — | — | — | — | — | — | — | — | — | — | — | — | — | principal |
| *Adalia bipunctata* | — | (0) | — | 0 | — | — | ++ | 00 | — | 00 | — | — | — | — | — | — | — | — | — | — | — | 00 | | principal |
| *Coccinella septempunctata* | 0 | — | — | 00 | — | — | ++ | — | — | — | — | — | — | — | — | (0) | — | 0 | — | — | — | 0 | — | principal |
| | | | | | principal | | princ. | | | | | | | | | | | principal | | | | | | |
| *C. quinquepunctata* | ⧻ | ++ | 0 | 00 | 000 | — | ⧻ | ++ | — | — | — | 00 | — | — | — | 00 | — | 000 | 00 | — | — | — | — | principal |
| *C. distincta* | 00 | — | — | — | — | — | — | — | — | 0 | 0 | — | — | — | — | — | — | 00 | — | — | — | 000 | — | principal |
| *C. hieroglyphica* | — | — | 0 | — | — | — | — | — | — | (0) | — | — | — | — | — | — | — | 00 | (0) | — | — | — | — | principal |
| | | | | | principal | | | | | | | | | | | | | principal | | | | | | |
| *Synharmonia conglobata* | ⧻ | ⧻ | ⧻ | 00 | 000 | 00 | — | — | — | 000 | — | 00 | — | — | | | — | ⧻ | ⧻ | 000 | 000 | — | — | principal |
| *Myrrha octodecimguttata* | 0 | — | — | — | — | (0) | — | — | — | — | — | — | — | — | — | — | — | 0 | — | — | — | — | — | principal |
| *Calvia quatuordecimguttata* | (0) | (0) | 0 | — | — | (0) | — | — | — | — | — | — | — | — | — | — | — | — | — | — | — | — | — | principal |
| | principal | | | | principal | | | | principal | | | | | | | | | | | | | | | |
| *C. quinquedecimguttata* | ⧻ | ⧻ | ⧻ | — | 000 | 000 | (0) | 0 | | | | | | | | | | | | — | — | — | 0 | |
| *Anatis ocellata* | 0 | 00 | 000 | — | — | (0) | — | — | — | — | — | — | — | — | — | — | 00 | — | 00 | — | — | — | — | principal |
| *Exochomus quadripustulatus* | — | — | ⧻ | — | — | — | 0 | — | — | — | — | — | — | — | 00 | — | — | — | — | — | — | — | — | principal |

[1]) accepted — *000* and ⧻: in abundance; *00* and ++: rather amply; *0* and +: rarely; *(0)* very rarely; observed — *0* in laboratory; + also in the open.
[2]) E — eggs, L — larvae, P — pupae.
[3]) Principal food is marked for comparison, according to Kanervo (1940).

The strange finding that the larvae of *Coccinella undecimpunctata* can complete their development when fed solely on fresh manure (Hawkes and Marriner 1927) has been questioned by recent authors such as Benham and Muggleton (1970). They believe that in making the observation, Hawkes and Marriner overlooked cannibalism. Also with *C. undecimpunctata*, Harpaz (1958) in Israel achieved the completion of larval development on *Aphis pomi*.

## 6.13 Cannibalism

It is even possible for individuals of the same or of other coccinellid species to serve as alternative food in the absence of other prey. Mainly eggs or soft individuals which have just moulted are consumed. This habit has been observed in the majority of papers dealing with coccinellids and is one of the main obstacles preventing simple mass production of coccinellids. In order for cannibalism to be eliminated the larvae, as soon as they hatch, must be isolated and be cultured in isolation. A good and more practical alternative is to provide an increased travelling space in the cages by filling them with fluffed-up wood shavings ("excelsior") (Shands et al. 1970).

The value to newly hatched larvae of cannibalising other eggs of the same batch is discussed in 6.31. Cannibalism increases the chances of survival of the larvae only when there is a very low density of prey.

In general, cannibalism can be considered as having a selective value for the species; however, when late instar larvae feed on pupae of their own species, there is probably no such advantage to the species (Way in Hodek 1966 p. 91).

Koide (1962) obtained complete larval development of *C. septempunctata* on an exclusive diet of eggs of their own species. This diet was less favourable for the larvae than were aphids: the rate of development decreased by 38 % and the larvae were 15 to 25 % smaller. *Coleomegilla maculata* larvae can also develop when reared on eggs of their own species, and consume on average 76.3 eggs during the whole development. The duration of larval plus pupal development (21.1 days at 26.7°) was slightly longer than when the larvae were fed on pig's liver (20.6 days) and increased markedly compared with that of larval feeding on *Hyphantria cunea* eggs (18.2 days) (Warren and Tadić 1967). In contrast to these two findings is the older observation on the coccidophagous species *Exochomus flavipes*. The larvae generally developed in a shorter time when fed on the eggs of their own species than when fed on their normal food, *Dactylopius opuntiae* (Geyer 1947).

## 6.14 Artificial diets

As the prey can be collected in the field only for part of the season and as the breeding of aphids or coccids requires a large investment in both space and man power, attempts have been made to feed predaceous coccinellids on substances other than live prey.

The first attempts were rather primitive. Hawkes (1920), for example attempted to rear the larvae of *Adalia bipunctata* on cooked or raw hens' eggs and on pounded dates. Considerable progress has since been achieved, even though successfull breeding, especially on chemically defined diets, is still rather a rarity.

The simplest way to compensate for a shortage of aphids from the field is to use them dried (Smith 1965a) or frozen (Shands et al. 1966). Smith (1965a) succeeded in rearing *Anatis mali*, *Adalia bipunctata*, *Hippodamia tredecimpunctata* and *Coleomegilla maculata* on dry powdered aphids, but failed with *Coccinella spp*. When fed quick-frozen aphids, development of larvae and adults of *Coccinella septempunctata* was the same as when fed living aphids (Shands et al. 1966). Similar evidence is quoted by Hagen (1962) for *Hippodamia convergens.*

There are two approaches for developing artificial foods for insects. The by far more complicated method is to prepare a mixture of chemically defined substances (holidic diet). This has been achieved with only a few insects, as the demands of the insect on its food are very complex and are not yet completely understood. Not only must the diet contain all necessary nutrients, but also the food must have the correct mechanical properties to enable the intake of food to occur (e. g. sufficiently small particles for lst instar coccinellid larvae — Smith 1960) and furthermore gustatory properties to induce feeding. In many insects the complex nature of feeding behaviour must also be considered as e. g. in the development of artificial food for the curculionid *Anthonomus grandis* (Vanderzant and Davich 1961), where the boll weevil will only accept food at the tip of a vertical stick.

Usually, however, an easier technique for preparing artificial foods for insects is to mix different natural substances (meridic diets). Substances such as honey, yeast, royal jelly, are known to be rich in most of the necessary nutrients, including carbohydrates, proteins, essential aminoacids, minerals, lipids and sterols. In some cases, some of these nutrients may need to be added in the chemically pure form. As yet, all attempts to breed coccinellids artificially have been based on meridic diets.

When it is necessary to transport coccinellids, droplets of honey mixed with agar have often been successfully used to increase survival. The task of composing an artificial diet, which can completely substitute for the essential food, is much more difficult. The best results have been achieved with *Coleomegilla maculata.* This is probably a reflection of the extensive polyphagy or rather pollinivory of this species (see 6.111 and 6.12). Several authors have succeeded in rearing this ladybird on laboratory diets. Szumkowski (1952) successfully (93 % development) used mammalian liver enriched with vitamin C for rearing this species. Later he found that a mixture of minced liver with an added 30—50 % honey solution was the best diet for oviposition, but the results were also satisfactory if liver was substituted by meat or fish and the honey by sugar. A mixture of protein hydrolysates and sugar solution was not so successful, and caused an increase in mortality of the beetles and a reduced oviposition (Szumkowski 1961a). For larval development, however, a mixture of fresh yeast (30—60 %) with 2—10 % glucose or sucrose solution was superior to the liver diet. Warren and Tadić (1967) used Szumkowski's first method (pig's liver with 1 or 2 drops of a liquid vitamin mixture — Deca-Vi-Sol) with success for ovipositing *C. maculata* females.

These diets although they had proved satisfactory for *C. maculata,* failed to support oviposition of females and development of larvae in *Cycloneda sanguinea* and *Hippodamia convergens* (Szumkowski 1961b). Hukusima and Sakurai (1963) tested a series of diets of which the least unsuccessful was quite a simple diet. This simple diet was composed of: grasshoppers (1 g), yeast (0.3 g), agar (0.7 g), honey (3 g) and sugar (8 g). This concentrate was made up with a mixture of water (25 cc) and milk (25 cc) with vitamins added in the form of a multivitamin preparation (0.3 g). When second instar-larvae of *Coccinella septempunctata bruckii* were fed on this food, about 10 % completed larval development and became adults. When lst instar larvae were put on the diet, they died in the 3rd instar.

Similar lack of success with a series of 25 diets is reported by Tanaka and Maeta (1965). The diets had value only as alternative food for all three species tested (*Harmonia axyridis, C. septempunctata bruckii* and *Chilocorus kuwanae*). The females could not mature or oviposit, and the larvae could not develop from lst instar to adult.

Smith (1965a, b) tested several meridic, mostly powdered diets. A diet based on brewer's yeast and sucrose, with other growth factors added (Tab. 6.12), enabled a higher percentage of *C. maculata* larvae to complete larval development (86 %) and at a higher developmental rate (21.3 days) than if they were reared on corn pollen (52 %, 22.9 days) (Smith 1965a). This diet was, however, not satisfactory for oviposition by adults; for this purpose, the yeast had to be replaced by a liver preparation (Smith 1965a). The diets were prepared by dehydrating an aqueous mixture of the ingredients and grinding the residue to a fine powder. Three other similar dry foods and a banana diet were tested for adults of 13 species (Smith 1965b). All species accepted at least some of the diets tested, but only 8 species oviposited (Tab 6.13). In these latter tests, Smith did not compare fecundity on artificial foods with that on natural foods as he had done for *C. maculata* in his earlier tests (Smith 1965a). For the remaining 5 species, among them all three *Coccinella spp.*, the mixtures served only as alternative food, and merely increased longevity without promoting oviposition. On three diets oviposition by *Hippodamia convergens* was achieved. Hagen (1962, in de Bach, 1964) also obtained some eggs from *Hippodamia convergens* adults that had overwintered, after feeding them with a fluid consisting of water, fructose, enzymatic protein hydrolysate of yeast, choline chloride, cholesterol, vitamins and standard nutritional salt mixture.

Atallah and Newson (1966) formulated and tested 16 experimental diets, based on chemical analysis of aphids and calf liver, for rearing *Coleomegilla maculata.* Six diets were satisfactory for larval development and oviposition. The final and most satisfactory of the diets[1]) (Tab. 6.12, No. 16) differed from all the others in two respects. Double the amount of liver preparation was included, but a more important addition was an extract from 50 cotton leaves (carotenoids and sterols). Beetle copulation was increased when the diet also included 0.2 ml vitamin E. Eight generations were reared on diet no. 16 without any reduction in viability after the 2nd generation. The diet failed, however, to support other coccinellids tested (*Coccinella novemnotata, Cycloneda spp., Hippodamia convergens, Olla abdominalis*).

The above paragraphs tell a story of little success in rearing species other than *C. maculata* on artificial diets, and no complete successes with *Coccinella spp.* Good results were reported by two authors who added a small amount of natural food to their artificial diets; these achievements could not, however, be repeated by other authors (Tanaka and Maeta 1965, Iperti pers. comm.). Smirnoff (1958) claimed good results with 18 species (aphido-, coccido-, acaro-, and mycophagous); he added a small amount of pulverized dry prey to a medium otherwise made from water, agar, cane sugar, honey, alfalfa flour, yeast, and royal jelly. He believed that royal jelly was essential to the formula because of its content of growth stimulating factors and aseptic qualities. This diet was reported as excellent for adults, and was supplemented with beef jelly for larvae. Both rate of development and longevity were even higher than obtained on natural food, fecundity was not recorded. Another medium with added prey has been reported very successful in the rearing of *Cryptolaemus montrouzieri* by Chumakova (1962). The medium was composed of water, agar, sucrose, honey, casein, peptone, corn oil, vitamin E, cystine, yeast autolysate, and cotton oil. She even achieved an increased oviposition on a similar diet; no coccids were added, and peptone was replaced by nucleic acid. The larvae were reared on a more complicated diet, but, nevertheless suffered from a high mortality.

[1]) The technique for preparing the diet:
1. Half the amount of water was mixed thoroughly for 3—4 min. with all the ingredients except agar.
2. The other half of the water was heated to boiling, then the agar was added and carefully dissolved. The agar solution was then added to the mixture and blended at high speed for 1,5 min.
3. The diet was poured while still hot into the containers, where it solidified.

Tab. 6.12 Artificial diets for *Coccinellidae*

| | Smith 1965a[1]) | | Smith 1965b[2]) for adults | | | | Atallah and Newsom 1966[3]) for both larvae and adults | | |
|---|---|---|---|---|---|---|---|---|---|
| | for larvae | for adults | banana diet | dry diet 1 | dry diet 2 | dry diet 3 | diet 13 | diet 14 | diet 16 |
| water | — | | | | | | 150 ml | 200 ml | 200 ml |
| salt mixture | 1.5 % | | | | | | | | |
| Wesson' s salt | | | | | | | 1.25 g | 1.25 g | 1.25 g |
| brewer's yeast | 40 % | | | 30 % | 10 % | 10 % | | 1.0 g | 1.0 g |
| Difco egg-meat | | | | 40 % | | | | | |
| Difco liver | | 40 % | | | | 40 % | | | |
| liver extract 1—20 | | | | | | | | 3.0 g | 3.0 g |
| liver fraction L | | | | | | | 3.0 g | | 3.0 g |
| fresh banana | | | 90 % | | | | | | |
| pollen | | | | | | | 3.0 g | | |
| casein | | | 9 % | | 40 % | | 7.0 g | 6.0 g | 6.0 g |
| soybean hydrolysate | | | | | | | 3.0 g | 3.0 g | 3.0 g |
| sucrose | 55.0 % | | | 30 % | 50 % | 50 % | 3.0 g | 2.0 g | 3.0 g |
| dextrose | | | | | | | | 1.0 g | 1.0 g |
| glycogen | | | | | | | 2.1 g | 2.0 g | 2.0 g |
| cholesterol | 0.3 % | | | | | | | | |
| butter fat | | | | | | | 2.5 g | 2.0 g | 2.0 g |
| ribonucleic acid | 1.5 % | | | | | | | | |
| corn oil | | | | | | | 2.5 ml | 2.0 ml | 2.0 ml |
| wheat germ oil | 1.5 % | | 0.9 % | | | | | | |
| wheat germ | | | | | | | 4.0 g | 4.0 g | 4.0 g |
| choline chloride | < 0.1 % | | | | | | | | |
| inositol chloride | < 0.1 % | | | | | | | | |
| sorbic acid | | | | | | | 0.1 g | 0.1 g | 0.2 g |
| ascorbic acid | | | | | | | 0.4 g | 0.4 g | 0.5 g |
| carotenoids + sterols extr. from cotton leaves | | | | | | | | | from 50 leaves |
| vitamin solution[5]) | | | | | | | 3.0 ml[5]) | 2.0 ml[5]) | 3.0 ml[5]) |
| B-vitamins[4]) | [4]) | | c. 0.1 % | | | | | | |
| tegosept (methyl p-hydroxybenzoate) | | | | | | | 0.1 g | 0.2 g | 0.2 g |
| aureomycin | | | | | | | 0.06 g | 0.06 g | 0.06 g |
| agar | | | | | | | 5.0 g | 4.0 g | 4.0 g |

[1]) [3]) For results with these diets see Tab. 6.13.
[4]) All three diets enable both larval development and egg laying in *Coleomegilla maculata*; No. 16 is the most satisfactory.
[3]) B-vitamins, for each 10 g of diet approx.:

| | |
|---|---|
| niacinamide | 5.0 mg |
| Ca-pantothenate | 2.5 mg |
| thiamine-HCl | 1.5 mg |
| riboflavin | 1.2 mg |
| pyridoxine-HCl | 1.2 mg |
| folic acid | 0.1 mg |
| biotin | 0.02 mg |
| vitamin B 12 | 0.002 mg |

[5]) Vitamin solution:

| | |
|---|---|
| distilled water | 100 ml |
| niacinamide-HCl | 100 mg |
| Ca-pantothenate | 100 mg |
| thiamine-HCl | 25 mg |
| riboflavin | 50 mg |
| pyridoxine-HCl | 25 mg |
| folic acid (crystal.) | 25 mg |
| biotin (crystal.) | 2 mg |
| vitamin B 12 | 0.2 mg |

Tab. 6.13 Laboratory diets (table 6.12) tested on several *Coccinellidae* (Smith 1965 b) (A — accepted, O — promoted oviposition)

| Coccinellid species | Diet | | | | | Longevity |
|---|---|---|---|---|---|---|
| | fresh liver | banana diet | dry diet 1 | dry diet 2 | dry diet 3 | |
| *Adalia bipunctata* | A | O | A | | O | 66.4 ( 55 — 88) |
| *A. frigida* | | A | A | O | A | 121.4 ( 45 — 173) |
| *Anatis mali* | O | O | A | A | O | 52.6 ( 26 — 87) |
| *Coccinella novemnotata* | | A | | | | 21.8 ( 17 — 36) |
| *C. trifasciata perplexa* | A | A | A | A | A | 55.1 ( 43 — 72) |
| *C. transversoguttata richardsoni* | | A | | | | 66.8 ( 27 — 112) |
| *Coleomegilla maculata lengi* | O | O | O | A | O | |
| *Cycloneda munda* | A | O | | | | 53.7 ( 33 — 126) |
| *Hippodamia convergens* | O | O | A | O | | 134.3 ( 76 — 167) |
| *H. glacialis* | | O | A | | | 120.6 ( 78 — 198) |
| *H. parenthesis* | A | A | A | A | A | 40.8 ( 29 — 75) |
| *H. tredecimpunctata tibialis* | A | O | A | A | A | 55.3 ( 36 — 108) |
| *Mulsantina sp.* | | A | | A | A | 134.3 (106 — 191) |

In view of the current position that artificial foods for the majority of typically carnivorous coccinellids have proved comparative failures, it is understandable that a number of experimenters have endeavoured to use as food insects which are easier to rear than are the typical prey (like aphids, coccids, mites). Thus Sundby (1966) succeeded in rearing *C. septempunctata* on eggs of *Ephestia kuehniella* and Kesten (1969) alternated *Musca* larvae with bananas daily, and frequently achieved a completion of larval development and pupation in *Anatis ocellata.* The adults, however, died in the process of emerging from the pupa or shortly afterwards. Surprisingly enough, the females oviposited for months if fed only on bananas by which they were strongly attracted. A claim that is very interesting is that continuous breeding of *Harmonia axyridis* has been achieved on frozen or lyophilized larvae or young pupae of drone honey-bees (Okada 1970). Ten generations were bred without any apparent decrease in vitality. If this food proves as 'essential' also for other species — particularly for the difficult *Coccinella spp.* — there may here be an important step forward in the study and utilization of coccinellids. The lack of a suitably practicable food is a barrier not only to making mass rearing of coccinellids for release an economic proposition, but also to providing sufficient specimens for experiments so that significant results can be obtained. It is no exaggeration to claim that the discovery of such food is a prerequisite for further qualitative progress in coccinellid studies.

In the event that this aim cannot be reached through some fortunate discovery involving a natural substitution (as mentioned above this discovery may already be with us), it will be necessary to search for it systematically by analysing natural food and by elimination experiments; a process which has already been carried out for a series of phytophagous insects. The study by Atallah and Killebrew (1967) of the amino-acid requirements of adult *Coleomegilla maculata* is perhaps the only work of this kind as yet done with coccinellids. Instead of eliminating the individual acids from the diet, the authors used a faster and more elegant method with labelled compounds. They fed the beetles with a metabolic precursor, $^{14}C$-labelled sodium acetate. After 4 hr, the beetles were homogenized and the activity of the individual amino-acids (separated chromatographically) was measured. Since non-essential amino-acids can by definition be synthesized by the animal, the labelled precursor will be incorporated in them. Glycine, serine, aspartic acid, glutamic acid, proline and lysine showed high activity. These must therefore be synthetized in vivo and thus are non-essential. In contrast, threonine, phenylalanine, isoleucine and valine had a negligible activity, and are therefore apparently essential or derived from essential precursors.

## 6.15 Food of phytophagous Coccinellidae

Food specificity has also been described in phytophagous coccinellids. In Japan, *Epilachna vigintioctomaculata* prefers potato plants and its larvae cannot develop on eggplant, in contrast to *E. sparsa orientalis* for which evidently both plants are essential food (Iwao 1954). For *E. pustulosa*, which some workers regard as only a biological race of *E. vigintioctomaculata*, thistle (*Cirsium spp.*) is the normal food. However, a certain proportion of the beetles, which differs in various localities, accept potato as food. Within any one population the individual specimens even show a variable and apparently inherited response to potato. Tab. 6.14 shows a positive correlation between the acceptance of potato by parent females and the ability of their larval progeny to develop on this food. Moreover, the acceptance of potato by the young adults is related to the food they receive as larvae; only 31.3 % of individuals which

anb. 6.14 Development of *Epilachna pustulosa* on thistle (*Cirsium microspicatum* var. *kiotoense*) Tad potato (Iwao and Machida 1961)

| Parent female | Larval progeny | | | |
|---|---|---|---|---|
| | Food | Completion of development [%] | Duration of larval development [days][1]) | N |
| rejected potato (even in absence of other food) | thistle | 69.0 | 18.2 | 58 |
| | potato | 43.1 | 16.0 | 65 |
| accepted potato (only in absence of other food) | thistle | 75.3 | 18.5 | 89 |
| | potato | 50.5 | 17.2 | 97 |
| accepted potato (in presence of thistle) | thistle | 73.7 | 18.2 | 57 |
| | potato | 60.0 | 16.6 | 60 |

[1]) 27 °C 16 hr light.

had developed on thistle accepted potato as adults while 74.2 % of those which had developed on potato did so. However it is not certain, whether this was due to pre-imaginal "conditioning" or to negative selection, i. e. there may have been a high larval mortality of those individuals which have a low ability to sustain themselves on potato. The percentage of beetles which accept potato, can be markedly increased in another way; they can be "conditioned" in the first days of adult life (Iwao and Machida 1961).

Klausnitzer (1965) failed to obtain eggs from *Epilachna argus* which had been collected on *Bryonia alba* and subsequently fed on *Bryonia dioica*. Dissection of dead females revealed an abundance of matured eggs.

*Epilachna vigintioctopunctata*, which feeds on *Luffa aegyptiaca* in India, prefers to feed on the corolla of the flower of this cucurbitaceous plant if under illumination, but makes no distinction between flowers and leaves in the dark (Sinha and Krishna 1969). A pest of cucurbits, *Henosepilachna elaterii* ( = *Epilachna chrysomelina*) has been recorded also feeding on aphids (El Khidir 1969).

As well as the above species (all *Epilachninae*) which feed on vegetative parts of plants, there is an isolated genus *Bulaea* in the *Coccinellinae* which has obviously specialised on a different form of phytophagy. Capra (1947) reported that *B. lichatschovi* and other closely related species of the genus were "predominantly if not exclusively" pollinivorous both as larvae and adults, with a preference for *Chenopodiaceae*. *B. lichatschovi* is also considered as phytophagous by other authors (Dyadechko 1954, Bielawski 1959, Savoïskaya 1966, 1970a) and in the two last mentioned papers pollinivory is reported from Kazakhstan on various plants (*Euphorbia*, *Artemisia*, *Eurotia*, *Atriplex*, *Nitraria*, *Tamarix*, *Clematis*). In the last mentioned paper, various types of phytophagy are reported for this species from central Asia: feeding on pollen (*Tamarix*, *Euphorbia*, *Artemisia*, *Eurotia ceratoides*, *Atriplex*), on nectar (*Nitraria*, *Clematis*) and on leaves (sugar-beet, young apple trees).

There is much less information on the genus *Verania* for which pollinivory has also been reported (quoted by Goidanich 1943). Pollinivory in *Coleomegilla maculata* has already been discussed in 6.111 and 6.12.

## 6.16 Food of mycophagous Coccinellidae

The coccinellids comprising the mycophagous genera *Halyzia*, *Vibidia* and *Thea* have long been mistakenly considered as carnivorous. This mistaken idea is evident both in observations listed by Schilder and Schilder (1928), Börner and Heinze (1957), Fulmek (1957), and from the taxonomic placing of these three genera as subgenera of the genus *Halyzia* together with the other subgenera of this genus (*Sospita*, *Calvia*, *Myrrha*, *Propylaea*) in spite of the fact that morphologically they are very clearly distinguishable by the shape of their mandibles (see 1.1/Head). The history of how the ecological and morphological differentiation of the members of the tribus *Psylloborini* ( =*Halyziini* sensu Bielawski 1959) came to be realised and the consequences for taxonomy were discussed by Strouhal (1926) who also gave a detailed comparison of adult and larval mouthparts in the different species and a survey of observations on mycophagy to that date.

Since then the members of the three genera mentioned above as well as of other genera of this tribe (for example *Illeis*/Liu 1950/) have been reported as feeding on lower fungi, particularly the powdery mildews (*Erysiphaceae*); no experimental work has yet been done on their food. Strouhal's (1926) original concepts remained basically unchanged in later discussions of the topic by Goidanich (1943) and Capra (1945) who added further observations.

*Tytthaspis sedecimpunctata* is classified into the tribe *Coccinellini* and is usually reported as aphidophagous (Dyadechko 1954, Sem'yanov 1965a, Klausnitzer 1966), although Dauguet already in 1949 wrote: "serait mycophage". No precise information on its food had been published until Turian (1969) reported on observations and preliminary experiments demonstrating that *T. sedecimpunctata* feeds also on *Erysiphaceae*, like the *Halyziini*. According to Turian, *T. sedecimpunctata* adults have no specificity as to various species of mildew but this has only been established with respect to food intake. Turian observed *Thea vigintiduopunctata* larvae and adults also feeding on various *Erysiphaceae* and suggests the term micromycetophagy to describe feeding on lower fungi (*Micromycetes*).

# 6.2 Food consumption

A great deal has been published about the amount of food consumed by larvae and adults of individual coccinellid species. It does not seem particularly helpful to list here all the accumulated numerical data since they can hardly be adequately compared. Even within a group with the same prey-relations (aphidophagous, coccidophagous, acarophagous) the individual findings are based on different species and developmental stages of prey and have been obtained under a variety of physical conditions. Thus, for any one coccinellid species, figures quoted for the amount of food vary enormously.

Moreover, there are several different ways of measuring food consumption. The base-line mortality of prey may be checked in cages identical to experimental ones, with the same amount of prey but without predators. The base-line value thus obtained is subtracted from the number of dead aphids in the experimental cages, using the formula (Hodek 1956)

$$f = n - (s_{24} + m_{24})$$

where $f$ is the daily feeding rate; $n$ is the daily supply of prey, $s_{24}$ is number of prey survived after 24 hr, $m_{24}$ is mortality of aphids within 24 hr in the check cages (base-line mortality). Alternatively, Kaddou's method (1960) may be used. He felt it possible to identify the aphids which have died from natural causes and eliminate them from the calculations. Each prey killed is then classified arbitrarily by the proportion of it which has been consumed, and these fractions are summed up. Kaddou quotes as separate figures the number of prey killed and the units consumed. Most authors have not specified how they obtained their measure of feeding rate. A promising approach for collecting data in a more defined and more easily comparable way, is to express the quantity of prey consumed in terms of biomass, i. e. in weight units (Blackman 1967a). However, this approach requires the precaution that changes in weight by evaporation must be taken into account.

To give some idea of the variability in figures quoted for food intake, Tab. 6.15 presents some data for *Coccinella septempunctata*. In aphidophagous *Coccinellidae* the daily feeding rate of adults usually amounts to about 100 aphids (the females, particularly if ovipositing, feed more than males) and the larvae consume several hundreds of aphids (mostly 200—600) during their entire development. Usually, about 60—70 % of the total food intake is consumed during the 4th instar (Tab. 6.16). Naturally, the smaller coccinellids take less food than the larger ones.

Tab. 6.15 Feeding rate of *Coccinella septempunctata* (Hodek and al. 1965)

| Author | Food intake (in number of aphids) by larvae during the whole larval development | by adults (daily average) | Aphid species |
|---|---|---|---|
| Smee | 305 | — | *Microlophium urticae* |
| Moroshkina | — | 243 | *Schizaphis graminum* |
| Jöhnssen | 624; 846 | — | *Aphis fabae* |
| | | | *Brevicoryne brassicae* |
| | | | *Acyrthosiphon pisum* |
| | | | *Hyalopterus pruni* |
| Bodenheimer | 629; 1930 | — | *Aphis frangulae* |
| | | | *A. asclepiadis* |
| Volkov | 200 | | *Dactynotus sonchi* |
| | 297 | | *Rhopalosiphon padi* |
| | 316 | | *Brevicoryne brassicae* |
| | 373 | | *Schizaphis graminum* |
| Dunn | 100 | | *Macrosiphum euphorbiae* |
| Blattný | up to 1000 | 92 | |
| Dyadechko | 665 (max. 668) | — | |
| Hodek | 216; 275 const. temp. | — | *Aphis fabae* |
| | 506 fluct. temp. | — | *Aphis fabae* |
| Weismann. Valló | 239; 229; 339 const. temp. | — | *Aphis fabae* |

## 6.21 Effect of physical factors on feeding rate

Naturally, the daily feeding rate both of larvae and adults increases with temperature within the favourable range. There is, however, a certain discrepancy between the data of individual authors (Tab. 6.17) concerning the effect of temperature on total food

Tab. 6.16 Proportion of food intake in individual larval stages

| Coccinellid species (conditions) | Author | Food intake in larval instars [%] I | II | III | IV | V[1]) |
|---|---|---|---|---|---|---|
| *Coccinella undecimpunctata* | Ibrahim | 6.4 | 12.1 | 16.7 | 64.6 | |
| *aegyptiaca* | 1955b | 6.1 | 11.8 | 19.8 | 62.3 | |
| *Coccinella septempunctata* | Asgari 1966 | | | | | |
| (15° C/darkness) | | 4.5 | 17.0 | 29.2 | 49.3 | |
| (20° C/darkness) | | 3.2 | 6.5 | 9.6 | 80.6 | |
| (20° C/8 hr light) | | 4.2 | 13.3 | 25.5 | 57.0 | |
| (20° C/12 hr light) | | 5.3 | 13.4 | 24.2 | 57.1 | |
| (20° C/16 hr light) | | 4.1 | 10.4 | 31.7 | 53.5 | |
| (25° C/darkness) | | 5.8 | 7.8 | 9.7 | 76.7 | |
| *Coccinella septempunctata bruckii* | Koide 1961 | 8.0 | 5.6 | 14.5 | 71.9 | |
| *Adalia bipunctata* | | | | | | |
| (*Aphis pomi*) | Sem'yanov | 8.2 | 26.8 | 29.2 | 45.7 | |
| (*Hyalopterus pruni*) | 1970 | 7.3 | 15.3 | 29.6 | 47.8 | |
| *Cycloneda limbifer* | Zelený 1969 | 1.1. | 4.7 | 23.5 | 70.6 | |
| *Leis conformis* | Moursi, Kamal 1946 | 4.9 | 7.0 | 18.5 | 69.6 | |
| *Chilocorus bipustulatus* | Yinon | 10.8 | 15.6 | 28.6 | 44.7 | |
| | 1969 | 6.2 | 8.9 | 16.3 | 25.5 | 43.0 |
| *Exochomus flavipes* (own eggs) | Geyer 1947a | 4.0 | 5.6 | 16.9 | 73.5 | |

[1]) Some proportion of laboratory cultured larvae of *C. bipustulatus* revealed 5 instars see 4. 11.

consumption during the entire larval development. Whereas Hodek (1957b), Kaddou (1960 — only his data for the lower temperature range below 27 °C) and Sundby (1966) found a slight increase of total consumption with increased temperature, Kaddou (1960 — at temperatures above 27 °C), Asgari (1966), Ellingsen (1969a), Gurney and Hussey (1970) reported the reverse. With the exception of the data of Gurney and Hussey (1970), however, the differences are rather small, so that the general impression is that the total larval food consumption of *Coccinellidae* is more or less stable irrespective of temperature. Kehat (1968a) reported that the total feeding capacity of larvae of *Pharoscymnus numidicus* was not affected by temperature. This would be in accordance with Rubner's hypothesis and with results obtained with other insects [Rubner 1908 (in Allee et al. 1949)].

The inconsistent food consumptions referred to above may have been subject to various inaccuracies arising from the methodological difficulties discussed earlier. That this is so is partly indicated by the rather great inconsistencies in data obtained at the same temperature but in different photoperiods (Asgari 1966 — Tab. 6.17).

While the total larval food consumption seems to be similar at different constant temperatures, it was found to be doubled when larvae of *C. septempunctata* were kept under naturally fluctuating temperatures in summer (Hodek 1957b). When Sundby (1966) alternated two constant temperatures, 16 and 21 °C, food consumption remained unaffected by the alternation. In *Adalia bipunctata*, Ellingsen (1969b) found an increase (though only by 10.6 %) of total larval food consumption if the larvae were reared under an artificial diurnal cycle of temperature ranging between 8 and 28 °C with

Tab. 6.17 Total food consumption of larvae in relation to temperature

| Author | Prey species | Coccinellid species | Temperatures [°C] 15 | 16 | 17 | 18 | 19 | 20 | 21 | 22 | 23 | 24 | 25 | 26 | 27 | 28 | 29 | 30 | 31 | 32 | 33 | 44 | 35 |
|---|---|---|---|---|---|---|---|---|---|---|---|---|---|---|---|---|---|---|---|---|---|---|---|
| Jöhnssen 1930 | several | *Coccinella septempunctata* | | | | | | | 846 | 624 | | | | | | | | | | | | | |
| Bodenheimer 1943 | several | *Coccinella septempunctata* | (629[1]) | | | | (1930)[1]) | | | | | | | | | | | | | | | | |
| Hodek 1957b | *Aphis fabae* | *Coccinella septempunctata* | | | | 216 (const. light) | | | | | | | | 276 (const. light) | | | | | | | | | |
| Kaddou 1960 | *Macrosiphum pisi* | *Hippodamia quinquesignata* | | 190 | | | | | | | | | | | 241 | | | 216 | | 158 | | | |
| Sundby 1966 | *Myzus persicae* | *Coccinella septempunctata* | | 399 | | | | | | 420 | | | | | | | | | | | | | |
| Asgari 1966 | *Aphidula pomi* | *Coccinella septempuncatata* | 1550 (dark?) | | | | | 1236 (dark)<br>1063 (8hr light)<br>1465 (12hr light)<br>991 (16hr light) | | | | | 1010 (dark?) | | | | | | | | | | |
| Ellingsen 1969a | *Myzus persicae* | *Adalia bipuncatta* | | | | 188 | | | | | | | | | | 168 | | | | | | | |
| Gurney and Hussey 1970 | *Myzus persicae* | *Coleomegilla maculata* | | 652 | | | | | 449 | | | 359 | | | | | | | | | | | |
| | | *Cycloneda sanguinea* | | 320 | | | | | 292 | | | 205 | | | | | | | | | | | |

[1]) the data were achieved in different generations in the open and are here plotted according to the aver. of temperatures.

an average of 18 °C. That fluctuating temperatures have a stimulating effect is also indicated by the fact that pupae of *C. septempunctata* derived from larvae reared under the above mentioned artificial cycle were heavier (Sundby unpubl., in Ellingsen 1969b).

There are some indications (Hodek et al. 1965) which suggest that at lower relative air humidities food intake may be increased, particularly when the prey is aphid. As lower humidities are usually combined with higher temperatures, it is difficult to separate both effects. It seems quite probable, however, that at lower humidities the coccinellids may — in the absence of water — consume more of a prey with a high water content to make up for the decrease in water content they suffer from increased transpiration.

As far as the effect of light conditions on food consumption is concerned, only the results of Asgari (1966) are available, from which no consistent conclusions can be drawn.

## 6.22 Response to prey density

Within certain limits, food consumption seems to be positively correlated with the population density of the prey. This follows from the findings on adult *Coccinella septempunctata* (Popov 1960) (Tab. 6.18) and larvae of *Harmonia axyridis* (Mogi 1969)

Tab. 6.18 Effect of prey density (*Aphis gossypii* and *A. fabae*) on feeding rate of *Coccinella septempunctata* adults (Popov 1960)

| Number of aphids offered to 5 beetles | Daily average number of aphids consumed by 1 beetle |
|---|---|
| 125 | 21.1 |
| 250 | 40.2 |
| 375 | 59.8 |

(Fig. 6.05). Thus, coccinellids can act as a density-dependent mortality factor for aphids, although such an application of the above laboratory results to the field may not be fully justified. To emphasize this latter point, field observations do not confirm for *Adalia bipunctata* the dependence ascertained in the laboratory for *Coccinella septempunctata*. The proportion of *Drepanosiphum platanoides* taken by *A. bipunctata* on sycamore trees did not increase as the prey population density rose. Thus *A. bipunctata* does not regulate populations of *D. platanoides* (Dixon 1970).

The increase in the number of aphids killed at high aphid density may be the result of two different phenomena operating simultaneously. Firstly it has been observed that hungry coccinellids completely devour the first few prey they tackle but exploit subsequent prey with a gradually decreasing efficiency. Thus, with an increase in prey population density, an increasing proportion of partially consumed prey may be left. Secondly, when prey is abundant, the larvae may also consume considerably more than the minimum required for their development (see below). An ability to adjust to variable resources of food is very important if predators are to have an effective impact on their prey.

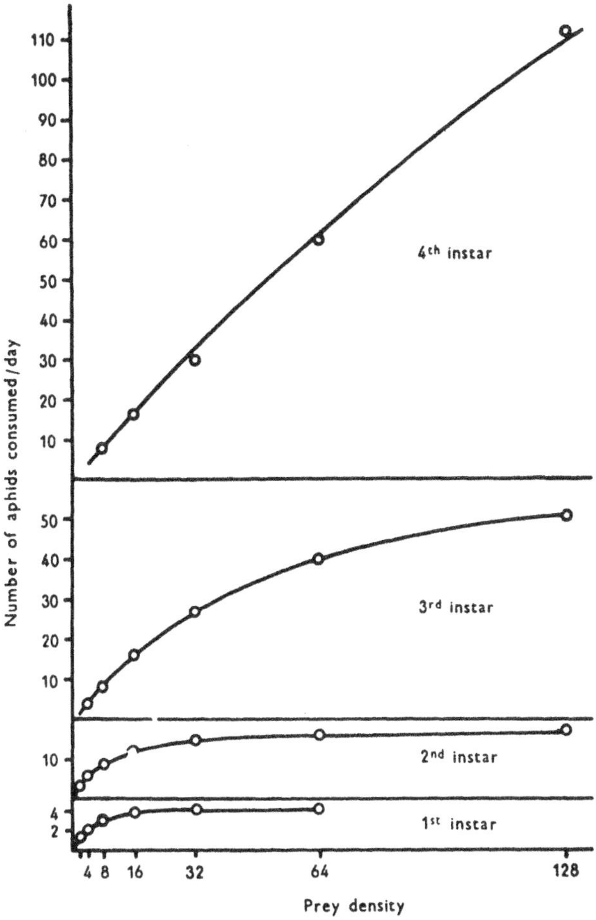


Fig. 6.05 Effect of prey density (*Aphis craccivora*) on the feeding rate of larvae of *Harmonia axyridis* (Mogi 1969).

## 6.23 Minimum food requirement

Larvae of *Coccinella septempunctata* complete their development even when their total food consumption is artificially reduced to 55 or even 40 per cent by limiting their daily supply in the 4th larval instar to 30 or even 10 aphids (*Aphis fabae*) (Tab. 6.02) (Hodek 1957a). Sundby (1966) verified this by showing that the larval period of *C. septempunctata* can be completed on only 7 aphids (*Myzus persicae*) per day. This feeding rate was only one-third of the normal value. Most of the larvae pupated, but the emergence of adults was low (only 3 out of 8). These adults were smaller and produced less eggs.

Similar observations have also been made with other coccinellids. The females of *Coccinella undecimpunctata aegyptiaca* began to oviposit with only 30 aphids per day and until 65 aphids were supplied per day there was a positive correlation between the consumption of food and egg-production. Longevity remained low up to 45 aphids per day (Fig. 6.06) (Ibrahim 1955b). The minimum supply on which larvae could complete their development was 15 *Aphis durantae* per day. An increase to 45 aphids per day shortened the duration of development and any further increase merely resulted in an increase in the size of adults produced (Ibrahim 1955b). Kaddou (1960) reported analogous results on larvae of *Hippodamia quinquesignata* (Tab. 6.19).

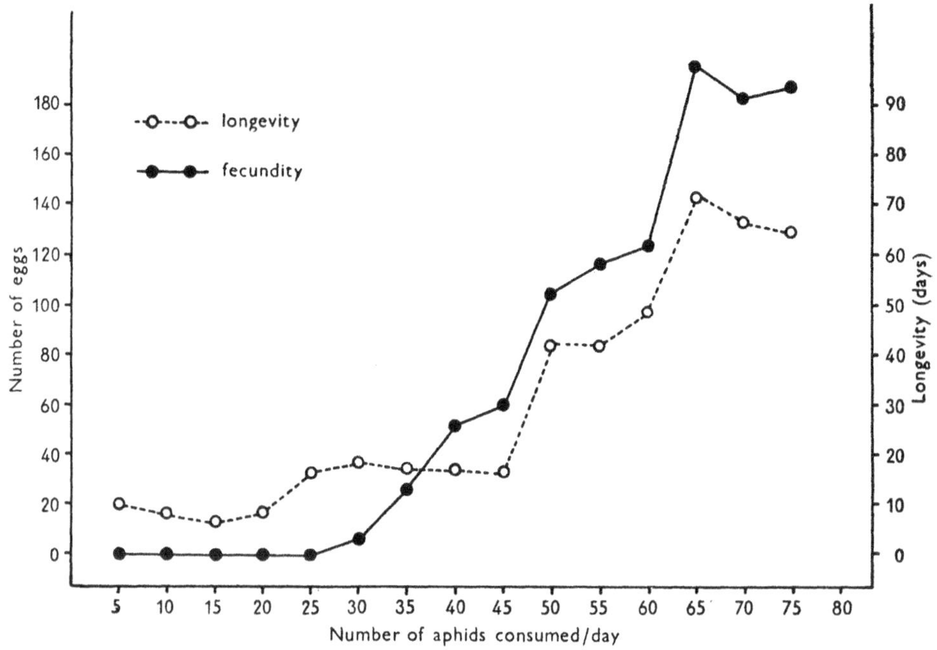


Fig. 6.06 Effect of daily feeding rate on the fecundity and longevity of *Coccinella undecimpunctata aegyptiaca* (Ibrahim 1955b).

Tab. 6.19 Effect of reduced feeding in larvae of *Hippodamia quinquesignata* (Kaddou 1960)[1])

| Total larval aphid consumption [aphids][1]) | Duration of stages larval [days] | pupal [days] | Weight of pupae [mg] | adults [mg] | N |
|---|---|---|---|---|---|
| 32.3 | 27.0 | 4.0 | 5.4 | 4.5 | 3[2]) |
| 78.4 | — | — | 10.7 | — | 4[2]) |
| 131.7 | 8.5 | 4.0 | 17.6 | 14.1 | 8 |
| 191.8 | 8.2 | 3.8 | 20.5 | 16.6 | 5 |
| 215.6[3]) | 8.1 | 3.1 | 23.1 | 18.7 | 10 |

[1]) Food: 2nd instar larvae of *Macrosiphum pisi*: temperature: 30° C.
[2]) Only 3 and 4 larvae completed development out of an initial 32 and 11, resp. (in other replicates mortality is not given).
[3]) Check replicate supplied with surplus of aphids.

The minimum daily feeding rate for individual larval instars and its effect on oviposition (Tab. 6.20) have also been measured for the acarophagous *Stethorus bifidus* (Collyer 1964). Larvae of *Stethorus picipes* require a minimum of one fifth of their maximum diet in order to complete the life cycle successfully (Fleschner 1950).

Smith (1965c) expressed minimum food consumption in terms of prey weight. The dry weight and size of *Coleomegilla maculata lengi* adults increased with the quantity of larval food, but only up to about 10 mg of the prey (dried *Acyrthosiphon pisum*). Additional food consumed contributed nothing to the weight of the adult (Fig. 6.07). In *Anatis mali* an excess of larval food, unimportant for survival, nearly doubled

**Tab.** 6.20 Fecundity of *Stethorus bifidus* on different prey-density (*Panonychus ulmi*) (Collyer 1964)

| Prey density (mites per day) | Fecundity | |
|---|---|---|
| | total number of eggs | daily number of eggs (av. per female) |
| 40–60 | 308<br>333<br>438 | 6 |
| 20 | 24<br>64<br>67 | 2 |
| 10[1]) | 2<br>19<br>27<br>30<br>33 | 1 |

[1]) Some of the females ceased oviposition prematurely, although they continued to live.

adult weight (Tab. 6.21). When the food supply was increased, developmental time decreased significantly in *A. mali*, but remained unchanged in *C. maculata*. *A. mali* was more flexible than *C. maculata* in this ability to adjust developmental time and adult weight to changes of food available (Smith 1965c).

A similar flexibility is found in *Pharoscymnus numidicus*. The 4th instar larvae feeding under conditions of food scarcity showed a decreased rate of development and increased mortality; also small adults developed (Fig. 6.08). The larvae could complete their development on 1/7 of

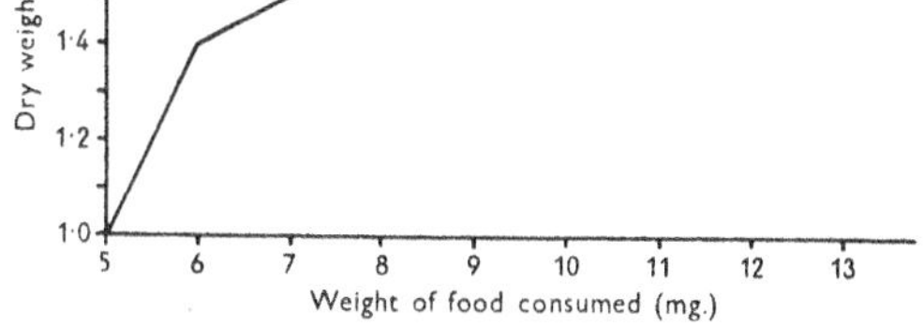


Fig. 6.07 Dependence of the adult size of *Coleomegilla maculata* on the amount of food consumed during larval life (Smith 1965c).

the normal food consumption. Thus *P. numidicus* has both a high tolerance to food scarcity and the ability to increase voracity when prey is abundant (Kehat 1968a).

Tab. 6.21 Effect of food intake (*Acyrthosiphon pisum* — dried) on *Anatis mali* (Smith 1965c

| Amount of food [mg] | Length of development from egg to adult [days] | Survival [%] | Adult dry weight [mg] |
|---|---|---|---|
| 5 | — | 0 | — |
| 15 | 24.7±0.1 | 30 | 4.6±0.3 |
| 30 | 19.2±0.1 | 90 | 6.7±0.4 |
| 45—55 | 20.9±0.3 | 85 | 12.8±0.3 |

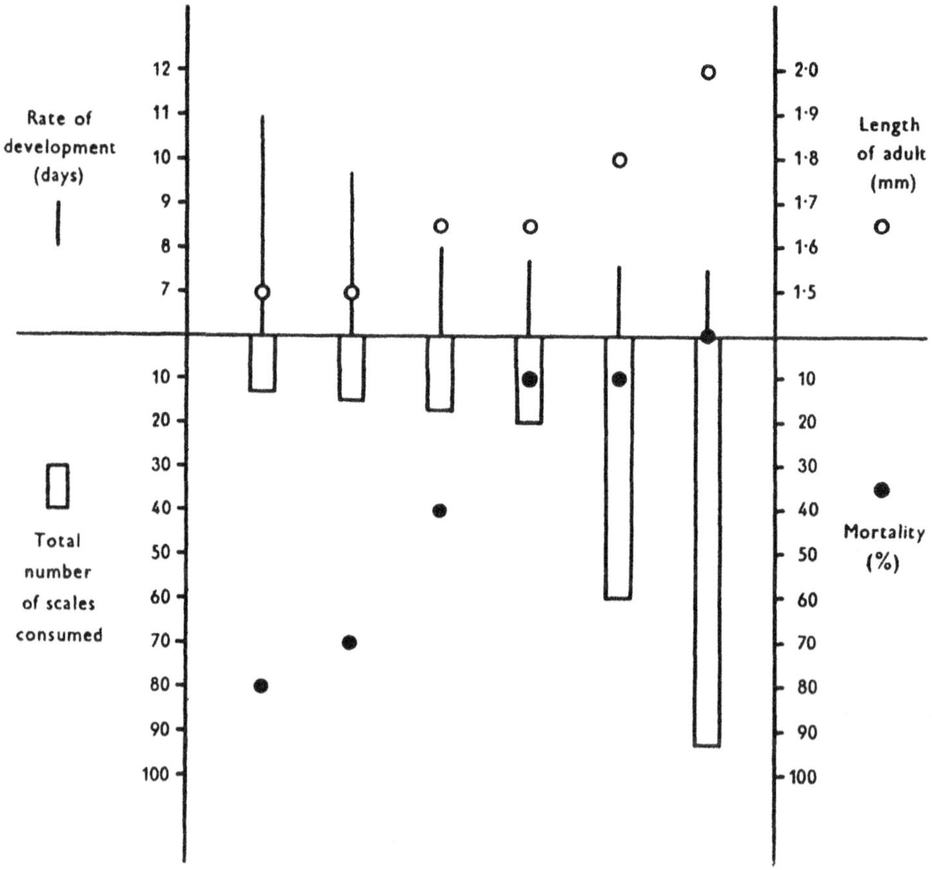


Fig. 6.08 Development and feeding of 4th instar larvae of *Pharoscymnus numidicus* under conditions of food scarcity (*Parlatoria* females as food) (Kehat 1968a).

## 6.3 Behaviour related to food

### 6.31 Searching behaviour

Thompson (1951) wrote that coccidophagous *Coccinellidae* "have a full complement of sense organs and there is no doubt that they can perceive their hosts at a distance". Further he believed that "certainly there seems no reason to think that insect predators are so casual in their food relations that their behaviour can be described in any sense as random action". This distinguished specialist on entomophagous insects made this statement on the basis of observation in the absence of experimental evidence, and now that such is available, these statements can no longer be regarded as valid, particularly for aphidophagous and acarophagous coccinellids.

Nearly all the experimental work on searching behaviour was published after Thompson wrote these words. Detailed experiments with acarophagous and aphidophagous coccinellids (also one coccidophagous species included) have now shown that neither optic nor olfactory orientation operates in prey searching behaviour (Fleschner 1950 — *Stethorus picipes*, Putman 1955 and Moter 1959 — *S. punctillum*, Banks 1954b, 1957 — *Coccinella septempunctata*, *Adalia bipunctata*, *Propylaea quatuordecimpunctata*, Dixon 1959 — *Adalia decempunctata*, Kaddou 1960 — *Hippodamia quinquesignata*, Bänsch 1964, in Hodek 1966 p. 123 — *C. septempunctata*, *A. bipunctata*, Kehat 1968a — *Pharoscymnus numidicus*, Kesten 1969 — *Anatis ocellata*).

No discovery of prey is made by the coccinellid until actual contact occurs. The prey may be missed if only a few millimeters away (Banks 1957), even when wind carries the smell to the coccinellids (Bänsch 1964, in Hodek 1966 p. 123, experiments on larvae). Fleschner (1950) quotes the value of 0.5 mm as the area of perception. Kesten (1969) found out that the maxillary palps are perhaps the most important sensory organs through which both adults and larvae of *Anatis ocellata* recognize the prey.

However, the movement of coccinellid larvae and adults is not completely haphazard. Certain regularities of behaviour, which make the encounter of predator and prey more probable, have been observed. (1) Such encounters are made more likely, at least between aphidophagous and acarophagous species and their prey, by a correspondence of tactic responses: like coccinellids, both mites and aphids show positive phototaxis and negative geotaxis (Fleschner 1950, Dixon 1959, Kesten 1969). (2) Searching efficiency may be increased further by the tendency of coccinellids to crawl along an edge or a raised surface. Thus on leaves, veins influence the direction of movements, and it is near veins that colonies of aphids occur most often (Banks 1957, Dixon 1959, Bänsch 1964, in Hodek 1966 p. 123). (3) In addition, the change in search pattern that the coccinellids show after finding prey has an advantage for predators of prey which forms colonies. After feeding, the larva increases the thoroughness of its subsequent searching by slower, winding movements in the vicinity, often turning frequently in a short distance (Banks 1957, Dixon 1959, Kaddou 1960, Kesten 1969) (Fig. 6.09).

Apart from these three regularities in behaviour, crawling is haphazard and the larvae revisit places already searched and neglect other areas. So, for example, a lst--instar larva of *Propylaea quatuordecimpunctata*, moving for three hours on a clump of 14 bean stems spent 52 % of the time searching on leaves visited more than once and only 12 % on leaves visited only once (the rest of the time was spent on stems or stationary) (Banks 1957).

To study the movements of predators on the stem and twigs of plants, Bänsch (1964, in Hodek 1966 p. 123) used a model tree 50 cm high. When moving upwards on the

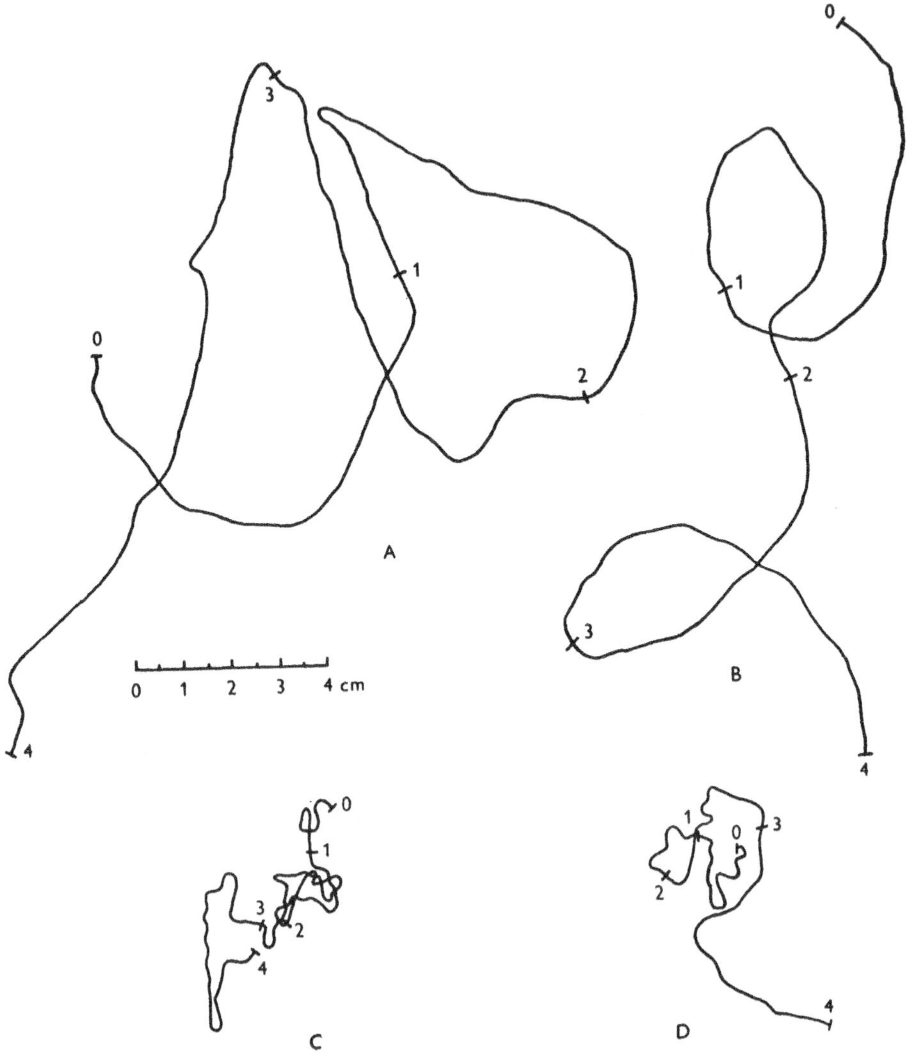


Fig. 6.09 Tracks of an unfed 4th instar larva of *Adalia bipunctata* on paper, before (A, B) and after (C, D) feeding on an aphid. (On the tracks there are indicated the passage of 4 successive intervals of 15sec.) (Banks 1957).

stem, coccinellids stop moving at bifurcations, but soon continue walking to follow the main axis. After an unsuccessful search at the top the larvae return down the stem, but their behaviour changes and they now follow each bifurcation. If also unsuccessful on a twig, they return to the stem and move to the top again. So, on the model tree, the larvae are more or less "trapped" between the apex and the tops of branches (Fig. 6.10). If the larvae eventually reach the ground, they do not move upwards again until they have covered a certain horizontal distance on the ground. Larvae of *C.*

*septempunctata* which have found aphid colonies remain close to them for long periods (e. g. 18 hr) (Banks 1957). The adults differ in their searching behaviour in that they soon fly away after an unsuccessful search.

In contradiction to all previous observations which have rejected the possibility of sense orientation by coccinellids towards their prey, Colburn and Asquith (1970) found that *Stethorus punctum* adults were attracted to their prey by smell in a special cage with four vials attached. Air was blown through the vials towards the beetles at the centre, but only some vials contained prey.

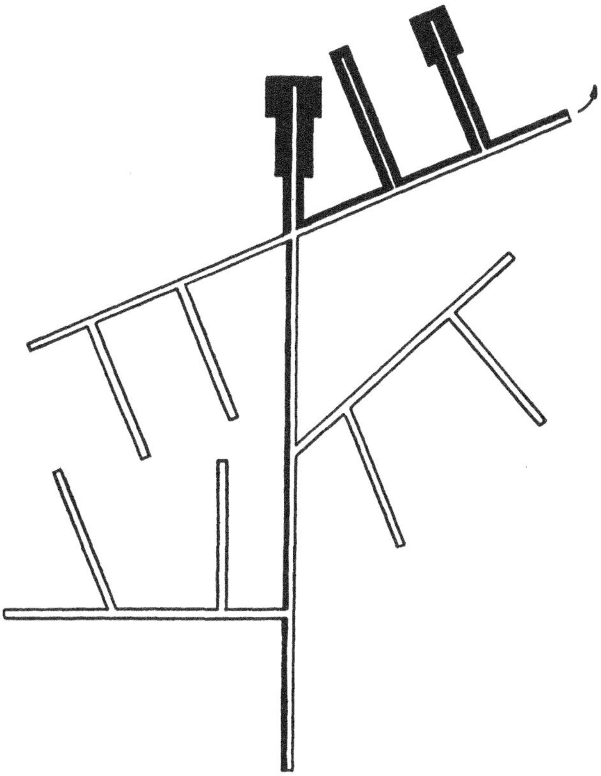

Fig. 6.10 Search pattern of one *Adalia bipunctata* adult on the model tree until flying away (arrow) after 23 min. (Bänsch 1964). The thickness of the line indicates the searching frequency of the predator.

Two behaviour studies which support Thompson's (1951) view have recently been reported and concern two large coccinellids preying on coleopterous and lepidopterous larvae. Savoïskaya (1970b) reports that the later instar larvae of *Aiolocaria mirabilis*, a specialized predator of chrysomelids, are very mobile and aggressive and actively pursue their prey. A similar behaviour is shown by the adults of the imported *Anatis ocellata*, which in Michigan preys on larvae of the tortricid *Choristoneura pinus*. Although the beetles search at random, they show a characteristic attack behaviour. "Within a distance of 1/2 — 3/4 in. (= 1.3 — 1.9 cm) from the prey . . . . the hungry adults stopped momentarily before moving forward and quickly snatching it in their mandibles". Physical contact was not required to arouse the attack and visual stimuli seemed to play a role (Allen et al. 1970).

From the point of view of economic impact of coccinellids, searching at random has its advantages and disadvantages. Such behaviour may improve the predator impact in particular situations (see 9.121).

However, random searching behaviour evidently markedly increases the mortality of newly hatched first instar larvae when prey density is low. According to Banks (1957) the first instar larvae of *Propylaea quatuordecimpunctata* spend 12—24 hours on the empty egg shells before they disperse and when food is absent then search actively for 25—35 hours. Activity then ceases, and death follows within 24—28 hours. They therefore need to find food within 1—1.5 days. During this time they can crawl for 75—300 cm. This distance is rather short for survival because, in searching randomly, the larvae may cover the same ground several times. Moreover, the movements of first instar larvae become impeded by honeydew and by hairs (Tab. 6.22). Therefore, when ladybird eggs are laid remote from aphid colonies, there is only a poor chance that the newly hatched larvae will find prey. The immature stages therefore suffer a high mortality (Banks 1954b). Laboratory observations of searching behaviour of *Adalia bipunctata* larvae agree with observations over 8 years in the field that under field conditions, the first instar larvae would be unable to survive unless the population density of young aphids on sycamore leaves exceeds 2 aphids per 100 cm² (Dixon 1970).

With first instar larva of *A. decempunctata* (and presumably generally), the chance of capturing prey increases once the first aphid has been eaten (Dixon 1959). Also cannibalism on eggs may prolong survival (see 6.13). Banks (1954b) found that the consumption of just a single egg nearly doubled the time a larva can survive and considered that the searching ability of newly hatched larvae was increased by feeding on eggs of the same egg batch. This has been queried by Pienkowski (1965), who found that cannibalism on eggs makes a larva disperse later from the egg batch and then be less active and even after 4 days of feeding on aphids less successful in finding prey (Tab. 6.23). As the larvae even feed on viable but more slowly developing eggs, the value of egg cannibalism for the impact of coccinellids on their prey seems rather doubtful. Cannibalism only increases the chance of survival of the larvae where there is a very low density of prey.

Tab. 6.22 Mean rates of movements (mm/min) of first instar larvae of *Propylaea quatuordecimpunctata* on various surfaces (Banks 1957)

| | Paper | Clean bean leaf | Honey-dewed bean leaf | Hairy potato leaf |
|---|---|---|---|---|
| mm/min | 151 | 154 | 104 | 54 |
| m/1—1.5 day | | 225—300 | 150—200 | 75—100 |

Tab. 6.23 Influence of egg cannibalism on the predatory efficiency of first-instar larvae of *Coleomegilla maculata* (Pienkowski 1965)

| Average number of eggs eaten per larva | 0.3 | 0.1 | 0 |
|---|---|---|---|
| larvae observed attacking aphids | 34 % | 36 % | 51 % |
| larvae active after 3 days | 20 % | 34 % | 46 % |
| larvae active after 4 days | 38 % | 44 % | 67 % |

### 6.32 Prey capture

Even when an encounter occurs, the coccinellid may fail to capture the prey and feed on it. This section of prey/predator relationships has been largely neglected, although it can be of paramount importance for the impact of the predator (Varley and Gradwell 1970).

Although aphids are generally considered to be completely "helpless" (Imms 1947), their defensive and escape mechanisms have been observed several times. Recently two authors have studied such responses of aphids towards two *Adalia spp.* (both larvae and adults of *Adalia decempunctata* — Dixon 1958, 1959, and larvae of *A. bipunctata* — Klingauf 1967). The aphids show various defence responses, as e. g. kicking movements, movements of the body, pulling free the appendage seized by the coccinellid and "waxing" the coccinellid with an oily liquid appearing on the tip of the siphunculi. Moreover, the aphids may escape by walking away or dropping off the plant (Fig. 6.11). The intensity of response varies with the size and stage of the predator. If an *A. decempunctata* adult approaches, the response of aphids (*Microlophium evansi*) is more intense than with coccinellid larvae; adult coccinellids induce escape rather than defence responses of aphids (Dixon 1959). The success or failure of a predator to capture the prey depends on their comparative size (Fig. 6.12).

Older larvae of *A. bipunctata* (9 mm long) succeed in capturing 90—100 % of first instar and about 60—70 % of adult *Myzus persicae* or *Neomyzus circumflexus*, but only 0—50 % of the much larger *Acyrthosiphon pisum* (Klingauf 1967). This finding agrees with the observation that aphids with long appendages were more difficult for *Hippodamia quinquesignata* larvae to capture than the more compact and tenacious aphids

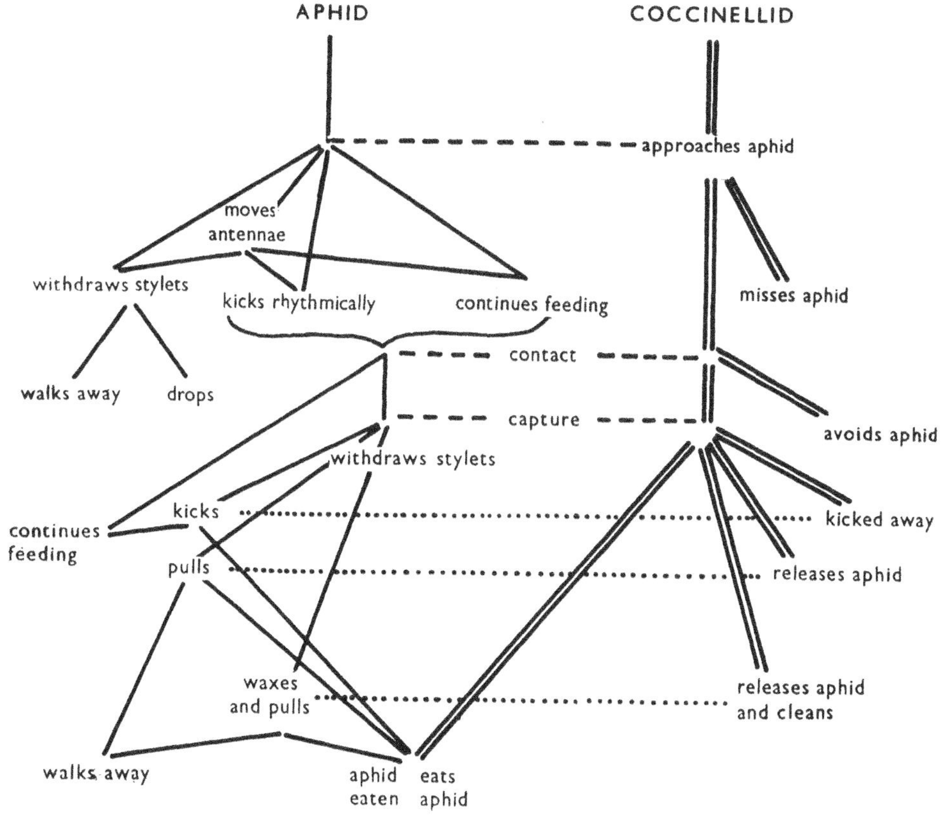


Fig. 6.11 Diagrammatic representation of the possible interactions between the behaviour patterns of *Microlophium evansi* and *Adalia decempunctata* during an encounter (Dixon 1958).

Instars of Adalia decempunctata

| | 1st instar | | | | | 2nd instar | | | | | 3rd instar | | | | | 4th instar | | | | | adult | | | | |
|---|---|---|---|---|---|---|---|---|---|---|---|---|---|---|---|---|---|---|---|---|---|---|---|---|---|
| A | 48 | 20 | 15 | 10 | 10 | 26 | 42 | 22 | 22 | 10 | 20 | 20 | 20 | 20 | 20 | 20 | 20 | 20 | 20 | 20 | 20 | 20 | 20 | 20 | 20 |
| B | 240 | 335 | 300 | 200 | 179 | 54 | 207 | 369 | 439 | 200 | 24 | 40 | 49 | 206 | 229 | 22 | 34 | 46 | 62 | 105 | 53 | 74 | 103 | 217 | 148 |
| C | 48 | 10 | 1 | 0 | 2 | 26 | 39 | 9 | 1 | 0 | 20 | 20 | 20 | 19 | 5 | 20 | 20 | 20 | 20 | 20 | 20 | 19 | 18 | 13 | 16 |

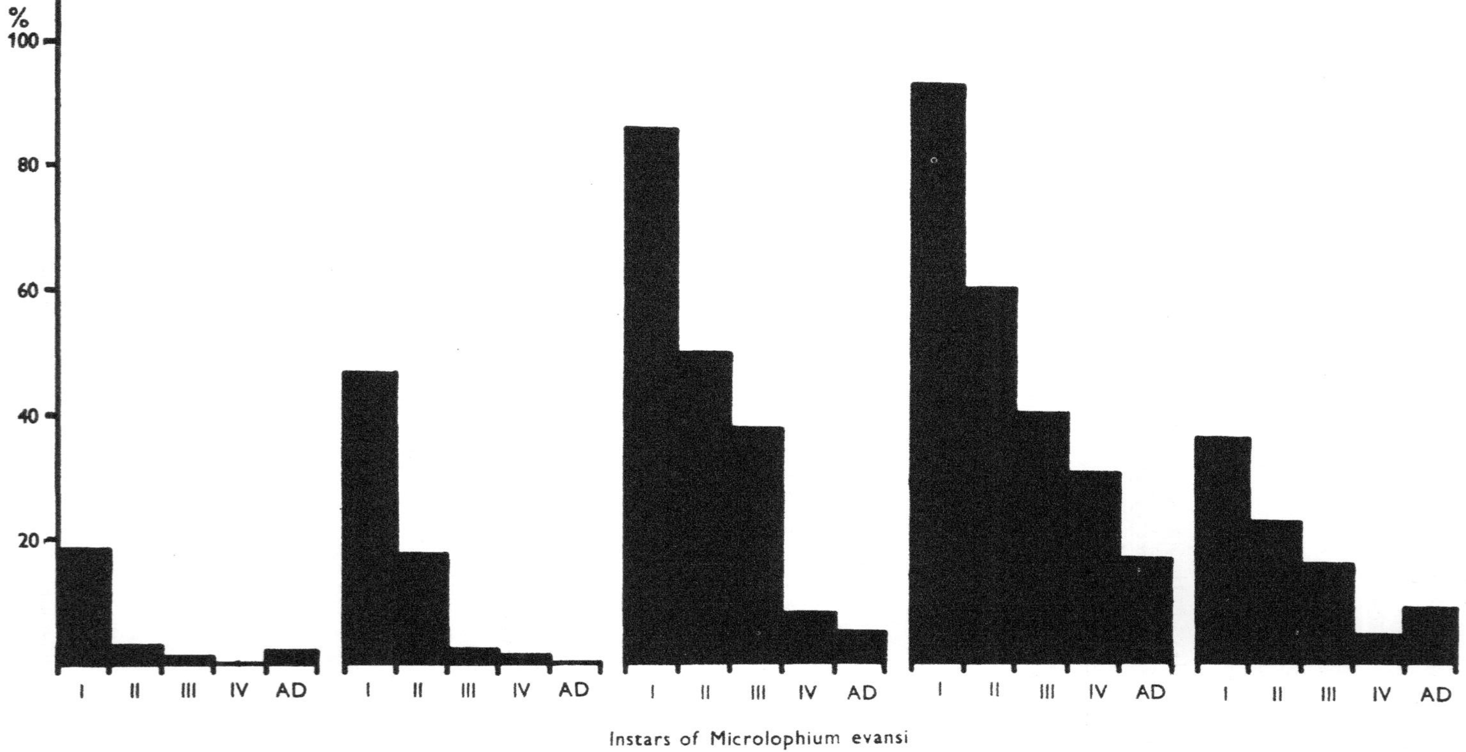


Fig. 6.12 The efficiency (% of encounters resulting in capture) of *Adalia decempunctata* in capturing *Microlophium evansi* (Dixon 1959).

with short appendages (Kaddou 1960). The aphids may apparently succeed in escape because they perceive the predator at a distance of several millimetres (*Acyrthosiphon pisum* at 4 mm, sometimes even at 10 mm — Klingauf 1967), while the predators perceive the prey only by contact. Thus the aphid attains a time advantage and may use this for escape. The aphid may also gain time when it is encountered by a small coccinellid larva which at first shows a fright response and only attacks the prey after a delay (Bänsch 1964). The intensity of the fright response depends on the size of the prey and is gradually suppressed once the predator has captured several prey (Bänsch 1964, Klingauf 1967).

Some aphids are protected against their predators by attendant ants (see 8.1). Some adult coccids are protected by their hard scale coverings to such an extent that only their larvae are preyed upon by coccinellids.

### 6.33 Food intake

The food of the *Epilachninae* consists of plant juices. The **phyllophagous** coccinellids scrape the parenchyma off the leaves and suck it, taking only fluid. The sucked parenchyma remains on the leaf leaving a characteristic feeding symptom (Butt 1951, Klausnitzer 1965).

The food of **aphidophagous** coccinellids consists not only of the sucked body fluid of aphids, but also of solid parts of the aphid body, e. g. legs, antennae, etc. (Butt 1951). The type of feeding evidently depends on the comparative size of prey and predator. When the prey is too large, it is merely sucked out and the skin with appendages is rejected (Butt 1951, Harpaz 1958, Hagen 1962[1]), Kesten 1969). This type of feeding behaviour is normal in younger larvae (1st and 2nd instar). Older larvae and adults may consume smaller prey completely (Butt 1951, Hagen 1962[1]), Kesten 1969).

Most coccinellids, particularly their larvae, show extra-intestinal digestion; they periodically regurgitate fluid from the gut into the chewed prey and suck back the pre-digested food. Some coccinellid larvae even perform extra-intestinal digestion through their perforated, sucking mandibles, like the larvae of *Dytiscus*, *Chrysopa*, etc. (Wigglesworth, 5th ed., 1953). This type of pre-digestion has been observed in larvae of *Stethorus spp.* (Blattný and Osvald 1949, Fleschner 1950, Putman 1955, Hagen 1962[1]) and in the unusually shaped larvae (Plate XXXIV) of *Platynaspis luteorubra*. These larvae were repeatedly observed sucking aphids hollow merely via one of the aphid's appendages; during the process the body of the aphid was alternately deflated and inflated (Hodek unpubl.).

[1]) A series of references, quoted by Hagen (1962) in his discussion of feeding behaviour, are not repeated here.

# 7 DORMANCY

## 7.1 The adaptive function of diapause

On most land surfaces, periods which are favourable for growth and morphogenesis alternate with unfavourable periods. In insects (as in other animals and plants) this has led to the evolution of appropriate annual life - cycles. The period of favourable environmental conditions (i. e. sufficient food and suitable physical factors) must be utilized for active life, i. e. for growth and development (of preimaginal stages and adults) made possible by adequate feeding. It is advantageous for a species to bridge the unfavourable period in a more or less inactive state in a stage which in some way has an increased survival potential. The presence of reserves and a reduced metabolic rate makes survival possible over a long period without feeding; different adaptations increase resistance to unfavourable physical conditions.

At first, workers mainly considered only the aspects concerned with resistance to drastic changes in climatic factors, especially the cold winter period in temperate climates. This is indicated by the introduction of general terms for hibernation and hibernation quarters in all languages (zimovka; overwintering; Überwinterung, Winterschlaf; l'hivernation). These terms are often misleading because they are commonly used for dormancy, which begins in late summer and ends in spring. Some coccinellid species, however, already enter their 'hibernation' quarters in late July! In this connection an assumption also arose that the main aim of diapause is to increase **frost-resistance,** This assumption has largely been abandoned. It appears in fact that gradual cooling gradually increases cold resistance, and that a series of arctic insects survive the winter successfully in a simple quiescence without entering diapause (Downes 1965).

Only more recently has there been a stronger emphasis on the need for a **synchronization** of the occurrence of the feeding stages with periods of sufficient food, although resistance to adverse conditions is a necessary corollary. This is the substantial purpose of diapause "during which the resistance to adverse climatic conditions is increased by physiological and sometimes morphological features fitting together into an 'adaptation syndrome' varying from species to species" (De Wilde 1970). Diapause is necessary so that morphogenesis (i. e. the maturation of ovaries in coccinellids, which almost always only have an imaginal diapause) shall be arrested in time, before there is a lack of food; diapause is therefore most necessary for the period during which climatic conditions alone would in fact allow morphogenesis to continue. In this sense the function of diapause is finished, for example, in temperate climates once the temperature has fallen below the level necessary for development and reproduction. In general, in those insects where this problem has been investigated, diapause has been found to be terminated in December or January.

By and large, the timely inhibition of morphogenesis (prevention of maturation in adults) in insects is governed by a series of signals from the environment. Decrease in temperature or the physiological ageing of host-plants is not repeated with such regularity year after year as are changes in day-length. Therefore, it is photoperiod[1]) which most often serves as the signal, often called 'seasonal token'.

## 7.2 Anatomical and chemical changes in dormant Coccinellidae

### 7.21 Anatomical changes

A greatly enlarged **fat body** is a conspicuous feature of diapausing coccinellids of both sexes, and this has been reported in all species observed, as well as the fact that the digestive tract becomes emptied of food. The **mid-guts** of adults of *Coccinella novemnotata* that were feeding on aphids were distended, thin walled, semitransparent, and packed with green aphid parts. In contrast, the mid-guts of beetles that had attained the stage of dormancy were reduced to whitish, opaque, thick-walled tube that contained a brown fluid (Mc Mullen 1967a).

The **ovaries** which in most cases do not develop beyond the stage of mere germaria, stop developing in diapausing females (Photos 17, 18). Such a situation has been recorded in *Coccinella septempunctata* (Hodek and Čerkasov 1961, Hodek 1962a, Hariri 1966), in *Semiadalia undecimnotata* (Hodek and Landa 1971), in *Adalia bipunctata* and *Propylaea quatuordecimpunctata* (Hariri 1966), in *Coccinella novemnotata* (McMullen 1967a) and *Hippodamia convergens*. The situation in the genus *Chilocorus* is rather different. For *C. renipustulatus*, Pantyukhov (1965, 1968a) describes that the majority of females diapause with ovaries in which development was blocked at quite an early stage (this condition is not defined); in about 7 % of the females development is stopped even earlier and follicles are completely missing. The females with no follicles have a substantially longer longevity, by 1 — 2 months in the laboratory, than those where some development has occurred. The situation is probably similar in *C. bipustulatus* and *C. geminus*. Zaslavskiĭ and Bogdanova (1965) report that in some diapausing females of these two species, the development of the ovarioles is inhibited at the stage where only one follicle is formed. The diameter of this follicle is, however, not larger than the diameter of the germarium, although yolk is deposited in the oocyte.

It is at the end of dormancy in spring, often still at dormancy sites while no food is taken, that previtellogenesis and the formation of the first oocyte occurs, e. g. in *A. bipunctata* (Hariri 1966) and in *S. undecimnotata* (Tab. 7.01) (Hodek and Landa 1971). In *C. septempunctata* and *P. quatuordecimpunctata*, Hariri (1966) states that this stage can only be reached after the beetles have emerged from hibernation and have fed on aphids.

The females of *Stethorus punctillum* often live for 2 years and enter a second dormancy. The mature ovaries are largely resorbed and the metabolic reserves are re-accumulated in the enlarged fat body. This resorption of ovaries even occurs during the first hibernation of those females which have emerged from pupae early in the season (before late July) and have already oviposited in that season (Putman 1955). This latter

[1]) Duration of light period = photophase and dark phase = scotophase within 24 hours.

Tab. 7.01 Gonads in dormant *Semiadalia undecimnotata* (Hodek and Landa 1971)

| Date of sampling[1] | Ovaria without egg chambers | Ovaria with one oocyte | Spermatheca without sperm | Spermatheca with sperm | Testes[2] spermatocytes in testicular follicles absent | Testes[2] spermatocytes in testicular follicles present |
|---|---|---|---|---|---|---|
| 23.10.57 | 20 | — | (17) | (3) | 20 | — |
| 16.12.57 | 20 | — | 20 | — | 20 | — |
| 11.2.58 | 20 | — | 20 | — | 20 | — |
| 8.4.58 | 30 | — | 30 | — | 20 | — |
| 23.4.58 | 20 | — | 18 | 2 | 11 | 9 |
| 29.4.58 | 20 | — | 12 | 8 | 7 | 13 |
| 5.5.58 | 20 | — | 4 | 16 | — | 20 |
| 13. 5.58 | 10 | 10 | — | 20 | — | 20 |
| 19. 5.58 | 7 | 13 | — | 20 | — | 20 |

[1]) All samples collected in Louny-hills (northern Bohemia).
[2]) Vesiculae seminales always full of sperm.

situation may be expected to apply in every polyvoltine cycle (or at least partly bivoltine), as occurs in *C. septempunctata* (Hodek 1962a). It is also unlikely that a second hibernation is an exceptional event; it has been proved to occur in *P. quatuordecimpunctata* by Hariri (1966) in England, in *C. septempunctata* by Sundby (1968) in Norway, and in two species (*Harmonia axyridis* and *Aiolocaria mirabilis*) introduced to central Asia from the Far East (Savoiskaya 1970b, d).

In males, the activity of the follicular tissue in the **testes** already starts in the pupa, and, moreover, does not stop in diapausing individuals. In *S. undecimnotata* (Hodek and Landa 1971) both spermatogenesis and spermateliosis proceed during the first, warmer period of dormancy (Tab. 7.01) (Photos 19—22). Only during the cold period (November — March) are the testicular follicles mostly empty and there is only a cluster of spermatogonies ("apical cells") at their distal tip and there is a mass of sperm bundles (spermatodesms) in the proximal third of their lumen. In April, spermatogenesis is resumed. In the seminal vesicles, the viable spermatozoids are permanently stored. Copulation takes place before the beetles emerge from the dormancy sites, so that all females disperse with a reserve of spermatozoids in their spermathecae.

This process, although it has not been studied in detail in other species, seems to be similar in *C. septempunctata* (Hodek and Čerkasov 1961, Hodek and Landa 1971) and in *A. bipunctata* (Hariri 1966) where its passage is indicated by changes of dimensions: the diameter of testes is least during the winter months and increases in the spring. Mating activity just prior to dispersing from aggregation has also been reported in California for *Hippodamia quinquesignata punctulata*, *H. sinuata*, and *H. parenthesis* (Hagen 1962). Well developed testes in *Hippodamia convergens* in dormancy sites have been reported by Stewart et al. (1967).

For *Chilocorus bipustulatus* and *C. geminus*, Zaslavskiĭ and Bogdanova (1965) similarly did not ascertain any effect of conditions inducing diapause on the state of testes. Gametogenesis evidently ceases very early in the life of both these species because it was very difficult to find mitotic divisions in active or diapausing males older than 7—10 days.

The presence of spermatozoa in the spermathecae of some *C. septempunctata* females in autumn (Hodek and Čerkasov 1961) and particularly the fertilization of females by males kept under diapause conditions, observed in *C. septempunctata* by Bonnemaison (1964) and in *S. punctillum* by Putman (1955), both would suggest that the activity of the male accessory glands is also independent of diapause.

It has been proved experimentally in *S. undecimnotata* and *C. septempunctata* that the activity of the testicular follicles depends on the ambient temperature; under favourable temperature conditions it may be continued or resumed, at least if alternative food (e. g. honey solution) is provided. At a temperature around +12 °C the males can embark on a normal and gradual decrease in gonad activity (Hodek and Landa 1971).

## 7.22 Changes in chemical composition

### 7.221 Lipids

Diapausing insects typically show an increased fat content (Lees 1955, Hodek and Čerkasov 1963, Beck 1968). Their survival during dormancy largely depends on the amount of metabolic reserves they can accumulate before diapause.

In many coccinellids the deposition of reserves in the fat body is quite extensive. A hypertrophied fat body is always mentioned in diapausing coccinellids. Numerical estimates based on biochemical analysis of the amount of fat accumulated are, however, given only for 6 coccinellid species (Tab. 7.02).

The data for *C. septempunctata* indicate that in the laboratory slightly less fat is accumulated than in the field (Hodek and Čerkasov 1961). Females, sampled in the hibernation sites in early October, when an important proportion of the reserves must already have been utilized, still had 0.205 mg of lipids per mg total weight. Females, in which an "artificial" diapause had been induced (by 12 hrs light and a constant 18 °C) had 20 % less lipids — 0.164 mg.

The usual way in which lipid contents are stated as a percentage of dry weight is not very suitable for following changes during dormancy. It is better to express fat content (and also other components) in absolute values of weight per specimen. In this way, changes in individual components become mutually independent.

We can compare both these ways of expression in Tab. 7.02. While the fat content in *S. undecimnotata* in reality decreased substantially from late August to late October, e. g. in males from 6.4 to 3.6 mg, i. e. by 44 %, this decrease is hardly reflected in the relative expression: the initial 41 % fat of dry weight only fallsot 38 % dry weight over the same period. The substantial decrease of fats is masked by the decrease of other components of dry weight.

All aphidophagous coccinellids that have been studied enter hibernation with considerable amounts of fat. In *Semiadalia undecimnotata* 6 — 7 mg of fat per beetle were recovered, i. e. about 40 % of dry weight (Hodek and Čerkasov 1963). As the first sample for analysis was taken as late as one month after the arrival of the beetles in the winter quarters, it may be supposed that the initial fat content, especially before migration, was even substantially higher. The comparison of our results with those in the literature shows that *S. undecimnotata* has the highest relative fat content of all the insects which have been analysed. Hariri (1966c) ascertained similarly high values in 3 further species of aphidophagous coccinellids: 5 — 9 mg fat in *Coccinella septempunctata*, 1 — 3 mg in *Adalia bipunctata* and 1—2 mg in *Propylaea quatuordecimpunctata*. The fat content of these three species ranged from 37 to 48 % dry weight at the beginning of diapause. On the other hand, the coccidophagous *Chilocorus rubidus* only has 26.5 % fat of dry weight at the beginning of hibernation (Pantyukhov 1968b) and

Tab. 7.02 Fat content in dormant and active *Coccinellidae*

| Coccinellid species | Amount of fat in | Dormant beetles: Origin of sample | Dormant beetles: Changes in fat content | | | | | | | | | Active beetles: Origin of sample | Active beetles: Changes in fat content | | | | | | Reference |
|---|---|---|---|---|---|---|---|---|---|---|---|---|---|---|---|---|---|---|---|
| *Coccinella septempunctata* | mg/mg of total weight | natural hibernacula N. Bohemia | Date | 10/10 | 9/5 | | | | | | | | | | | | | | Hodek and Čerkasov (1961) |
| | | | ♂ | 0.21 | 0.11 | | | | | | | | | | | | | | |
| | | | ♀ | 0.21 | 0.11 | | | | | | | | | | | | | | |
| | | labor. culture (12hr, 18°) | ♀ | 0.16 | | | | | | | | labor. culture (18hr. 18°) | ♀ | 0.07 | | | | | |
| | mg/beetle | artificial hibernacula S. England | Date | 17/9 | 17/11 | 2/3 | 5/5 | | | | | feeding on aphids | Date | 12/6 | | | | | Hariri (1966c) |
| | | | ♂ | 5.3 | 4.4 | 3.3 | 2.7 | | | | | | ♂ | 1.8 | | | | | |
| | | | ♀ | 9.4 | 7.6 | 5.8 | 3.7 | | | | | | ♀ | 2.5 | | | | | |
| *Semiadalia undecimnotata* | mg/beetle | natural hibernacula N. Bohemia | ♂ | 6.4 | 5.7 | 4.2 | 3.6 | 2.2 | 1.6 | 1.5 | 1.2 | | | | | | | | Hodek and Čerkasov (1963) |
| | | | ♀ | 7.2 | 6.1 | 5.4 | 4.1 | 2.9 | 1.9 | 2.4 | 1.7 | | | | | | | | |
| | % of dry weight | | Date | 29/8 | 11/9 | 25/9 | 22/10 | 2/4 | 16/4 | 29/4 | 9/5 | | | | | | | | |
| | | | ♂ | 41 | 47 | 36 | 38 | 30 | 25 | 21 | 17 | | | | | | | | |
| | | | ♀ | 38 | 51 | 44 | 37 | 34 | 25 | 27 | 18 | | | | | | | | |
| *Adalia bipunctata* | mg/beetle | artificial hibernacula S. England | Date | 15/8 | 14/9 | 13/11 | 9/2 | 7/4 | 4/5 | | | feeding on aphids | Date | May | | | | | Hariri (1966c) |
| | | | ♂ | 2.0 | 1.4 | 1.2 | 0.9 | 0.7 | 0.5 | | | | ♂ | 0.7—0.9 | | | | | |
| | | | ♀ | 3.1 | 2.1 | 1.5 | 1.2 | 0.9 | 0.4 | | | | ♀ | 0.8—1.1 | | | | | |
| | | natural hibernacula S. England | Date | 16/10 | 22/12 | 28/3 | 23/4 | | | | | | | | | | | | |
| | | | ♂ | 1.9 | 1.0 | 0.8 | 0.4 | | | | | | | | | | | | |
| | | | ♀ | 2.1 | 1.8 | 1.1 | 0.9 | | | | | | | | | | | | |
| *Propylaea quatuordecimpunctata* | mg/beetle | artificial hibernacula S. England | Date | 17/9 | 17/11 | 2/3 | 15/5 | | | | | feeding on aphids | Date | 10/6 | 8/8 | | | | Hariri (1966c) |
| | | | ♂ | 1.4 | 1.1 | 0.9 | 0.4 | | | | | | ♂ | 0.7 | 1.6 | | | | |
| | | | ♀ | 2.0 | 1.5 | 1.2 | 0.8 | | | | | | ♀ | 0.8 | 2.0 | | | | |
| *Chilocorus renipustulatus* | % of dry weight | labor. beetles in artificial hibernacula | Maikop | 20 | 18 | 17 | 17 | 16 | 15 | 13 | | labor. culture | Maikop | 10 | 16 | 19 | 11 | | Pantyukhov |
| | | | Date | 30/10 | 28/11 | 30/12 | 30/1 | 28/2 | 28/3 | 29/4 | | | Age [days] | 1 | 15 | 40 | 75 | | |
| | | | Leningr. | 20 | 19 | 17 | 16 | 16 | 15 | 13 | | | Leningr. | 10 | 16 | 19 | 9 | | (1965) |
| | | labor. culture | Age [days] | 21 | 30 | 40 | 50 | 75 | | | | | | | | | | | |
| | | | | 23 | 24 | 25 | 24 | 21 | | | | | | | | | | | (1968a) |
| *Chilocorus rubidus* | % of dry weight | natural hibernacula (Leningrad) | Date | 15/9 | 30/10 | 30/11 | 30/12 | 28/2 | 30/3 | 20/4 | | labor. culture | Age [days] | 1 | 15 | 30 | 45 | 60 | Pantyukhov (1968b) |
| | | | | 27 | 25 | 24 | 23 | 22 | 21 | 18 | | | | 12 | 18 | 20 | 27 | 27 | |

*C. renipustulatus* 25 % (Pantyukhov 1968a). The latter results come from individuals reared in the laboratory. Earlier experiments with this species yielded even less fat (20 %) (Pantyukhov 1965).

**Rate of decrease.** — The decrease in the substantial fat reserves during hibernation is also substantial: during the eight months of dormancy the absolute amount of fat decreases to a mere quarter in *S. undecimnotata*. The decrease is half as great in the pentatomid bug *Eurygaster integriceps* and a quarter as great in the chrysomelid *Leptinotarsa decemlineata*, which are both dormant in the soil and thus partly isolated from temperature changes. The proportions of the initial fat content consumed throughout the whole hibernation period in *C. septempunctata*, *A. bipunctata* and *P. quatuordecimpunctata* were, for males and females respectively, as follows: 49 and 61, 75 and 87, and 71 and 60 %. Therefore, except in *P. quatuordecimpunctata*, males consumed a greater proportion of their fat than did females (Hariri 1966c). *Chilocorus rubidus* decomposed 32 % of the initial amount of fat during dormancy.

The rate of decrease in fat reserves in *S. undecimnotata* is evidently related to the effect of ambient temperature on locomotion and metabolism. The rapid decrease in fat content commences in late summer and early autumn. In September 1957 the average decrease in absolute weight amounted to 0.082 mg of lipids per day, in October 1958 the decrease was reduced to 0.022 mg per day. During the cold period of November till March the decrease of fat was 0.009 mg per day in both seasons studied. In April and early May the utilization of fat reserves again increased, and varied in the three springs studied due to different temperatures: the decrease of fat in *S. undecimnotata* males amounted to 0.018 mg, 0.045 mg and 0.027 mg per day in 1957, 1958 and 1959, respectively (Hodek and Čerkasov 1963).

Hariri's findings (1966c) completely agree with the above. He calculated a close positive correlation between the rate of fat consumption and temperature (the regression coefficient was 0.002 ± 0.0004 mg of fat per day per °C ($P = 0.02$). Also in coccidophagous coccinellids of the genus *Chilocorus* (*C. rubidus*, *C. renipustulatus*) it was found that the metabolism of fats was at a low level during the winter months (Pantyukhov 1968a, b).

By keeping *A. bipunctata* and *C. septempunctata* for six months at temperatures below zero (—1 to —3 ° or —2 to —10 °C) Hariri found out that negligible amounts of fat, but considerable amounts of glycogen, were consumed. At these temperatures glycogen is probably used directly as an energy source.

In Hariri's experiments, the coccinellids were stored in artificial hibernacula (zinc cages with dead leaves and pieces of corrugated cardboard, covered with polythene sheeting to keep out rain). Here they consumed fat, but not glycogen, much faster than those hibernating in the field.

Just before emerging from hibernation, *S. undecimnotata* males contained 17 — 19 % and females 18 — 22 % fat of dry weight (Hodek and Čerkasov 1963), *A. bipunctata* 14 %, *P. quatuordecimpunctata* 19 % and *C. septempunctata* 23 % fat of dry weight (Hariri 1966c).

**Sexual differences.** — On most occasions, *S. undecimnotata* (Hodek and Čerkasov 1963) females had a higher absolute fat content (by 0.1 — 1.2 mg) than males. Taking the amount of fat in males as 100 %, the amount of fat in females on ten occasions then represented 100 — 120 %, on twelve occasions 120 — 150 %, and on two occasions 150 — 160 %. Values less than 120 % were more common in the first period of dormancy, the higher values occurred in the second period and particularly at the end of dormancy. In the three species studied by Hariri (1966c), females contained 1.4 to 1.7 times more fat than males at the start of hibernation, but as the females consumed fat faster, both sexes contained about the same amount towards the end of hibernation. This differs from the findings for *S. undecimnotata.*

In addition to recording changes in fat content, Pantyukhov (1968b) also observed changes of the iodine number in the course of hibernation of *C. rubidus.* Although the iodine number before the onset of frosts in September and October amounted to 88 — 89, and to 86 at the end of March, in January and February this index of the proportion of fatty acids in fat which are unsaturated rose to 93 — 94. As the unsaturated acids have a lower freezing point than the saturated ones, their rise may increase cold resistance. In *C. renipustulatus* (Pantyukhov 1965) similar changes in the iodine number can be observed: there is an increase in winter from 59 — 60 to 62 — 64 and a decrease in spring to 58 — 59.

### 7.222 Glycogen

Like lipids, glycogen reserves are also accumulated during prediapause. Glycogen is unstable in comparison with the more stable fat, thus both absolute and relative glycogen content have been found greatly to fluctuate in the course of dormancy in *S. undecimnotata* (Hodek and Čerkasov 1963). Also a comparison of glycogen content during similar periods in different years revealed marked differences[1]).

The amount of glycogen in dormant coccinellids was roughly ten times less than the amount of fat (Tab. 7.03). In late August *S. undecimnotata* adults had about 0.4 mg of glycogen per beetle which represented about 2—3 % of dry weight. Moreover the amount of glycogen presumably decreased due to migration and a one-month stay in the dormancy site (Hodek and Čerkasov 1963). Values ascertained by Hariri (1966c) for *Adalia bipunctata* in mid-October are one tenth of those for *S. undecimnotata* — 0.04 mg of glycogen per beetle. The relative glycogen content is also substantially less, ranging from 0.4 to 1.2 % of dry weight. On the other hand, coccidophagous coccinellids of the genus *Chilocorus* appear to contain much more glycogen (10—12 % of dry weight) at the beginning of diapause (Pantyukhov 1965, 1968b).

**Rate of decrease.** — In all coccinellid species studied, a decrease in glycogen during dormancy has been observed. In *S. undecimnotata* the rate of decrease (as for fat content also) depended on temperature. In late summer and early autumn the decrease was very steep so that the daily decrease averaged 2.5 and 2.8 $\gamma$ in males and females respectively. During the winter and early spring (November — March) the decrease was slow — 0.8 and 0.7 $\gamma$ per day. In the last 37 days in the dormancy site the fall was about the same in males as during winter (0.9 $\gamma$ per day), but became much steeper in females (2.0 $\gamma$ per day). In a second year, when the emergence from winter quarters was more delayed, the decrease was much steeper in males (1.7 $\gamma$ per day). In the females, however, the amount of glycogen increased within the last week before dispersion, and so the total average decrease was only 1.2 $\gamma$ per day. On the whole, therefore, during the 8 months of dormancy in *S. undecimnotata*, glycogen decreased in males to about 1/5 (22 %) and in females to 1/4 of the initial amount (Hodek and Čerkasov 1963).

Hariri's (1966c) data for *C. septempunctata*, *A. bipunctata* and *P. quatuordecimpunctata* show that, at the end of hibernation the insects contained 11 to 35 % of the amounts of glycogen they had contained initially. In overwintering *C. septempunctata* and *A. bipunctata* which had been kept experimentally for 6 months in glass vials at subzero

[1]) The amount of glycogen is reported in different works, even for one species of insect, on the basis of different methods of analysis or in different units (surveyed by Hodek and Čerkasov 1963). It is possible that the variability of the results is to some extent due to such differences.

Tab. 7.03 Glycogen content in dormant *Coccinellidae*

| Coccinellid psecies | Amount of glycogen in | Origin of sample | Changes in amount of glycogen content | | | | | | | | | Reference |
|---|---|---|---|---|---|---|---|---|---|---|---|---|
| *Semiadalia undecimnotata* | mg/beetle | natural hibernacula (N. Bohemia) | ♂<br>♀ | 0.388<br>0.432 | 0.310<br>0.357 | 0.302<br>0.378 | 0.252<br>0.283 | 0.120<br>0.176 | 0.095<br>0.135 | 0.096<br>0.131 | 0.085<br>0.104 | Hodek and Čerkasov (1963) |
| | | | Date | 29/8 | 11/9 | 25/9 | 22/10 | 2/4 | 16/4 | 29/4 | 9/5 | |
| | % of dry weight | | ♂<br>♀ | 2.5<br>2.3 | 2.5<br>3.0 | 2.6<br>3.1 | 2.7<br>2.6 | 1.6<br>2.1 | 1.5<br>1.8 | 1.4<br>1.5 | 1.2<br>1.1 | |
| *Adalia bipunctata* | mg/beetle | natural hibernacula (S. England) | Date | | | | | 16/10 | 22/12 | 28/3 | 23/4 | Hariri (1966c) |
| | | | ♂<br>♀ | | | | | 0.043<br>0.045 | 0.016<br>0.042 | 0.014<br>0.019 | 0.008<br>0.007 | |
| *Chilocorus renipustulatus* | % of dry weight | laboratory beetles in artificial hibernacula | Maikop | | 11.7 | 10.5 | 9.9 | 9.2 | 8.0 | 7.4 | 6.2 | Pantyukhov (1965) |
| | | | Date | | 30/10 | 28/11 | 30/12 | 30/1 | 28/2 | 28/3 | 29/4 | |
| | | | Leningrad | | 11.5 | 10.7 | 9.8 | 9.3 | 8.5 | 7.8 | 6.9 | |
| *Chilocorus rubidus* | % of dry weight | natural hibernacula (Leningrad) | Date | 15/9 | 30/10 | 30/11 | 30/12 | 30/1 | 28/2 | 30/3 | 20/4 | Pantyukhov (1968b) |
| | | | | 9.6 | 9.1 | 7.9 | 8.3 | 6.8 | 5.0 | 5.3 | 6.5 | |

temperatures, Hariri (1966) found that negligible amounts of fat, but considerable amounts of glycogen, had been consumed. In the field, however, the results for these species paralleled those of Hodek and Čerkasov (1963); glycogen content decreased faster during the autumn and spring, and more slowly during the winter.

Before emergence from the dormancy site, *S. undecimnotata* contained 50 — 85 γ and 90 — 105 γ of glycogen in males and females respectively which equalled 0.7 — 1.3 % of dry weight. In *A. bipunctata* it was 5 — 8 γ in males and 6 — 9 γ in females and equalled about 0.2 % of dry weight.

**Sexual differences.** — A markedly higher absolute glycogen content was found in the females of *S. undecimnotata*. If the content in males is taken as 100 %, then in 7 cases the amount of glycogen in females represented 100 — 120 %, in 14 cases between 121 — 180 %, in 2 cases 181 — 200 % and 2 exceeded 200 % (Hodek and Čerkasov 1963). At the start of hibernation, *C. septempunctata* females contained very much more glycogen, and *A. bipunctata* females considerably more glycogen than did the males. Males and females of *P. quatuordecimpunctata* contained almost equal amounts of glycogen initially. Later the differences were negligible in all three species (Hariri 1966c).

According to the concept that glycerol and other polyols are possibly responsible for frost resistance, both fat and glycogen must be considered not only as a source of energy, but also as precursors for polyols (see 7.24).

#### 7.223 Water

Hagen (1962) found that the water content in monthly samples of *Hippodamia convergens* taken from Sierra Nevada aggregations remained remarkably constant. Also in *S. undecimnotata* there was no general tendency for a decrease or increase in absolute water content during the eight months dormancy (Hodek and Čerkasov 1963). The water content usually amounted to 10 — 12 mg per male and to 11 — 14 mg per female. The continual increase in the relative water content from about 50 — 55 % to about 60 — 63 % which has been observed during dormancy is simply caused by the fall in dry weight, particularly of fat.

In *Chilocorus renipustulatus* and *C. rubidus*, Pantyukhov (1965, 1968b) found a slight decrease in the relative water content from 58 — 59 to 54 — 55 % and from 62 to 59 %, respectively, from October to January, and again a further increase till April to 64 and 68 %.

The body water content may be increased by drinking or by the production of metabolic water when splitting fat. The loss of body water by transpiration is affected by air humidity. The possibility that the intensive splitting of fat contributes to the increase in water content has been supposed (Hariri 1966c). He kept the coccinellids (*C. septempunctata*, *A. bipunctata*, *P. quatuordecimpunctata*) in artificial hibernacula covered with polythene sheeting to keep out rain. As the beetles were not able to drink, he related the relative increase in water content to the production of metabolic water. However, no explanation really need to be searched for the increase in water relative to dry matter. Only in one sample of *A. bipunctata* was there a small increase of 0.9 or 0.8 mg (males and females) in absolute water weight. In a second season no increase at all was found in *A. bipunctata*, and in both *C. septempunctata* and *P. quatuordecimpunctata* there was a gradual decrease in absolute water content.

In our studies of *S. undecimnotata* increases in water content usually coincided — especially after a period of drought — with rainfall and high humidity of the air. In contrast, decreases in water content coincided with drought. Hagen (1962) assumes, for *H. convergens*, that the water balance is maintained by imbibing water. If the beetles from aggregations are kept in a refrigerator for a month or so, in the absence of litter, there is a distinct water loss. When these beetles are then exposed to water, they drink avidly.

### 7.23 Changes in metabolic rate

The onset of diapause is invariably associated with a striking fall in the level of metabolism (Lees 1955), e. g. in *Leptinotarsa decemlineata* a drop to 15 — 20 % of the normal respiratory rate was observed in dormant beetles (De Wilde 1969).

Although rarely measured, a decrease in oxygen consumption has also been demonstrated for diapausing coccinellids. When measuring the respiratory rate at 18 °C in diapausing *Hippodamia convergens*, Stewart et al. (1967) obtained the value of 12 μl $0_2$/beetle/hr immediately after the arrival at the aestivation quarters in Pinnacle mountains (Arkansas, USA) in late June. Later the oxygen consumption decreased somewhat further, so that during aestivo-hibernation, from July to February, it fluctuated between 5 and 9 μl. On the other hand, in reproducing beetles the oxygen consumption in the lowlands in March was 29 μl. Thus during diapause the level of metabolism was reduced to about 1/3 — 1/6.

Pantyukhov (1968a, b) measured a decrease to c. 1/8 of active rate in *Chilocorus rubidus* (1968b) and to c. 1/2 in *C. renipustulatus* (1968a) (see Tab. 7.04). Oxygen

Tab. 7.04 Oxygen consumption in coccinellids of the genus *Chilocorus* (μl $0_2$/mg weight/hr)

| Species | Reproducing adults | | Diapausing adults | |
|---|---|---|---|---|
| | (30 days old) | (40 days old) | (40 or 45 days old) | (50 or 60 days old) |
| *C. renipustulatus*[1]) | 1.25 | 1.12 | 0.52 | 0.43 |
| *C. rubidus*[1]) | 0.80 | — | 0.10 | 0.09 |
| *C. bipustulatus* (females)[2]) | 2.04 | | 1.22 — 1.25 | |

[1]) Measured at 25 °C — Pantyukhov (1968a, b).
[2]) Measured at 32 °C — Tadmor and Applebaum (1971).

consumption by diapausing *C. bipustulatus* females reared in the laboratory was only a little more than half that of non-diapausing individuals (Tadmor and Applebaum 1971). In southern Japan (Aichi Prefecture), where *Coccinella septempunctata bruckii* Mulsant breeds during the winter but aestivates, Sakurai (1969) ascertained a decrease in respiration rate from 2 — 3 μl $0_2$ /mg/hr in active beetles to 0.4 — 1.3 μl $0_2$ during aestivation, i. e. a fall to 1/2 — 1/5 of the active rate (measured at 30 °C). Hibernating *Semiadalia undecimnotata* have a respiration rate around 0.8 μl/mg/hr (measured at 25 °C), but if re-activated in the laboratory the beetles show a doubled oxygen consumption, so that in ovipositing females it amounts to 1.6 (1.1 — 2.0) μl/mg/hr (Hodek and Čerkasov 1958, Hodek 1970).

## 7.24 Changes in cold-resistance

Studies concentrated on the effect of low temperatures are extremely rare in coccinellids, although such data are of utmost importance especially if a species is to be introduced to a different climate for biological control. The scarce data available add some evidence to the recent concept of frost-resistance. In contrast to the original idea that diapause and the correlated changes in chemical composition are primarily connected with increased frost tolerance or resistance, the recently accumulated evidence is convincing that frost-resistance is low at the beginning of diapause and that it then only gradually rises under the influence of decreasing temperature.

Further evidence for this latter idea comes from the considerable increase of cold-resistance in *Chilocorus rubidus* between September and January (Pantyukhov 1968b, Fig. 7.01). In spring the cold-resistance decreases again. Similar changes in resistance have been noticed in measurements of the super-cooling points (Table 7.05)[1]. The

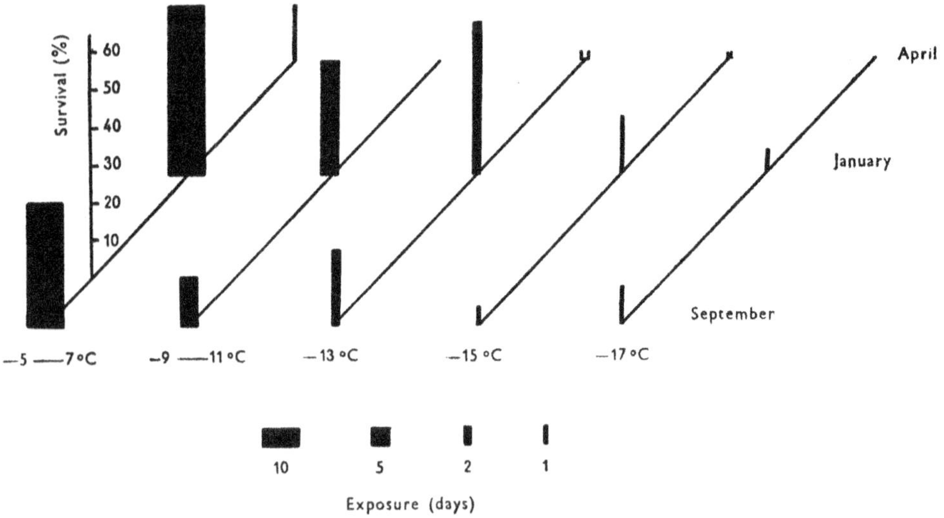


Fig. 7.01 Changes in frost-resistance during hibernation of *Chilocorus rubidus* (Pantyukhov 1968b). The width of columns indicates the length of exposure to cold.

Tab. 7.05 Supercooling point and freezing point in *Chilocorus rubidus* (Pantyukhov 1968b)

| Months | N | Supercooling point [°C] | | | Freezing point [°C] | | |
|---|---|---|---|---|---|---|---|
| | | aver. | min. | max. | aver. | min. | max. |
| IX | 41 | − 8.2 | −2.7 | −14.2 | −2.6 | −0.4 | −5.2 |
| XII | 57 | −13.2 | −4.9 | −17.4 | −3.2 | −0.7 | −6.8 |
| I | 50 | −12.9 | −4.3 | −17.2 | −3.2 | −0.8 | −6.7 |
| II | 40 | −13.5 | −5.6 | −17.8 | −3.4 | −1.1 | −6.9 |
| IV | 32 | − 3.4 | −0.2 | − 5.5 | −0.8 | −0.1 | −2.7 |

[1]) As the temperature falls, supercooling occurs which is followed by freezing of the tissues accompanied by a rapid rise in internal temperature. The external temperature at which this occurs is defined as the supercooling point. For the general theory of frost-resistance and definitions of terms see Salt (1961, 1964).

lowest temperature that could still be survived in January by a considerable number of beetles (40 %) was —13.5 °C for 2 days. *C. rubidus* hibernates in the litter and if this is additionally covered by a sufficient layer of snow, the beetle is not exposed to extreme low temperatures. In the Leningrad region Pantyukhov (1968b) measured the minimum temperature on the earth under the snow; in the winter 1964/65 this was —2.5 °C and —7 °C in the winter 1965/66; the survival of the beetles was 67 — 82 % and 52 — 70 %, respectively. By contrast, in Alma-Ata where the minimum temperature (without snow) was —10 °C, only 12 % of the beetles survived due to alternations of temperature. In the Far East region the survival was 21 % in a microhabitat where the temperature was not measured (Pantyukhov 1968b).

The more common *C. bipustulatus* also has low cold-resistance: most individuals have their super-cooling point at —8 to —9 °C, the limit for survival being —10 to —12 °C (Pantyukhov 1965). A longlasting decrease of temperature down to —5 to —6 °C at the soil surface in the hibernation sites causes rather high mortality. Considerable mortality is only to be expected where there is a thin snow cover or where the spring temperature fluctuates.

The results achieved with coccinellids suggest conclusions valid for most insects. Resistance against temperatures below zero is rather high in the middle of dormancy, after 'hardening' due to the gradual decrease in temperature. Early or late frosts, however, can be very damaging.

The degree of resistance naturally varies in different species, in connection with their type of hibernation. Species which hibernate in litter are certainly more sensitive than the more exposed species which hibernate in crevices. This is indicated by a simple comparison of winter mortality in the species.

A much higher resistance to low moisture and larger extremes of temperature was found by Novák and Grenarová (1967) in *A. bipunctata* when compared with coccinellids hibernating in litter. When placed under the roof of an open hut or into the crown of *Picea*. *A. bipunctata* suffered only 13, 26 or 40 % mortality, while *C. septempunctata*, *C. quinquepunctata* and *Exochomus quadripustulatus* died out completely or by more than 90 % within 4.5 months (mid-October — early March). There were no differences, however, in the winter survival in the protected grass environment among the four species tested. The different tolerance of *A. bipunctata* is obviously connected with its hibernation in the drier habitat of bark crevices where it is not protected from ambient extremes.

*H. convergens* is obviously not very frost resistant, as Latta (1928) reported 100 % mortality caused by a sudden decrease in temperature from about —2 °C to —12 °C in December.

## 7.3 Behaviour patterns in dormant Coccinellidae

Coccinellids which have been induced to diapause change their behaviour to a greater or lesser extent. Some species make lengthy migratory flights, form large (dormancy) aggregations and exhibit a long dormancy. This behaviour is most pronounced in two *Hippodamiini* species from warm temperate and subtropical regions, *Hippodamia convergens* in America and *Semiadalia undecimnotata* in the Palaearctic region. In general, these features of flight, aggregation and dormancy developed as an adaptation in those species which depend on "ephemeral" prey. Consequently they very often occur in aphidophagous coccinellids of the tribes *Coccinellini*, *Hippodamiini* and *Anisostictini*. By contrast, they are missing or occur only rarely in coccinellids of the tribes *Hyperaspini*, *Scymnini*, *Psylloborini* and *Chilocorini*, which for the most part are coccidophagous or mycophagous.

Although ample literature exists on the behaviour of coccinellids, the causal relations remain by and large obscure. The vast majority of the literature only contains isolated observations accompanied with speculations based on insufficient evidence.

A complete review of the literature would therefore not be of much value and thus no such review is presented. What is required is a series of comparative observations of laboratory and field phenomena, the field observations carried out at a number of hibernation sites for several years. Unfortunately, such data are still extremely scarce.

There is the danger of providing misleading generalisations by basing them on the few well-documented experiments that have been carried out. This is especially because the behaviour of dormancy is very dissimilar in different species of coccinellids. Therefore the individual stages in the behaviour associated with dormancy will be briefly described, and then their specific variations in most common species discussed.

## 7.31 General account of the behaviour

### 7.311 Pre-diapause feeding

There is some controversy regarding pre-diapause feeding activity. Some coccinellids accumulate the reserves for hibernation as adults before the migration to hibernation quarters (*C. septempunctata*), other continue to feed after the main phase of migration (*H. convergens*) and in some it is even presumed, that the adults do not feed at all (*S. undecimnotata*).

In *C. septempunctata* it seems that the onset of diapause is determined in late pre--imaginal and early imaginal life (for details see chapter 7.411). The adults in which the "points" have been set in direction of diapause feed voraciously on aphids for a relatively short period. The ingested food is used for building of large metabolic reserves in the fat body instead of for reproduction; these reserves are mainly fat and glycogen, and serve as a source of energy during the long time without food that follows (see 7.22). It can be assumed that it is this accumulation of a sufficient amount of reserves that forms the stimulus for the beginning of migration.

In *Hippodamia convergens* it is presumed that diapause is induced mainly by food (Hagen 1962). When aphids are lacking, the coccinellids feed on alternative plant food (nectar of flowers or pollen — see 6.111) and this is why they accumulate reserves instead of reproducing. They migrate into the mountain forests, without having built up any great degree of fat. There they feed and deposit sufficient fat for aestivo-hibernation. Then they move to lower altitudes for the actual overwintering (*H. convergens* — Hagen 1962).

In *S. undecimnotata* the "switching of the points" apparently must already take place in the larva, for it is reported that the adults do not feed at all after they emerge from the pupa and manage during dormancy with the reserves accumulated by the larva (Yakhontov 1962).

It is possible that the controversy is to some extent the result of a difference in the way the various species have been studied and that further study will remove the difference which is currently thought to exist. Any comparison is at present made difficult by the fact that no publications with numerical data are available for *H. convergens* and *S. undecimnotata*.

Pre-diapause feeding of adults has been ascertained for many other species, e. g. *Chilocorus geminus* and *C. bipustulatus* (Zaslavskiĭ and Bogdanova 1965, Zaslavskiĭ 1970), and *Coccinella novemnotata* (Mc Mullen 1967a, b).

### 7.312 Migration

According to the latest concepts of migration (Johnson 1969) both the flights to and from the dormancy sites are migration. This type of displacement fully fits in migration class III., defined by Johnson as follows: "Some long-lived females fly, while sexually immature, from the site where they became adult to hibernate or aestivate at sites which may be either within or beyond the areas where climate limits seasonal breeding. The males may or may not accompany them. Next season, after diapause, they reinvade breeding places and oviposit there, at least some of them more or less retracing the emigration route".

Johnson's view is to some degree in disagreement with the Hagen's older conception of coccinellid migration. Hagen (1962) argues that the flight toward aggregational sites can be considered as migration since it is directional and under partial control of the beetle. The disbanding of aggregations is, however, a simple dispersal flight. Johnson (1969) counters that the direction of both displacements is greatly affected by wind. The contrasting ultimate results of migration (aggregation or dispersal) are not caused by differences in the flight behaviour during displacement, but by responses at the concluding phase of migration.

Although this disagreement might seem only semantic, it actually concerns the veritable crux of the problem of migration which is still quite unresolved: do coccinellids fly to hibernation sites by a directional flight, or are they brought there more or less passively by wind currents? No direct evidence is available to favour either of these possibilities. There is, however, some circumstantial evidence for both viewpoints.

The aggregations for dormancy are often formed at prominent features of the landscape. For those coccinellids that make lengthy migratory flights (*S. undecimnotata*) these may be peaks of hills, large rocks or posts on the slopes. There may, however, be a forest edge, a wave in the field, a shrub, a tree, or a post in a flat landscape for those coccinellids that aggregate within the breeding area but usually in a different habitat or microhabitat (*C. septempunctata*). These sites are the same year after year if the relief remains the same; but new sites are occupied if the relief is changed by the disappearance or formation of a prominent object. The creation or removal of triangulation posts, forest fences and huts has provided the author with "unintentional" experiments which show how often the aggregation sites of *S. undecimnotata* can change (Hodek unpubl.).

It seems improbable that coccinellids would always be transported passively to the same places or that wind currents would be so drastically changed, for example by a post of 15 cm or less in diameter, that the coccinellids are carried elsewhere. Also direct observations on the top of a hill (Raná in northern Bohemia or Kováčov-hills in southern Slovakia) have shown that *S. undecimnotata* landed actively e. g. on a shifting figure that constituted the most prominent object (Hodek unpubl.). Any air currents existing were too weak to be noticed by the observer.

Therefore an analogy has been sought between the aggregation of coccinellid at prominent objects and the swarming habits of pairing ants and other insects or the flight to the feeding sites in *Melolontha* hypsotactic orientation has been suspected (Hodek 1960a, Hagen 1962).

Some authors have observed that coccinellid adults move against the prevailing wind in the direction of mountain peaks (Mani 1962).

By contrast, coccinellids have also been observed to make use of air currents, especially in mountain valleys. Savoĭskaya (1966) reported that in the Zailiĭskiĭ Alatau mountains (Kazakh. SSR) the coccinellids fly to their hibernation places up the valleys with the help of a steady breeze blowing upwards in the daytime. According to Hagen's hypothesis (1962, in Hodek p. 135, 1966) on the migratory flight behaviour of *Hippodamia convergens*, this species is also transported to the hibernation site and back by air currents.

It is possible that migration behaviour may be different in individual species. For instance Hagen (1962) clearly distinguishes 'hypsotactic' and 'climatotactic' aggregators. However, it seems more probable that for any one species (at least for most species) behavioural responses of both types apply (Hodek 1967), even if one type may dominate in certain species. Johnson (1969) judges that 'the hypothesis that postulates an appetence by a migrant for some perceived but far-distant habitat or hibernation site is at present unacceptable'.

In this context the mechanism for long distance migrations by coccinellids which Hagen (1962, in Hodek p. 135, 1966) hypothesized for *H. convergens* (see 7.324) seems most probable. It is only towards the end of the long migration that the hypsotactic response may be shown and that an optically oriented directional flight can guide the coccinellids to prominent features. This may even apply to the aggregations of *H. convergens* on mountain tops which have been reported many times (recently by Stewart et al. 1967), and the hypsotactic response may also be involved when coccinellids choose places of fairly large clearings on the slopes of hills for winter aggregation (Hagen 1962). Sem'yanov (1965b) states that the tendency to hibernate on at least slightly elevated ground may have the survival advantage that in some lowland regions extensive parts of the plains may be flooded. Many other authors have speculated on the adaptive significance of hibernation in the mountains and in aggregations (summarized in Hodek 1960a). It is quite possible that mass hibernation facilitates contact between beetles of the opposite sex which will be important for less abundant species and leads to crossings between those beetles which have developed under different climatic conditions. Hibernation in high mountains in areas with warm climates may retard the beginning of spring activity of the beetles till aphids have multiplied sufficiently. A similar function may be attributed to hibernation among the litter in plains, under mosses, and in other humid microhabitats where cooling is brought about by the evaporation of water.

### 7.313 Aggregations

The formation of aggregations or 'assemblages' in the sense of Mani (1962) and Kapur (1954), or the non-random distribution (Banks 1954) is a very specific behavioural feature connected with dormancy in many *Coccinellidae*.

The behavioural responses induce the coccinellids to aggregate in two ways. Some of the responses act **indirectly** — the beetles guided by them are passively brought to the same place. Hypsotactic responses (dealt with in 7.312) cause the beetles to concentrate on a larger spot, a portion of a habitat, as e. g. surrounding a post or a rock on a hill-slope, or in a clearing surrounded by high trees, etc. After arriving there, the beetles are influenced by their responses to physical factors (hygrotaxis, thermotaxis, phototaxis), and by geotaxis and thigmotaxis so that they hide into small places, in microhabitats, as e. g. the space under a stone, crevices in a rock, in a grass tussock, in tree bark, etc.

The beetles hibernating in the litter, for instance, have apparently been guided by negative phototaxis and positive geotaxis (possibly also positive hygrotaxis) to hide on or near the ground after a period of sitting on the plants. It can be assumed that during this period of 'waiting' the behavioural patterns gradually change from those which led the beetles to the hibernation habitat during migration.

As regards the hygrotactic responses, these are perhaps more specific for individual species. In *H. convergens* and *Coleomegilla maculata* moisture availability is an important requirement during hibernation (Hagen 1962), and in *Coccinella novemnotata* in aestivation (McMullen 1967a), whereas in *S. undecimnotata* and in other species with a similar type of hibernation ventilated crevices are preferred and seem to be the most favourable (Hodek 1960a, Yakhontov 1962, Iperti in Hodek 1966 p. 137,) as they apparently reduce the danger of mycosis (Iperti 1964, in Hodek 1966 p. 189).

Few experimental studies have investigated the reactions to light, temperature and moisture, also geotaxis. Pulliainen (1963, 1964) studied the responses of *Myrrha octodecimguttata* hibernating in bark crevices in the final period of dormancy, and reactivated by temperature in the laboratory. The beetles showed a strong hygronegative reaction, which only changed after a prolonged desiccation. They reacted indifferently to long-wave light and photonegatively to short-wave light. After desiccation, they became strongly photonegative to long-wave light also. The optimum relative humidity is about 30 — 40 %. Novák (1966) found that the georeaction was only very slightly positive in dormant adults of *Coccinella septempunctata* collected on trees during their 'waiting' period before they hid in the litter. In humid environments, the response changed to be slightly geonegative.

Surprisingly enough, although there are many indications (e. g. renewed dispersal in spring) that responses must change during dormancy, no studies on this have been carried out.

The **direct** stimuli which result in aggregation may — apart from thigmotaxis whose function in this context is unquestionable — also include chemotaxis i. e. attraction by a specific odour. The specific odour of coccinellids is penetratingly noticeable to man, and long ago speculations were published on its function both as a protective agent (see 8.2) and as an agent enabling the formation of aggregations. Unfortunately there is absolutely no experimental verification of this function and thus we must limit ourselves to mere hypotheses. Of recent works, particularly Savoïskaya (in Hodek 1966 p. 139) assumes an aggregative function for odour following her observations on the coccinellids of the genus *Adalia* (see 7.323); she reports that the odour is stronger during migration. It may be that the smell attracts the later migrants to those beetles which have arrived earlier. Yakhontov (1962) supposed that the dead bodies remaining from the previous hibernation had an attraction, but this has been denied by Iperti (pers. comm.), and Novák (1965) has proved experimentally that the dead beetles are more likely to act as a repellent.

Both monospecific and heterospecific aggregations of coccinellids may be formed. As a general rule, whenever several species spend their dormancy in the same habitat, they tend to form monospecific clusters (McMullen 1967a) though the presence of few adults of other species has been reported.

Rarely (in two cases) mixed clusters of *S. undecimnotata* with *A. bipunctata* have been observed. It is more usual to find heterospecific aggregations between *C. septempunctata* and *S. undecimnotata* in grass tussocks on the tops of hills (Hodek 1960a and unpubl.), especially where rocks are missing. Pulliainen (1966) has reported that *Scymnus suturalis* and *Aphidecta obliterata* were admixed to aggregations of *Myrrha octodecimguttata*.

#### 7.314 Emergence from dormancy sites

During the closing stages of dormancy the beetles gradually change their behaviour in relation to changes in temperature in the converse way as happens in the introductory phase. The inactivity in the hiding places changes to slight mobility, sometimes already long before dispersion becomes noticeable. This closing stage is conspicuous and has often been observed in coccinellids hibernating in litter at the forest edges; such coccinellids stay for a prolonged period on the trees (esp. on young pines) in the dormancy site before dispersion (Bielawski 1961 etc. — see also 7.322).

In all species studied, dispersal from dormancy sites has been found to continue over a period of several weeks (Hodek 1960a, Bielawski 1961, Savoïskaya 1965, McMullen 1967a). The onset and progress of dispersal is regulated by the ambient temperature, as the second phase of dormancy has the characteristics of a simple quiescence (see 7.412). Usually a prolonged increase of average temperature over 10 °C induces the beetles to emerge (Hodek 1960a).

The different species of beetles, even if they have hibernated at the same site, do not disperse simultaneously. In Bohemia (Hodek 1960) *Adalia bipunctata* is the earliest species, followed by *Coccinella quinquepunctata*, *Propylaea quatuordecimpunctata* and *Tytthaspis sedecimpunctata*, then by *C. septempunctata* and *Adonia variegata*. *Semiadalia undecimnotata* seems to be very late. A similar, but not identical picture is reported by Bielawski (1961) at a hibernation quarter under young pines in Poland (Warszawa — Bielany). *A. bipunctata*, *A. decempunctata* and *C. quinquepunctata* leave about a month earlier than *C. septempunctata* and *P. quatuordecimpunctata*. Whereas these two observations agree that *A. bipunctata* is an early and *C. septempunctata* a late dispersor, Banks (1955) reports an opposite observation from England: *C. septempunctata* appeared on nettles one month earlier than *A. bipunctata* and *P. quatuordecimpunctata*.

### 7.32 Behaviour of individual species

#### 7.321 Semiadalia undecimnotata

The migratory and aggregational behaviour of this species is suprisingly uniform in all the three regions, in which aestivo-hibernation has been studied although they differ in many respects; the three areas are: the typically continental central Asian region of the USSR (Dobzhansky 1925, Radzievskaya 1939, Yakhontov, in Hodek 1966, p. 107), Czechoslovakia at the northern limit of the distribution area (Hodek 1960a, 1967) and the maritime region of south-eastern France — Drome, Vaucluse and Basses-Alpes (Iperti in Hodek 1966, p. 137).

The dormancy sites are always situated on hills or mountains, (see photos 23—29, 36—42) at prominent features (large rocks, heaps of stones, shrubs or other plants, or artificial structures, such as posts, triangulation points, and only very rarely buildings or ruins of castles). It is assumed that such sites are chosen as the result of hypsotactic responses. Coccinellids prefer rock cracks, especially those exposed to wind, where their mortality is lower. They only hibernate in vegetation when there is an absence of suitable shelter among rocks on the hill; then a high percentage often die of mycosis. The petrography of the site is not crucial, hibernation quarters can be found both on limestone and on igneous rocks (e. g. basalt). The altitude of the site (above see level) is also in no way crucial, in central Asia it is naturally much higher (> 2000 m) than in Czechoslovakia (400 — 900 m) or in France (usually above 700 m). The great majority of dormancy sites are situated on the south-western upper slopes of the hills or mountains. According to Iperti (in Hodek 1966, p. 137) particular attractivity may be attributed to the afternoon insolation of these sites.

The coccinellids arrive at the hibernation places over a period of several weeks on warm calm days and leave again over the same period under similar conditions. At the onset and end of hibernation they creep around close to the shelter during the day, forming numerous small clusters during the night or colder spells. In the course of the central cold part of dormancy they are aggregated in one or a few large clusters.

Their spring emergence seems to be dependent on the ambient temperature. In the Louny-hills (northern Bohemia) emergence occurs after the mean temperature has remained above 12 — 14.5 °C for some time (Hodek 1960a). Yakhontov (1962) regards the relationship between emergence from hibernation quarters and temperature as secondary in that it influences the maturation of gonads; a certain degree of the gonad maturation then becomes the signal for emergence. This does not agree with the findings from Czechoslovakia, where females emerged at different times in different years depending on meteorological conditions, but with ovaries at different stages of maturation. When the females leave the hibernation sites early, they have not yet reached the onset of previtellogenesis (i. e. formation of the first oocyte). If they disperse later, this stage is already attained at the dormancy site (Hodek and Landa 1971). In males any dependence of emergence from hibernation quarters on the maturation of testes seems even more improbable for the tissue of the testicular follicles is fully active already 2 — 4 weeks before the emergence and the males fertilize the females (Hodek and Landa 1971).

Year after year *S. undecimnotata* uses the same dormancy sites if the silhouette of the hill has not changed.

The east Siberian *Harmonia axyridis* seems to have the same type of dormancy behaviour as *S. undecimnotata*, at least as far as the short description by Telenga and Bogunova (1936) is considered.

### 7.322 Coccinella septempunctata and other species dormant in the litter

Attention should be drawn to the fact that those references to hibernation of *C. septempunctata* in the literature which describe mass assemblages between rocks on mountain summits, most probably do not concern this species at all but refer to *S. undecimnotata*. (The seven spotted form of *S. undecimnotata* is often confused with *C. septempunctata*, even by professional entomologists.) Iperti (unpubl.) has verified that such confusion occurred in Fabre's data (1879) from Mont Ventoux and St. Amand. It is possible that the same confusion has occurred elsewhere: Poulton (1904) from Spain, Benner and Franz (1905) from Poland, Werner (1913) from Turkey, Semenov-Tyanshanskiĭ (1911) from Italy (Vesuvius), Reineck (1918) from Germany, Radzievskaya (1939) from central Asia, and Benkevich (1958) from Crimea.

Even if these observations are ignored, the dormancy behaviour of *C. septempunctata*, in contrast to that of *S. undecimnotata*, is extremely heterogeneous even within one geographical area. The only safe generalisation is that *C. septempunctata* always hibernates on the ground: under stones, in litter, in holes in the soil surface, near the base of plants, in grass tussocks [never in the cracks of tree bark or walls — contrast *A. bipunctata* (see 7.323)]. The species forms only small or medium size aggregations, not exceeding tens of beetles. *C. septempunctata* often hibernates near its breeding sites. If any higher area, however small, is to be found near the breeding habitats, the beetles prefer to hibernate there. In lowlands the usual dormancy sites are situated at forest edges, clearings, or in wind breaks. If no such site is close by, *C. septempunctata* hibernates near isolated shrubs or other plants, or on slight unevennesses of the terrain, as often occurs at the road, etc. The types of site mentioned above may be combined: the beetles are very often found at forest edges or clearings on the slopes of hills. No special differences seem to occur between individual countries in this rather wide choice of habitats and microhabitats used as dormancy sites by *C. septempunctata:* information is available from England (Banks 1954a), Germany (Klausnitzer 1967), Poland (Bielawski 1961), the Ukraine (Dobrzhanskiĭ 1922a,b, Telenga 1948, Dyadechko 1954), Belorussia and in the Leningrad region of the Soviet Union (Sem'yanov 1965a, b), also Czechoslovakia (Hodek 1960a).

Savoïskaya (1960, 1965,) who distinguishes 5 types of hibernation quarters of coccinellids in the Kazakh. SSR, states that *C. septempunctata* utilises no less than four. The most characteristic is hibernation under small stones in the mountains, both along the river beds in groups of about 50 individuals (up to 900 — 1400 m) and on the slopes in the alpine zone (2600 — 3300 m). Here they form groups of 2 — 5 up to 30 individuals under small stones. *Harmonia axyridis* and *Aiolocaria mirabilis* also hibernate in this way in the Far East region of the USSR (Telenga and Bogunova 1936, Savoïskaya 1970b). In the alpine zone, *C. septempunctata* also hibernates at the base of grass tussocks and shrubs, but in smaller numbers. In the lowland of Kazakhstan the dormancy sites of *C. septempunctata* do not differ from the microhabitats described for several countries above. Thus the species hibernates in the upper layers of the soil or in litter or holes at the base of plants. The latter sites are used also by *Propylaea quatuordecimpunctata.*

After their gradual arrival at the dormancy site the coccinellids remain clustered on the plants close to the hiding places (during August and September in Bohemia) and only hide by degrees as the air becomes cooler. They can very often be found at the ends of twigs of young pines (Hodek 1960a, Bielawski 1961, Sem'yanov 1965a, b, Klausnitzer 1967). In a similar way the coccinellids re-appear and stay on the vegetation near the hibernation site before their dispersal in spring.

Of the common coccinellids, *C. quinquepunctata* is another species with a hibernation behaviour similar to that of *C. septempunctata;* it is, howeverfar, more abundant than the latter on young pine trees at forest edges (Hodek 1960a, Bielawski 1961, Sem'yanov 1965a, b, Klausnitzer 1967). It can also be found among small stones on hills (e. g. Louny-hills in north Bohemia), used as hibernation quarters by *S. undecimnotata* and *C. septempunctata* (Hodek 1960 and unpubl.).

Some other species are usually present in similar hibernation sites in the plain (the litter or upper soil-layer at forest edges). Such species are *Coccinula quatuordecimpustulata*, *Propylaea quatuordecimpunctata*, *Adonia variegata* (in dry places), *Hippodamia spp.* (if moist habitats are near), the phytophagous *Subcoccinella vigintiquatuorpunctata*, the mycophagous *Thea vigintiduopunctata* (especially in rather moist places), and *Tytthaspis sedecimpunctata* (particularly in drier localities) (Hodek 1960a, Bielawski 1961, Sem'yanov 1965a, b, Klausnitzer 1967, Novák and Grenarová 1967). Litter also serves as the dormancy site for *Stethorus punctillum* (Putman 1955, Berker 1958), although some authors report its hibernation in the bark (Moter 1959).

Besides these species for which litter in the widest sense typically provides shelter, *Calvia quatuordecimguttata* and *Exochomus quadripustulatus* sometimes also occur there. These two species are equally abundant in bark crevices.

### 7.323 Adalia bipunctata

This coccinellid is well known for its frequent occurrence in buildings at the time of its dormancy, and appears in cracks in walls, in lofts, or behind windows and even in rooms (Hawkes 1920, Sem'yanov 1970). It can be assumed that the coccinellids are brought here by their hypsotactic behaviour. This behaviour is also indicated by the fact that a small aggregation of *A. bipunctata* (20 — 30 individuals) has been found close to a large aggregation of *S. undecimnotata* in a rock crevice on a hill-top (Hodek unpubl.).

However, a fair proportion of *A. bipunctata* apparently do not leave the habitats of orchards, parks or forests, and hibernate either in crevices of tree bark or in artificial objects on trees. Thus Speyer (1934) used to find *A. bipunctata* together with *Chilocorus bipustulatus* and *Stethorus punctillum* in grease bands.

Mass hibernation in the thick bark of old Tyan-shan spruces (*Picea schrenkiana*) is described by Savoïskaya (1965, in Hodek 1966, p. 139) from Kazakhstan (USSR). Every autumn, the beetles migrate in masses to the same trees in mountain valleys flying up the valley and a rate of 50 — 55 beetles per min. was observed in late September. It was noticed that they turned sharply back just behind the tree, swarmed in its shade and dropped rapidly onto the trunk. Up to 2000 beetles could be found on one tree. Similar sites on *Picea* are also used by *A. fasciatopunctata* and by *Synharmonia conglobata*. The latter species is most abundant up to 1500 m, while the two *Adalia* species prefer higher altitudes.

Large aggregations of up to 200 *A. bipunctata* adults have been observed by Smee (1922) on elm-trees (*Ulmus*) in England (as many as 1000 beetles on one tree). Smee was the first to succeed in attracting coccinellids to artificial hibernation quarters; he achieved colonisation by *A. bipunctata* of cages containing tree bark and arranged around the garden in mid-October.

*A. bipunctata*, being a polyvoltine species, is relatively late (in early October in Germany — Speyer 1934) to enter diapause and emerges very early in spring.

*A. bipunctata* is very often found together with *Chilocorus bipustulatus* (Speyer 1934, Bielawski 1961, Savoïskaya 1965).

### 7.324 Myrrha octodecimguttata and other forest species which hibernate in bark crevices

Not much is known about the hibernation of coccinellid species which are typical of forests (especially coniferous). According to recent observations, they do not seem to leave their breeding habitat.

*Myrrha octodecimguttata* which lives and breeds in the crowns of pine-trees (Klausnitzer 1968), hibernates in bark crevices of old pines; this has been established for Poland by Bielawski (1961) and for Finland by Pulliainen (1963, 1964, 1966). Pulliainen observed that the coccinellids preferred the lowest 10 cm of the trunk (93 %) and the south and east sides of the tree (62 %). On these sides tendency to aggregate is most pronounced; the aggregations were on average of 3.8 individuals, the largest one consisting of 14 individuals.

Bark crevices of old pines are used as dormancy sites by many other species, mostly by species also bound to the forest habitat for their breeding life: *Scymnus* (*Pullus*) *suturalis* (Pulliainen 1966; although Bielawski (1961) found this species more frequently in the bark crevices of deciduous trees), *Scymnus* (*Pullus*) *impexus* (Delucchi 1954), *Scymnus nigrinus* (Bielawski 1961), and *Aphidecta obliterata* (Pulliainen 1966). *Exochomus quadripustulatus* (Bielawski 1961) also uses bark crevices but utilizes crevices and litter as its dormancy site with about equal frequency (Klausnitzer 1967). Other conifers than pines may also be visited, e. g. cedar by *Hippodamia tredecimpunctata* (Thomas 1932). Some coccinellids prefer the crevices in the bark of deciduous trees, e. g. *Scymnus* (*Nephus*) *quadrimaculatus* prefers chestnuts (Bielawski 1961), and *Synharmonia conglobata* and *Harmonia quadripunctata* chestnut and poplar (Bielawski 1961, Hodek unpubl.).

### 7.325 Hippodamia convergens

The diapause of this species, which is one of the most common coccinellids in North America, has been thoroughly studied by Hagen (1962, in Hodek 1966, p. 131, 135) in northern California and three types of dormancy have been determined. The majority of the population in northern California has a univoltine cycle. This is the original pattern, as before irrigation was introduced, the species was dependent on spring-occurring aphids. The irrigated crops provide a later supply of food which enables exotic aphids to maintain themselves during the summer and autumn. *H. convergens* and some other *Hippodamia spp.* may react to this later abundance of aphids by summer reproduction and multivoltinism.

The multivoltine adults of *H. convergens* enter hibernation in the late autumn. Diapause is induced mainly by photoperiod and temperature. *H. convergens* often joins the other *Hippodamia spp.* in the valley aggregations.

Most univoltine *H. convergens* adults show a facultative diapause that appears to be largely nutritionally induced. In the laboratory, however, there were occasions when 10 — 20 % of the beetles entered diapause although the conditions were optimal and the beetles were supplied with an excess of essential aphid food. Some interaction of photoperiod and diapause has also been observed, but this is much less important than nutrition. The main factor inducing diapause is feeding upon non-insect food. The progeny of diapausing beetles emerge from pupae in early May. They do not rove much locally, but migrate long distances towards the mountains (in northern California to Sierra Nevada) if there are not enough aphids nearby.

Hagen (1962, in Hodek, 1966 p. 135) has proposed a hypothesis to explain the mechanism of migration, which is in harmony with the concepts of Johnson (1969). Migration is initiated by a take-off during the calm of morning; the subsequent vertical upward flight is apparently assisted by convection currents. This goes on until the temperature permits the beetles to fly. At about 11 — 13 °C flight is inhibited (Hagen calls this the temperature ceiling) and the beetles fall about 300 m to a warmer layer of air (13 — 18 °C) where flight is resumed. The beetles which are thought to undergo these vertical oscillations are simultaneously blown sideways (Fig. 7.02). Each day during the late morning in May and June, westerly winds develop which carry the beetles toward the mountains. The migration is terminated when the temperature ceiling for flight contacts the mountains.

If aphids are found near the landing place, the beetles produce eggs there. Usually, however, they feed on non-insect diet which is used for building up reserves. After a week or so the coccinellids exhibit secondary, directional flights near the ground, flying up and down mountain creeks in search of summer aggregation sites. In the summer aggregations the adults remain rather inactive until October, when they become mobile again, presumably disturbed by the first rains. The tertiary flight during warm periods usually leads them to lower parts of the creek where new, larger overwintering aggregations are found.

*H. convergens* adults are guided by a series of factors in their search for the aggregation sites. In contrast to *S. undecimnotata*, by and large they do not respond to prominent objects. Sometimes assemblages of *H. convergens* are indeed found at the top of mountains (Douglas 1930, Throne 1935, Sherman 1938, Stewart et al. 1967), but according to Hagen's (1962) hypothesis they would reach summits in a semi-passive way. Moisture and light are considered to be the most important factors influencing selection of aggregation sites. At first the coccinellids assemble on bushes and trees in large clearings on slopes near creeks. After several days they move closer to the creeks, and

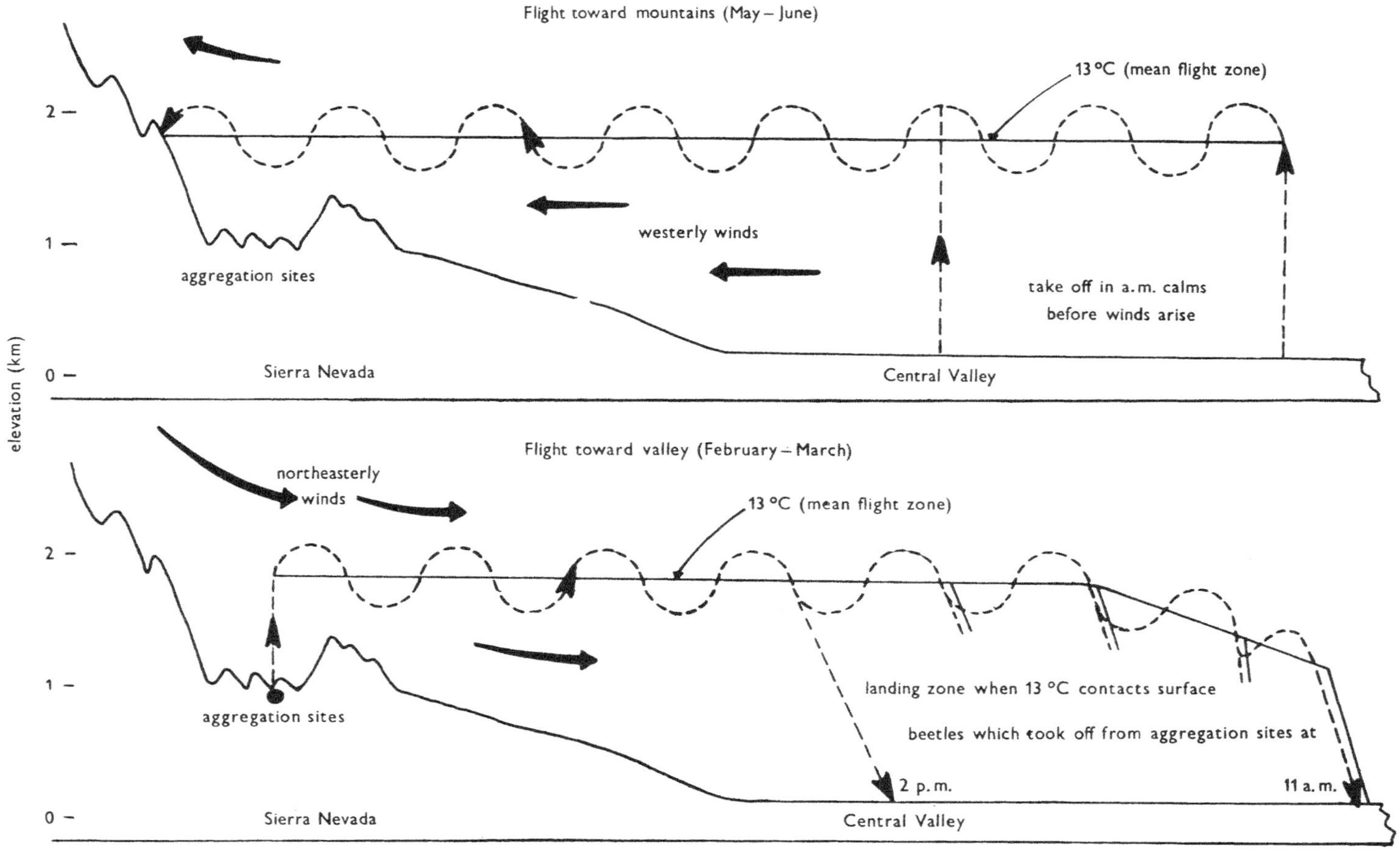


Fig. 7.02 Suspected temperature — controlled flight oscillations which *Hippodamia convergens* undergoes in its migratory flights (Hagen 1962).

settle down along the banks in the litter, often in spots exposed to the afternoon sunlight. As the beetles fly about, they concentrate in the sunlit spots until sunset. This may explain why the aggregations are formed in similar spots year after year. As new clearings are often selected, the assumption that dead beetles attract the aggregators (Yakhontov 1962) cannot apply here.

Hodson (1937) proved that the beetles exhibit a marked preference for litter with 20 % moisture. Hagen (1962) believes that *H. convergens* drinks water in order to maintain a constant water content and that a physiological requirement for free water may be the basic characteristic of "climatotactic" aggregators (*H. convergens*, *Coleomegilla maculata*), in contrast to "hypsotactic" species such as *S. undecimnotata*.

The large winter colonies (from one, about 40 million beetles have been collected) are formed by many non-contiguous small aggregations under leaves, at the bases of bushes, and on tree trunks. Snow often covers *H. convergens* aggregations for about three months at the higher altitudes. The majority, however, hibernate near the snow line in the Sierra Nevada, and during most of the winter they are not covered with snow.

Because the beetles aggregate in canyons, they are in the wind shadow. When they become heated to above 14 °C in the early spring, they take off vertically and are eventually caught up by the winds aloft (blowing in the opposite direction to that in summer) which "carry" them over ridges that lie to the west between the aggregation sites and the plains. The fall in air temperature in the evening forces the beetles to the ground, terminating the flight. Not all coccinellids, however, get so far. The beetles that aggregate during winter on the highest peaks of the Sierra are found after dispersal only to have reached the mountain valleys just below these peaks.

#### 7.326 Mycophagous and phytophagous species

Hibernation in aggregations does not occur only in entomophagous coccinellids. Three mycophagous species have been reported hibernating in litter, one in large aggregations. In mid-April Ruscinsky (1933) found four large aggregations of *Vibidia duodecimguttata* among fallen leaves at the base of trees on top of a hill in erstwhile Rumania (Bessarabia). Each aggregation covered about 1 m$^2$ and consisted of some 2500 individuals. Evans (1936) found a dormancy aggregation of *Tytthaspis sedecimpunctata* on a wooden post in Berkshire, England in December; Legay and De Reggi (1962) found aggregations (50 — 150 individuals per 100 m$^2$) of the same species under litter at the base of trees on a small rise overtopping the surrounding fields by only 20 m. The beetles left this hibernation site near Lyon, France in mid-April. This same site contained aggregations which included 1 % of *Thea vigintiduopunctata*. This latter species was often found in litter at moist dormancy sites at forest edges visited by *C. septempunctata* and related species (Hodek 1960a, Bielawski 1961) (see 7.322).

Aggregations have even been observed in phytophagous coccinellids in the subfamily *Epilachninae*. In Africa, Poulton (1936) observed aestivation aggregations of *Epilachna dregei*, once in Bechuanaland and once in northern Uganda. Both observations were of aggregations of immobile adults on a prominent feature (a hill, a termite-hill) and were found in mid-July, i. e. in the middle of the dry season in Uganda. In February, Kapur (1954) found four aggregations of several thousand individuals of *Epilachna bisquadripunctata* at an altitude of about 400 m in India (Chota Nagpur, Bihar). The aggregations were at the base of grass ca. 70 cm high, in the vicinity of an almost dry brook. The fact that the coccinellids had empty guts, a large fat body and quite unripe ovaries is evidence that they were in diapause. In both *Epilachna spp.* observed, "hypsotactic" aggregation may have been involved. Ghabn (1951) mentions winter migrations of *Epilachna chrysomelina* from the fields into the surrounding desert in Egypt without, however, describing the hibernation sites.

## 7.4 Ecological mechanisms of diapause

### 7.41 Development and current state of the problem in insects in general

This problem has been discussed with reference to coccinellids in only a few papers, so that a review of such papers cannot provide a general picture. This introduction, therefore, will give an overall informative survey of the developments in this field based on the whole range of insects. Nevertheless, the rapid developments in this field require that substantial changes need to be made in earlier hypotheses still frequently held.[1])

The old differential definition of diapause and quiescence by Shelford (1929) and Roubaud (1930) suffered from a lack of experimental evidence. The term diapause (la diapause vraie) was restricted to those instances where development is arrested spontaneously, whereas an interruption of development by unfavourable conditions was considered to be quiescence (la pseudodiapause).[2])

In many cases diapause has been proved to be induced by environmental conditions, therefore, 25 years after its inception, Lees (1955) considered that the definition was inappropriate and modified it satisfactorily to that date. In general, however, he retained the concept that these two types of developmental arrest were opposed, though he recognized that no clear distinction could be made between them.

Lees considered quiescence (la pseudodiapause) was a state of direct inhibition of development, caused by the direct effect of ambient conditions (low temperature, lack of humidity), and was terminated immediately the conditions became favourable. Diapause (la diapause vraie) was thought to be caused by conditions which did not directly prevent development, but which were merely signals of seasonal changes ("seasonal tokens"). Thus diapause usually begins long before the onset of unfavourable environmental change.

Lees himself realised the limitations of these definitions and they indeed have recently lost some of their validity since photoperiodism has been more fully understood. The length of the photophase cannot be regarded as directly preventing development and thus these cases of diapause which are directly regulated by photoperiods were not really diapause in terms of the classical definitions. This dilemma has been circum--navigated by different approaches. It is possible to separate these problem cases from typical diapause (or eudiapause) by giving them different names: parapause, oligopause (Müller 1965, 1970). This solution has the advantage of emphasizing the existence of transitions between quiescence and diapause. A certain disadvantage lies in the fact

[1]) The survey presented here can only be a simplified one. Students of diapause in coccinellids are recommended to refer to the literature cited.

[2]) Attention must be drawn to the fact that this concept, which may be misleading, has been retained by some authors and appears in the current literature. In some French papers (e. g. Bonnemaison 1964), the terms "la diapause vraie" in the sense of obligatory diapause and „la pseudodiapause" in the sense of facultative diapause are sometimes used.

that this system of terms and definitions, due to a relative shortage of facts, cannot yet be sufficiently elaborated to encompass all the striking complexity of the mechanisms involved. There are species of insects where the same individual contains the potential for alternative ways in which arrest or resumption of development can be affected; this is clearly seen in the coccinellid *Chilocorus bipustulatus* (Zaslavskiĭ 1970) (see 7.422).

Therefore many authors still retain the term diapause in a broad sense for any adaptive arrest of development which is accompanied by behavioural, structural and biochemical changes in the insects (Danilevskiĭ 1961, Fuzeau-Braesch 1961, Way 1963, Hodek 1965). Beck (1968) defines diapause as "a genetically determined state of suppressed development, the manifestation of which may be induced by environmental factors".

Not only do these terminological difficulties exist, but also all authors are aware of the variability of the ecological mechanisms which govern the induction, maintenance and termination of diapause, and accept as the only common feature the adaptive function of diapause which is (1) to synchronize the development of active stages with favourable conditions and (2) to enhance the survival potential during unfavourable periods. It is, however, to be expected that the underlying physiological mechanisms will, at least in principle, be the same for any one developmental stage (e. g. adult) in different insects.

It is difficult to imagine that simple quiescence in the sense of unadaptive torpor (whether caused by cold or heat) could enable a long survival under temperatures above zero.

Whereas the word 'dormancy' is used by most authors (and also throughout this book) as a more general term for any developmental arrest (comprising both quiescence and diapause), some authors use it in a restricted sense to designate 'quiescence', as e. g. Beck (1968): "simple dormancy, such as heat — and cold torpor . . .".

Just as with quiescence and diapause, so there is contradiction in the definition of the two basic types of diapause. If the potentialities for diapause are not realized in each generation, diapause is termed 'facultative'. The onset of such developmental arrest can be induced or prevented by environmental conditions. By contrast, obligatory diapause is entered by virtually every individual in each generation regardless of the environment (Lees 1955). Somewhat later it became emphasized that obligatory diapause is not essentially distinct from facultative diapause but is its marginal case (Fuzeau-Braesch 1961).

Recently the strict definition of obligatory diapause has been weakened by experiments involving the alternation of photoperiods in studies on the ecological mechanism of diapause. Some insects cease development under steady conditions, whether long or short-day, and require an alternation characteristic for the species. Müller (1958) ascertained that the requisite for a complete vitellogenesis in the delphacid *Stenocranus minutus* was a period of short days followed by a period of long days. Thiele (1966) found the same alternation was necessary for the carabid *Pterostichus nigrita*, and recently Zaslavskiĭ (1970) has shown the same for *Chilocorus bipustulatus*.

In some insects the situation is less marked, the succesion of long days following on short days leads only to a substantially increased percentage of females ovipositing (in the carabid *Agonum assimile* — Thiele 1966 and in the pentatomid *Aelia acuminata* — Hodek and Honěk unpubl.).

A different procedure was successful for preventing the diapause of the deutostage in *Gryllus campestris* where this effect was achieved by slowing down the development of larval stages (Fuzeau-Braesch 1966). Alternation of different photoperiodic regimes has as yet been very rarely tried; we can expect that many diapauses regarded as

PLATE XX

1 — *Tytthaspis sedecimpunctata*; 2 — *Scymnus suturalis*; 3 — *Stethorus punctillum*; 4 — *Exochomus quadripustulatus*; 5 — *Platynaspis luteorubra*. (I. Kovář del.)

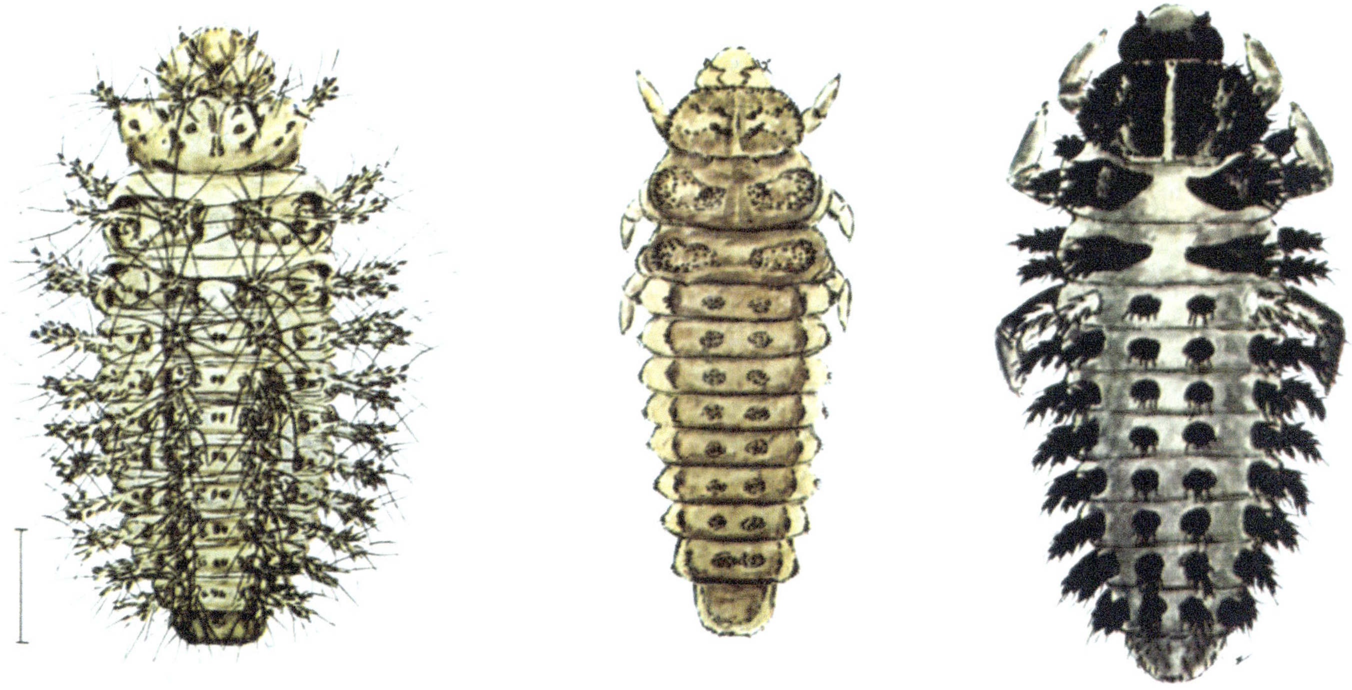

1 2 3

4th instar larvae. — 1 *Subcoccinella quatuordecimpunctata;* 2 — *Coccidula rufa;* 3 — *Exochomus quadripustulatus.* (J. Zelený del.)
(In the plates XXI—XXXIII, the length of the line represents 1 mm.)

PLATE XXII

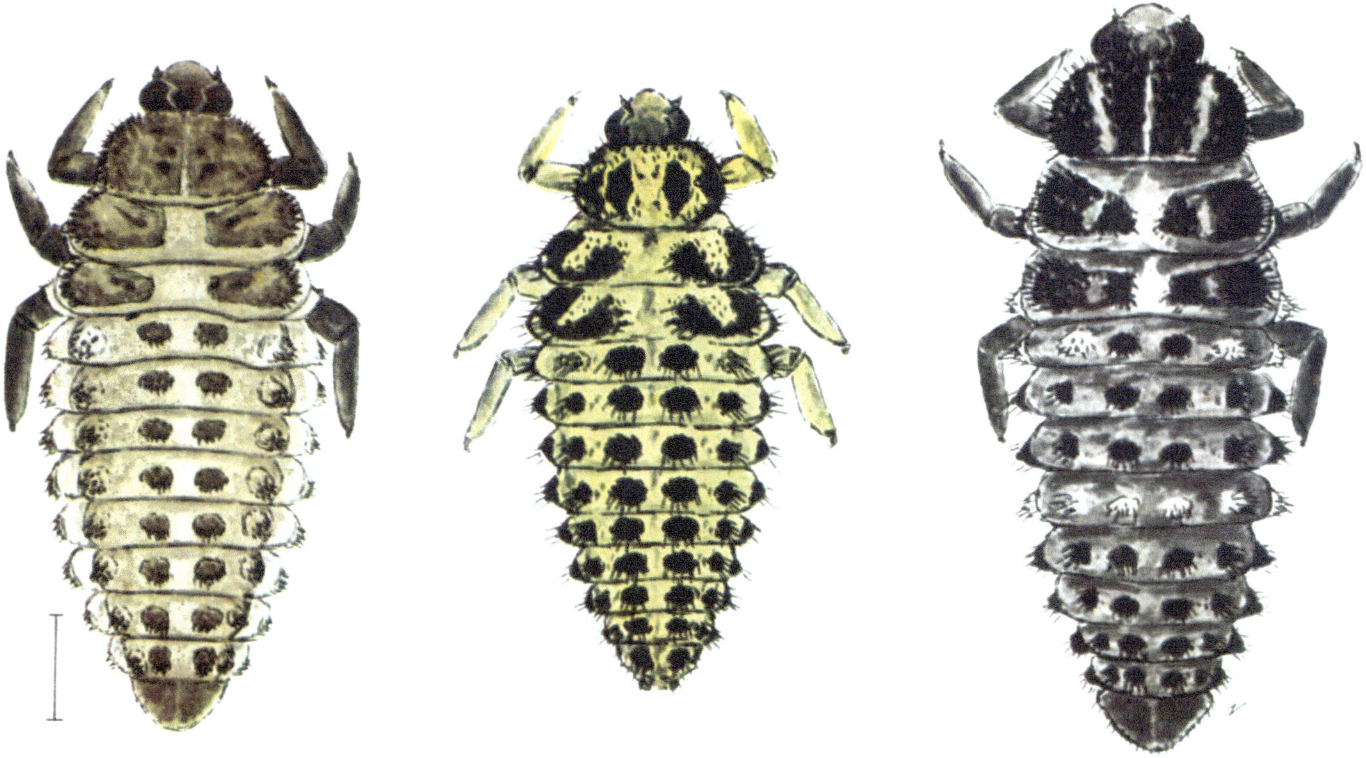

1 2 3

4th instar larvae. 1 — *Aphidecta obliterata;* 2 — *Thea vigintiduopunctata;* 3 — *Hippodamia tredecimpunctata.* (J. Zelený del.)

PLATE XXIII

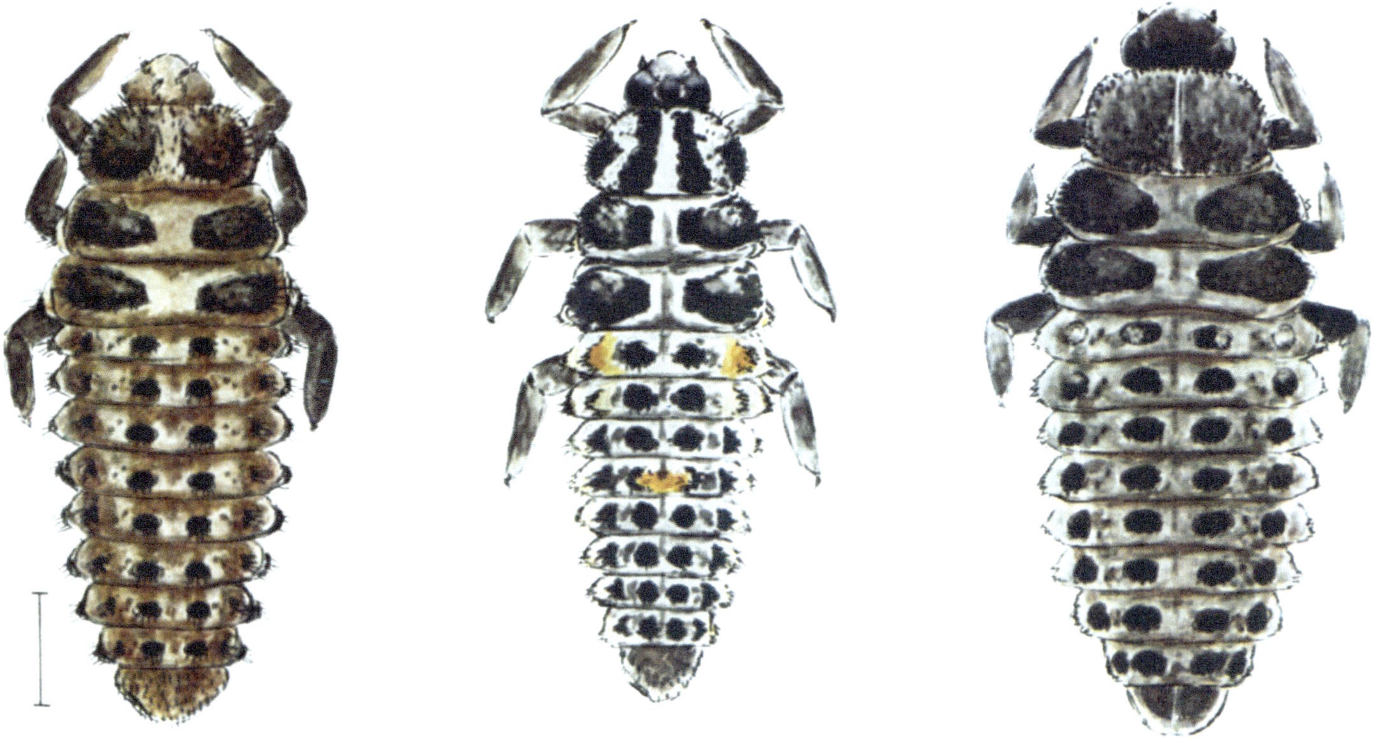

1 2 3

4th instar larvae. 1 — *Anisosticta novemdecimpunctata;* 2 — *Adalia decempunctata;* 3 — *Adalia conglomerata.* (J. Zelený del.)

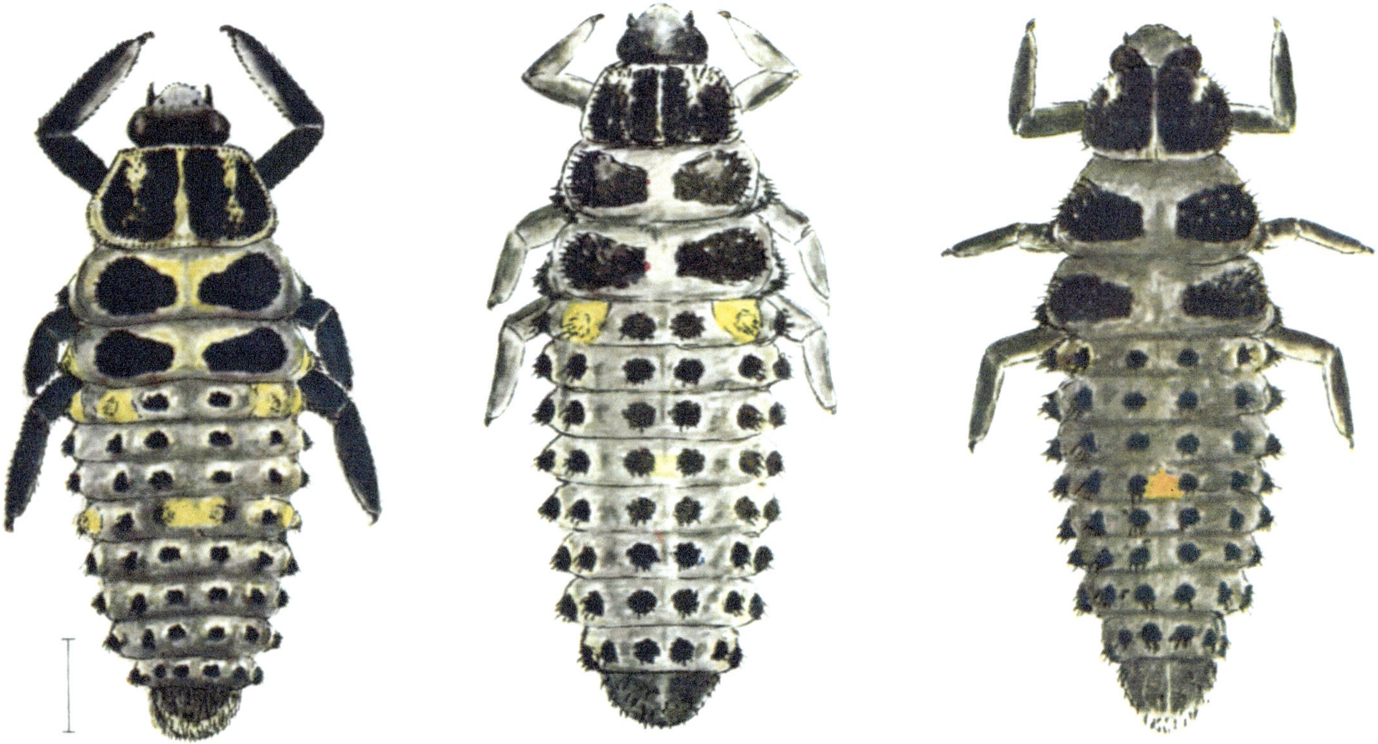

1 2 3

4th instar larvae. 1 — *Adalia bipunctata;* 2 — *Adalia bipunctata;* 3 — *Adalia bipunctata.* (J. Zelený del. )

4th instar larvae. 1 — *Adonia variegata*; 2 — *Coccinella septempunctata*; 3 — *Coccinella quinquepunctata*. (J. Zelený del.)

PLATE XXVI

1 2 3

4th instar larvae. 1 — *Propylaea quatuordecimpunctata;* 2 — *Synharmonia conglobata;* 3 — *Calvia quatuordecimguttata.* (J. Zelený del.)

PLATE XXVII

1 2 3

4th instar larvae. 1 — *Semiadalia undecimnotata*; 2 — *Harmonia quadripunctata*; 3 — *Cycloneda limbifer*. (J. Zelený del.)

4th instar larva of *Anatis ocellata*. (J. Zelený del.)

1 2 3

Pupae; 1 — *Exochomus quadripustulatus;* 2 — *Hippodamia tredecimpunctata;* 3 — *Calvia quatuordecimguttata.* (J. Zelený del.)

1 2 3

Pupae; 1 — *Semiadalia undecimnotata;* 2 — *Thea vigintiduopunctata;* 3 — *Cycloneda limbifer.* (J. Zelený del.)

Pupae; 1 — *Adalia decempunctata;* 2 — *Adalia bipunctata;* 3 — *Synharmonia conglobata;* 4 — *Propylaea quatuordecimpunctata.* (J. Zelený del.)

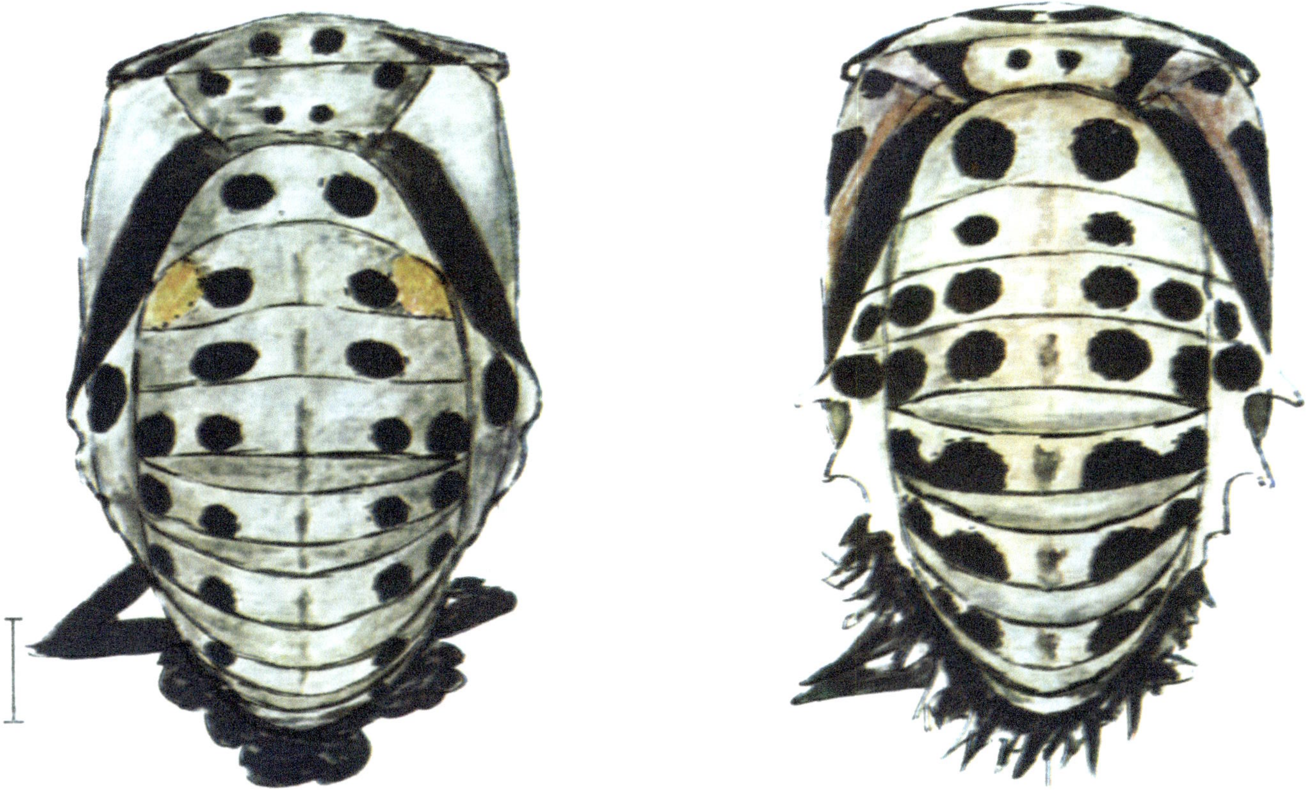

1 2

Pupae; 1 — *Neomysia oblongoguttata;* 2 — *Anatis ocellata.* (J. Zelený del.)

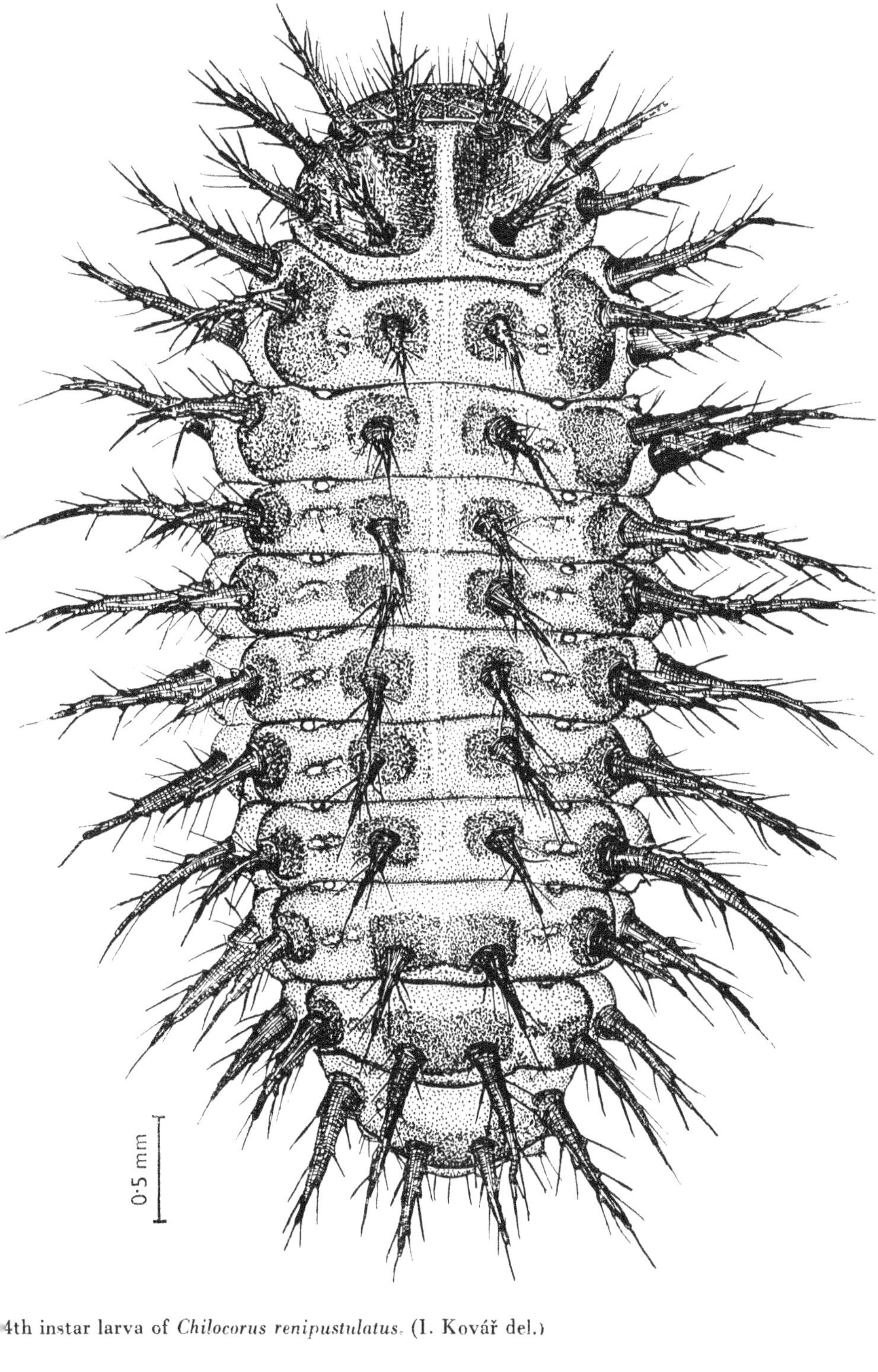


4th instar larva of *Chilocorus renipustulatus*. (I. Kovář del.)

PLATE XXXIV

1

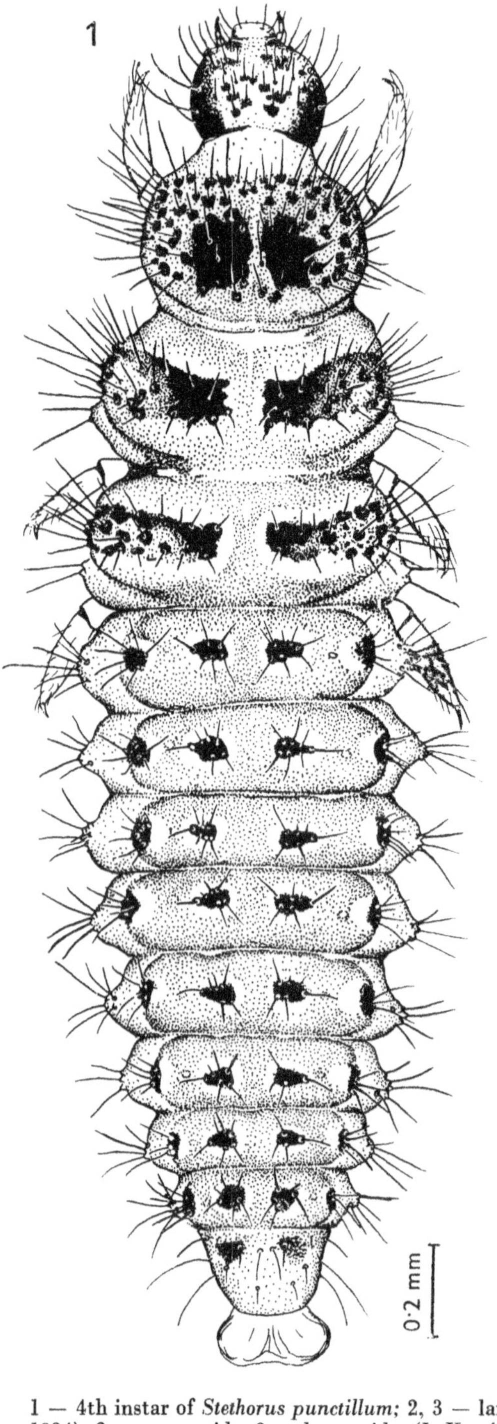


2

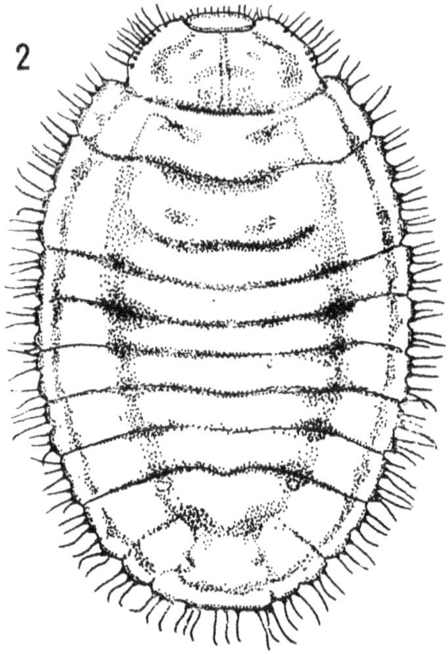

3

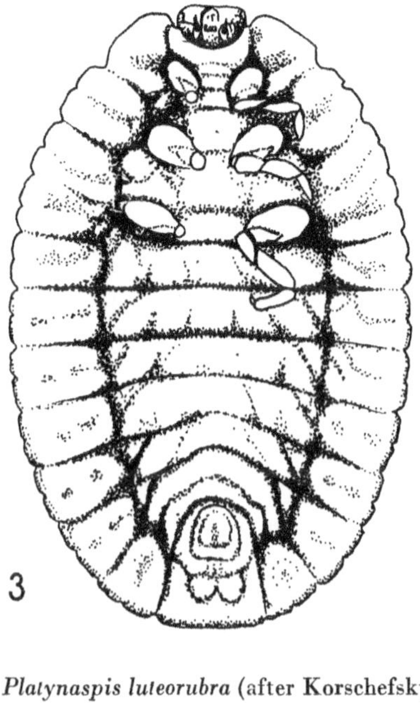

1 — 4th instar of *Stethorus punctillum;* 2, 3 — larva of *Platynaspis luteorubra* (after Korschefsky 1934); 2 — upper side, 3 — lower side. (I. Kovář del.)

"obligatory" could be eliminated from the life cycle by analogous procedures. It is therefore advisable not to restrict oneself to the use of constant photoperiods when studying the induction of diapause in coccinellids.

The ideas concerning the **termination of diapause** have also changed substantially during the short history of the study of diapause. As diapause had long been connected with hibernation, the classical concept assumed that the prerequisite for the termination of diapause was a certain period of chilling. This concept had two variations. The earlier variation contained the assumption that the inhibition of development was "broken" by the exposure to cold, whereas in the second variation Andrewartha (1952) hypothesized that at low temperature a gradual "diapause development" must be completed before normal development can be resumed. Obviously, "diapause development" is made possible by temperatures differing from those which enable normal development; most often the effective range is +5 — +10 °C.

This concept is retained in Müller's (1970) definition of eudiapause: "The further development is established not by the inducing factor, photoperiod, but by a completely different factor, temperature". When later on, instances of summer diapause — aestivation — were recorded, the necessity of "chilling" for the resumption of development began to be questioned.

Recently, however, photoperiodic termination has been achieved in many insects and examples are listed by Beck (1968), Danilevskiĭ (ed.) (1968), and Hodek (1971b). It is significant that photoperiodic and "chilling" termination may operate in the same insect species, and in a single population of a given species diapause may be terminated in either of the two ways. In temperate zones diapause usually covers only the first period of hibernation or aestivo-hibernation. In mid-winter, and in some insects already in autumn, the potential for resumption of development is recovered. Then, during post-diapause, development is inhibited only environmentally, and may be resumed in the laboratory if favourable conditions (including essential food) are provided. This period may be, therefore, considered a mere quiescence.

An important shortcoming in the study of termination of diapause by cold has been that any effect of mere passage of time has never been separated. While analysing the factors which cause the termination of diapause in some bugs (Hodek, unpubl.), it became evident that both factors — elapsing of time and the need for a cold period were combined and further complicated by the "shock" effect of an increase in temperature. It seems that imaginal diapause which is apparently less immutable than diapause in other stages (Norris 1964) responds especially easily to changes in the environment. Some preliminary results suggesting this have been achieved by Strübing (1963) in *Homoptera*.

As well as increase in temperature and prolongation of photophase, also disturbance and injury can both terminate diapause or at least accelerate its termination (Hodek unpubl.).

As is emphasized by Beck (1968), there is no explanation of how such different agencies (such as exposure to various ecological factors, their abrupt changes, simple time lapses or surgical and chemical treatments) can trigger events on a physiological level in the same way. Presumably, all these stimuli, in different ways (some by exteroreceptors, some by proprioreceptors), influence the neurosecretory cells of the brain which now appear to be the prime movers of the neuroendocrine system (see chapter 7.5).

It would therefore be very dangerous to adhere to the rather generally used and oversimple concepts of diapause termination when studying dormancy in *Coccinellidae*, and to underestimate the complex nature of the mechanisms involved.

### 7.42 Induction, maintenance and termination of diapause in coccinellids

Before any experiments to analyse the effect of factors regulating diapause were carried out, some hypotheses had been published based on mere observation. These hypotheses in so far as they were justifiable deductions from detailed observations are useful if only as working hypotheses.

Two short studies by Th. Dobzhansky (Dobrzhanskiĭ 1922a, b) have a special position among such hypotheses; they are surprising for their clear-sightedness and show, for that time, a modern attitude. For the first time, imaginal diapause in *Coccinellidae* was recognized and its adaptive significance for survival realized, i. e. that beetles in diapause can withstand a long period of aphid scarcity or absence. Around Kiev (the Ukraine SSR) Dobzhansky ascertained that *Coccinella septempunctata* had a bivoltine developmental cycle. The adults of the first generation emerge from the pupae between mid-June and mid-July and start ovipositing at the latest within two weeks. The females of the second generation, emerging from mid-July to September at the time of greatest heat, do not mature in the same year in spite of sufficient food, but only do so after hibernation. Having experimentally achieved the development of ovaries in the females of the first generation in spite of combined cold (13 °C) and starvation, the author came to the conclusion that maturation of ovaries is not influenced by external factors, but that a genetically fixed alternation of a generation with normal development with a generation entering diapause is involved. He confirmed this opinion by observing an anomalous situation in the year 1921 when overwintering females matured very late, in early July. In these unusual circumstances, the adults of the first generation emerged from pupae in August and early September — i. e. at a time when normally the second generation would have emerged. If diapause were caused by external conditions, it should then have been induced in the first generation in that particular year. However, it was not so, even so late the ovaries of the first generation females matured and oviposition took place. The emerged larvae then died in large numbers because of lack of food and later due to frost. (Surprisingly in the next March two live pupae were found which had successfully hibernated.) This data from the anomalous development of the beetles in 1921 seems to exclude the possibility that in a normal year diapause of the second generation could be induced by light conditions which Dobzhansky had not taken into consideration. At the same time he did not exclude high temperature (above optimum) as a potential inducing factor.

In over 30 other coccinellid species, where he found undeveloped ovaries during hibernation, he admits that environmental factors had influenced the induction of diapause.

In his paper of the same year, Smee (1922) wrongly states a 3 — 4 generation cycle of *C. septempunctata* in England, with the adults supposedly emerging from pupae in mid-May, early June, late July and in October!

In contrast to Dobzhansky's conclusions many later authors who merely speculated about diapause inducing factors (and did not quote his paper) emphasized the action of biotic or abiotic factors of the environment. Balduf (1935) did not actually hypothesize about diapause inducing factors in *Coccinellidae*, but about "causes of gregariousness" or of "flight from the place of their development". He mentioned three possibilities: high temperature, shortage of food or "state of obesity and satiety" due to the increase of fat content in the body. Yakhontov (1940) believes that shortage of food induces diapause in *Semiadalia undecimnotata* (which migrates without feeding after emerging from the pupa), whereas Telenga (1948) considers that a combination of low humidity and high temperature is the diapause inducing factor in *C. septempunctata.* Bodenheimer (1943) recognizes both biotic and physical factors in *C. septempunctata.*

Telenga and Bogunova (1936) drew attention to the fact that only those adults of *Harmonia axyridis* which emerge from pupae before mid-August can oviposit in the same season in East Siberia. Ul'yanova (1956) imported this coccinellid to a different climate of central Asia (Tashkent) where the beetles already emerged from hibernation in mid-February. The first adults of the first generation emerged from pupae in April, the second generation in June, the third in July, and the fourth in August. In the lst generation all females were reproductive, in the second a few females entered diapause, while in the third and fourth generations most females were diapausing. In both of these two papers there were not even guesses at causal relations made. Neither in the work on coccinellids of the Ukraine SSR (Dyadechko 1954), although very comprehensive as far as food ecology and distribution in habitats are concerned, is there included more than a simple statement of voltinism: "In the Ukraine, nearly all aphidophagous coccinellids reveal one complete and one incomplete generation".

### 7.421 Coccinella septempunctata

#### 7.4211 Induction of diapause

Surprisingly enough, the diapause of *C. septempunctata* remained experimentally neglected for about 40 years after the preliminary experiments by Dobzhansky. In a thorough ecological study of *C. septempunctata*, Jöhnssen (1930) records the univoltine cycle in central Europe; the acceleration by temperature of the development of all stages led him to admit the possibility of a 2nd generation under especially favourable climatic conditions. He expected this second generation to die out as eggs or larvae. The only experiment that he carried out was to rear adults emerged from pupae in August indoors on food, until their oviposition in early March the next spring. As neither the physical conditions of the experiment nor the food are given, no definite conclusions can be drawn. If we assume that these rearings took place in a heated room (which seems very probable), the results suggest no more than that "chilling" is not obligatory for the spring maturation of the females.

We tried to go a little further in understanding the factors which induce diapause in *C. septempunctata.* This species or at least its populations in Czechoslovakia proved to be complicated in this respect.

Although in some years aggregations of dormant *C. septempunctata* of both sexes may be found in their hibernation quarters from early August onwards, we can also find actively feeding coccinellids on vegetation with aphids (especially often on different weeds as *Carduus spp.* and *Umbelliferrae*) for the whole September and in early October. This agrees with observations by Telenga (1948) who also used to find, at the same period of time, that the population of *C. septempunctata* was partly dormant and partly feeding. We compared the physiological condition of our coccinellids by dissection and rearing (Hodek 1962a). Whereas the alimentary canal in dormant beetles is empty of food and there are no traces of vitellinization in the ovaries, we found that the digestive tract is full of food in more than half of the active adults and that 13 — 20 % of females possess one or more vitellinized oocytes or even eggs. The difference in the amount of fat body is not very great (Table 7.06). The difference between the dormant and active fractions of the population is much more striking, if the beetles of both fractions are reared for three weeks under long-day conditions, with a relatively high

Tab. 7.06 **Differences between two fractions of autumnal adults of *C. septempunctata* (Hodek 1962)**

| Date collected [1] [2] | Condition | N | Dissection [3] | Digestive tract | | Fat body | | | Ovarioles with | |
|---|---|---|---|---|---|---|---|---|---|---|
| | | | | empty | full of food | + | ++ | +++ | germaria only | at least one vitellinized oocvte |
| 18.9.[1]<br>27.9.[2] | active<br>active | 30<br>20 | after sampling | 4<br>9 | 26<br>11 | 3<br>5 | 8<br>3 | 19<br>12 | 26<br>16 | 4<br>4 |
| 18.9.[1] | active | 18 | after rearing[3] | — | — | — | — | — | 2 | 16 |
| 8.8., 6.9.[1] | dormant | 17 | after sampling | 17 | 0 | 2 | 5 | 10 | 17 | 0 |
| 8.8.[1] | dormant<br>dormant | 29<br>25 | after rearing[3] | —<br>— | —<br>— | —<br>— | —<br>— | —<br>— | 26<br>21 | 3<br>4 |

[1]) Collected in northern Bohemia, near Louny.
[2]) Collected in southeastern Slovakia, near Král'. Chlmec.
[3]) Reared for 3 weeks in long days, 19—22.5° C, fed with plenty of essential aphids.

temperature (19 — 22.5 °C) and are fed on plentiful essential aphid food. About 85 — 90 % of dormant females remain without any vitellinization, while about 90 % of the females collected on plants now possess vitellinized oocytes.

The summer dissections (mid-July) of females about a fortnight after emergence from the pupae indicated a strong tendency to univoltinism, and in the sugar-beet fields in central Bohemia about 84 — 93 % of females entered diapause. The incidence of diapause was still fairly high (about 65 — 80 %) when the coccinellids were reared indoors under conditions approaching those in the field (except for extremes) and with the natural light conditions of late June and early July. Similarly, an apparently very low percentage of ovipositing females was present in a simultaneous massive culture of more than 200 adults, as only 1 or 2 egg batches were found daily. In another simultaneous culture of 100 pairs most adults formed aggregations in corners of the cage from mid-July, only 5 — 10 beetles moving about, although the aggregation was disturbed purposefully every day from early August onwards. Plenty of essential food was supplied (field collected *Uromelan aeneus* and *Aphis fabae*), and the cage was heated with a lamp during colder days (during natural day-light only). Under these conditions only 1 or 2 egg batches were obtained per day, and by the end of October 75 males and 83 females still remained alive. Adults from the second generation, reared from the first batches, emerged from pupae from 17.8. onwards; they remained very active and feeding for about three weeks, and aggregated after four weeks (Hodek 1962).

By these crude experiments we proved that *C. septempunctata* adults feed before they enter diapause (in contrast to Yakhontov's (1962) finding for *S. undecimnotata*) and that the onset of diapause is not prevented by essential food being available (in contrast to Hagen's (1962) findings for *Hippodamia convergens*). The data showed no evidence of a possible diapause inducing function of decreasing day length because in some replicates the beetles entered diapause in spite of having emerged under the longest possible day length. The slight difference between the outdoor and indoor results indicated the possibility that temperature extremes may play some part in inducing diapause.

We tried to use a premature development of the 1st generation (the adults emerged from pupae as early as mid-May in spring 1968) to assess the potential for multivoltinism (or bivoltinism) of the species in a very warm region of Czechoslovakia — in southern Slovakia (Zohor near Bratislava). We reared the beetles in almost natural condititons, (except for extremes) by transferring them on cold days to the glass-house, or even to a constant 25 °C into the laboratory, into shadow during strong sunshine and indoors for the night. Diapausing females (i. e. females with ovarioles only consisting of germaria without any trace of vitellinization) ranged between 66 and 90 %, with the average of 79 % (N = 184) (Hodek unpubl.)

Notwithstanding the abnormally favourable conditions (a warm spring, a warm region) the tendency to a univoltine cycle persisted in a rather high percentage of *C. septempunctata* females in Czechoslovakia. As the last beetles were dissected on 21 June and therefore had lived through a period of still increasing daylength, the possibility that diapause was induced by the shortening of what were still long days was excluded in this experiment.

The next series of experiments was carried out under controlled light and temperature conditions (Hodek and Čerkasov 1960a, 1961, Hodek 1962a). Cultures were started in five successive seasons (1956—60) with beetles collected from the field after hibernation. As the eggs were collected at the end of the normally short pre-oviposition period, the eggs of any female with a delayed oviposition would have been excluded and thus a selection against the tendency to diapause was made. In this way, a gradual decrease in diapause incidence was achieved (Tab. 7.07). In spite of favourable conditions [i. e. long-day conditions (16 hr or 18 hr photophase in a 24 hr photoperiod), a constant 25 °C, an excess of essential aphid food], the incidence of diapause remained rather high in the 1st generation, and usually fluctuated between 60 and 90 %. These fluctua-

Tab. 7.07 Incidence of diapause in *C. septempunctata* in five laboratory cultures of successive generations[1]) (Hodek and Čerkasov 1961a, b, Hodek 1962)

| Culture (No. and year of its onset) | Generation | Percentage[2]) of diapausing females average | min.—max. (in replicates) | N | Emergence of adults (lst day) |
|---|---|---|---|---|---|
| I 1956 | 1 | 86 | | 22 | 3. 7. |
| | 2 | 64 | | 11 | 10. 8. |
| | 3 | 11 | | 9 | 30. 9. |
| | 4 | 0 | | 9 | 6.11. |
| | 5 | 0 | | 17 | 9.12. |
| | 6 | 0 | | 10 | 21. 1. |
| | 7 | 0 | | 10 | 3. 3. |
| | 10 | 0 | | 14 | 25. 7. |
| | 12 | 10 | | 10 | 27.10. |
| II 1957 | 1 | 63 | | 27 | 12. 7. |
| | 3 | 14 | | 14 | 26.10. |
| | 4 | 16 | | 19 | 3.12. |
| | 5 | 7 | | 30 | 20. 2. |
| | 6 | 17 | | 18 | 15. 4. |
| | 8 | 45 | | 11 | 18. 6. |
| | 9 | 10 | | 58 | 30.7. |
| III 1958 | 1 | 67 | — | 18 | 25. 6. |
| | 2 | 38 | (20 — 50) | 24 | 7. 8. |
| | 3 | 18 | — | 11 | 11. 9. |
| | 4 | 33 | (13 — 48) | 39 | 20.10. |
| | 5 | 6 | — | 16 | 24.12. |
| | 6 | 14 | ( 5 — 32) | 56 | 20. 1. |
| | 7 | 8 | — | 13 | 23. 2. |
| | 8 | 8 | ( 8 — 9) | 24 | 31. 3. |
| | 9 | 5 | ( 0 — 11) | 21 | 22. 4. |
| IV 1959 | 1 | 32 | (30 — 33) | 22 | 16. 6. |
| | 2 | 9 | ( 0 — 20) | 11 | 14. 7. |
| | 5 | 4 | ( 0 — 14) | 50 | 26.11. |
| | 6 | 5 | ( 0 — 9) | 22 | 1.12. |
| | 8 | 6 | ( 0 — 17) | 36 | 29. 1. |
| | 9 | 10 | — | 10 | 22. 2. |
| V 1960 | 1 | 81 | (46 —100) | 137 | 7. 3. |

[1]) 25° C ± 3° const.; long-day conditions (photophase 16 hr, from October 1958, 18 hr); essential food (*Aphis craccivora*).
[2]) Females with ovaries consisting of germaria only.

tions were perhaps related to varying conditions: e. g. the number of specimens used to start a culture, the locality where the beetles had been collected, the age of the females at the time of laying the eggs from which the next generation originated and the age of dissected females. In the subsequent generations a progressive decrease was sometimes obtained as in 1956 or in 1958 within the first three generations, or a trend of decreasing diapause with fluctuations.

To achieve incidences of diapause approaching or reaching zero in generations higher than the fourth or fifth it is essential that the rearings remain in perfect order (a steady excess of food, cleanliness, and a population density not exceeding 25 pairs in a cage of about 8 l). Otherwise, the percentage of diapause increases (Tab. 7.07 : 1957 8th gen., 1958 4th gen.). In the course of laboratory breeding under constant conditions the vitality of the beetles decreases (Tab. 7.08).

Tab. 7.08 Decrease in fecundity and longevity in generations three, four and six in the culture II (1957) of *C. septempunctata* (Hodek and Čerkasov 1961a, b)[1])

| Generation | No. of days on which oviposition occurred | Average longevity (days) | | Number | |
|---|---|---|---|---|---|
| | | females | males | females | males |
| 3 | 20.6 | 59.4 | 71.3 | 26 | 27 |
| 4 | 13.3 | 62.9 | 60.5 | 29 | 30 |
| 6 | 7.7 | 47.6 | 46.9 | 23 | 30 |

[1]) Kept as isolated pairs; the oviposition and the mortality in each cage were recorded daily.

That diapause incidence decreases in the course of culturing subsequent generations indicates the progress of a selection against a tendency for obligatory onset of diapause under long day conditions. The results for *C. septempunctata* are not exceptional in the literature on diapause (Hodek and Honěk 1970). It is assumed that multiple genes control the diapause tendencies.

In the later generations (beyond the 4th or the 5th), all or nearly all the beetles could oviposit under long-day conditions. Thus the way was opened for environmental factors to induce diapause and for this to be studied (Hodek and Čerkasov 1961). This was done gradually, with coccinellids obtained from different generations and cultures (Tab. 7.09). Up to the time that the larvae were transferred to the experimental conditions, they had been reared under the normal breeding conditions used for the negative selection of the diapause tendency. Although the material used was not the same in all experiments, the results are surprisingly consistent, with only one value (for III/7 at 12 hr) slightly discrepant. Two factors were operating in conjunction for the induction of diapause in these artificial conditions, i. e. photoperiod and temperature. The importance of photoperiod was dominant: diapause was prevented by long-day conditions (19 hr per 24 hr) even at low temperatures of 18 or 18.5 °C in respectively 87 — 96 % of females.

The constant temperature of 18 °C caused a high mortality of the larvae, particularly if they were transferred there at the early age of 2 — 3 days. One of the factors contributing to this high mortality was the evaporation of water from parts of the vegetation which carried the aphids; this produced excess of humidity in the incubator.

The response of diapause to short-day conditions was considerably modified by temperature (Tab. 7.09). At lower temperatures, corresponding to those used in the long-day experiments, the incidence of diapause reached about 90 % (with one exception, when it was only 50 %). An increase in temperature (within optimum) led to a marked decrease of diapause incidence so that at temperatures fluctuating between 24 — 25 °C (night) and 27 — 28 °C (day) it amounted to 10 % only. No significant differences were found between different lengths of photophase within the range of 'long' or 'short' day or between the exposure of younger or older larvae or of pupae to

Tab. 7.09 The proportion of *C. septempunctata* females entering diapause under different conditions of temperature and daylength (Hodek and Čerkasov 1961a, b)[1])

| Material culture/generation | Age of larvae at experiment onset[1]) [days] | Temperature [°C] | Photophase [hr/24hr] | Age at dissection [days] | N | Diapausing females [%] |
|---|---|---|---|---|---|---|
| I/7 | 3 | 17—18 (D)/20—22 (L) | 12 | 42 — 51 | 15 | 94 |
| I/7 | pupae | 17—18 (D)/20—22 (L) | 12 | 56 — 62 | 15 | 87 |
| I/9 | 5 — 8 | 17—18 (D)/20—21 (L) | 12 | 24 — 41 | 30 | 87 |
| III/4. 5 | 2 — 3 | 18 ± 0.5 | 12 | 26 — 43 | 20 | 85 |
| III/7 | 6 — 7 | 18.5 ± 0.7 | 12 | 20 — 37 | 38 | 50 |
| I/9 | 5 — 7 | 22 ± 0.5 | 8 | 33 — 35 | 10 | 70 |
| II/4 | 4 — 5 | 22 ± 1.0 | 12 | 24 — 34 | 42 | 60 |
| I/9 | 3 — 7 | 25 ± 0.5 | 12 | 30 — 33 | 40 | 33 |
| I/6 | eggs | 24—25 (D)/27—28 (L) | 8 | 24 — 26 | 20 | 10 |
| III/4, 5 | 2 — 3 | 18 ± 0.5 | 19 | 26 — 43 | 24 | 13 |
| III/7 | 6 — 7 | 18.5 ± 0.7 | 19 | 20 — 37 | 46 | 4 |

[1]) before the transfer to experimental conditions the insects were reared at 25°, 16 or 18 hr photophase; both before and during the experiment an excess of essential food was supplied

the experimental conditions. The critical photoperiod was not ascertained; it lay between 12 and 16 hr photophase.

These results obtained with beetles which had been cultured for several generations under a strong selective pressure against the diapause tendency, strikingly contradict the results using the progeny of hibernated beetles, i. e. the first generation. A possible explanation is that the central European population of *C. septempunctata* is very heterogeneous (Hodek 1962a). As this species is reported as univoltine in England and Norway (Banks 1954a, Sundby 1968) and obligatorily (Dobrzhanskiĭ 1922a, b,) or facultatively bivoltine in the Ukraine (Dyadechko 1954), it seems quite probable that Czechoslovakia lies in a transition zone (in the sense of Bodenheimer and Vermes 1957) between the distribution areas of uni-and multivoltine populations (Tab. 7.10).

Tab. 7.10 Hypothetical condition of *C. septempunctata* in early autumn (Hodek 1962)

| Activity | Ovaria | Voltinism | Type of diapause | Generation | Origin |
|---|---|---|---|---|---|
| Dormant at hibernation quarters | without vitellinization | univoltine | obligatory | 1. | (early batches) |
| Active (feeding) on plants | without vitellinization (primarily or retrogressed) | univoltine | obligatory | 1. | (later batches or slower development) |
| | | multivoltine | facultative | 1.<br>2. | (later batches or slower development) |
| | | | | 0. | (overwintered adults, not yet dead) |
| | reproductive | multivoltine | facultative | 1. | (early batches) |

This apparently holds for central Europe in general, for Jöhnssen (1930) mentions the possibility that extremely favourable conditions may increase the number of generations per year from the normal one to two. This assumption is supported by the experiments both under natural and almost natural conditions described above. Other explanations, however, have not yet been excluded by experimental evidence. Norris (1964), for instance, argues that environmental effects on the parents during hibernation may be responsible for the high tendency to diapause in the first generation. This possibility, however, though still open for *C. septempunctata*, has been excluded for the pentatomid bug *Aelia acuminata* by comparing the incidence of diapause in the progeny of adults collected before and after hibernation (Hodek and Honěk 1970). The Czechoslovakian results on *C. septempunctata* also leave room for another possibility that the induction of diapause has been partly influenced by the change from long to short day during larval life. This possibility is, however, rendered less likely by two replicates where the transfer was carried out very early in larval life (2 — 3 days old) and the incidence of diapause was nevertheless high (85 and 94 %).

There remains the very real possibility that obligatory univoltinism is induced by the absence of a special alternation of environmental conditions, e. g. the succession fo long-day conditions after short-day conditions (as has been found in some insects, see 7.41) which is a prerequisite for the complete maturation and oviposition. Then the change in response of the later generations could have been caused by the gradual negative selection for this requirement. This idea, however, is not in essence differing from our original interpretation.

Although the climate around Paris (Ile-de-France) is substantially different from that of central Europe, the results of a study of diapause of *C. septempunctata* in this region are fully consistent with ours (Bonnemaison 1964). 85 — 95 % of the progeny of hibernants enter diapause in spite of favourable conditions (long day of 16 or 18 hr, 22 °C, excess of aphids, Tab. 7.11) under which selection proceeds from 85 % by steps

Tab. 7.11 Incidence of diapause in the progeny of hibernating *C. septempunctata* in Paris region (Bonnemaison 1964)

| Photophase[1]) [hr/24 hr] | | Type of culture | Food | Diapausing females[2]) [%] |
|---|---|---|---|---|
| larvae | adults | | | |
| 16 | 16 | group | excess | 85 |
| 16 | 18 | group | limited | 95 |
| 16 | 18 | group | excess | 85 |
| 16 | 16 | isolated | excess | 90 |
| 16 | 12 | isolated | excess | 95 |
| 12 | 18 | group | excess | 85 |
| 12 | 16 | group | excess | 85 |
| 12 | 12 | group | excess | 95 |
| 12 | 14 | group | excess | 95 |

[1]) 22 °C.
[2]) dissected at the age of 50 days; in all replicates N = 20.

of 40, 20, 5 and 5 % diapausing individuals for the first 5 generations to a culture consisting solely of non-diapausing coccinellids by the sixth generation. Already by the second generation a substantially different percentage of diapause could be achieved with an unfavourable photoperiod: whereas 35 — 40 % diapaused with an 18 hr photophase and excess of food, the figure rose to 85 % with 12 hr light or a shortage of food with long day illumination. Bonnemaison also ascertained a considerable variability in diapause incidence and accepts the hypothesis of a mixture of uni- and multivoltine races. He considers that air movements and human transport regularly bring univoltines from colder regions and multivoltines from the meridional zone to the Paris area. The development of a 2nd generation near Paris is considered impossible by Bonnemaison, except during exceptional conditions. He considers that temperature and lack of aphids in late July and in August are limiting.

The tendency to a univoltine cycle seems equally pronounced in the French Riviera (Cote d'Azur) (Iperti pers. comm.). Iperti (1966) recognizes *C. septempunctata* as univoltine or partly multivoltine. The situation is apparently similar in Israel according to Bodenheimer (1943). In the French Riviera the first generation only oviposits under exceptional conditions (the peninsula of Giens) after an aestivation period and thus forms a second generation.

When rearing the introduced *C. septempunctata* for releases in potato fields, Shands et al. (1970) found that "food in the larval stage and quality of light in the adult stage" are important for diapause induction.

#### 7.4212 Completion of diapause

Only preliminary experiments have been conducted on the completion or termination of diapause in *C. septempunctata;* our ideas about the regulating mechanism involved are therefore still rather nebulous. "Artificial" diapause has been found to proceed to termination rather quickly; chilling is apparently not a prerequisite for further development (Hodek and Čerkasov 1961; Tab. 7.12).

Neither does "natural" diapause seem to be too firmly established (Tab. 7.13). After about three months of dormancy (17.11), all females can be activated within three weeks. Comparing this with the poor results obtained when activating the August samples (even when the beetles were exposed to a sufficiently high temperature of 24 °C for 8 weeks), indicates that the inhibition of maturation was lifted in the course of September, October (and early November) (Hodek 1962a). It was, however, not clear, whether merely some "time-keeping" factor is involved (as supposed by Hinton 1957

Tab. 7.12 Spontaneous termination of „artificial" diapause in *C. septempunctata* (Hodek and Čerkasov 1961)

| | | | | | |
|---|---|---|---|---|---|
| 12 hr, 22° / 5 weeks | 70 % diap. (N = 10) | 12 hr, 22° / 5 weeks | 36 % diap. (N = 39) | | |
| 12 hr, 17.5° — 20.5° / 3.5 — 6 weeks | 87 % diap. (N = 30) | 12 hr, 17.5 — 20.5° / 5 weeks / dark, + 4.5° | 12 hr, 25° / 3 weeks | (N = 7) | 0 % diap. (N = 11) |

Tab. 7.13 Attempts to terminate the „natural" diapause in *C. septempunctata* by transfer to the laboratory (Hodek 1962)

| Sample[1]) | Breeding[2]) Duration [days] | Breeding[2]) Temperature [°C] | Diapausing females [%][3]) | N |
|---|---|---|---|---|
| 8. 8. 61 | 22 | 17 ± 1/19.5 ± 0.5 | 90 | 29 |
| 30. 8. 61 | 18 | 20 ± 1.5 | 84 | 25 |
| 25. 8. 59 | 54 | 24 ± 1 | 78 | 18 |
| 4. 11. 60)[4]) | 20 | 24 ± 1 | 12 | 17 |
| 17. 11. 59 | 23 | 24 ± 1 | 0 | 23 |

[1]) Sampled at the dormancy sites.
[2]) Long day and excess of food.
[3]) Dissection before breeding revealed immature ovaria possessing germaria only.
[4]) Collected 18 Oct., stored in cold until 4 Nov.

and other authors) or whether the low autumn temperatures were effective, as is generally supposed to apply in insects.

A preliminary experiment with beetles collected in late August (i. e. after a maximum of 5 weeks in the dormancy sites) has not indicated any effect of low temperatures above zero: the highest temperature used (+ 12 °C const.) was most efficient in activation (Tab. 7.14). Zero was substantially less effective than other temperatures, even though the exposure lasted 9 weeks. At other temperatures, the activation rate was related to the duration of exposure (Hodek 1970). Thus diapause could be terminated in 60 — 100 % of the beetles after only about 14 weeks.

Tab. 7.14 Attempts to terminate the "natural" diapause in *C. septempunctata* by exposure to cold (Hodek 1970)[1]

| Exposure to cold[2]) | | Diapausing females [%] | N |
|---|---|---|---|
| Duration [weeks] | Temperature [°C] | | |
| 3 | 12 | 45 | 11 |
| 6 | | 0 | 8 |
| 3 | + 5/12[3]) | 67 | 21 |
| 6 | | 40 | 20 |
| 3 | + 5 | 67 | 30 |
| 6 | | 11 | 9 |
| 9 | 0 | 80 | 30 |

[1]) Sampled in late August.
[2]) After cold exposure the beetles were reared for 3 weeks (long day, excess of food, 18 °—D/23 °—L)
[3]) One increase and one decrease in temperature per 24 hr.

Near Paris, Bonnemaison (1964) found that diapause lasted 3 — 6 months in *C. septempunctata* females which had emerged from pupae in early August and were kept at 20 — 22 °C and a natural photoperiod. He failed to reactivate the diapausing females (collected in late July or early September) by an exposure to 8 ° (or 5 °) for 5, 9 and 13 weeks and by subsequent rearing at 20° and 16 hr for 15, 11 and 7 weeks to make a total of 20 weeks for each batch. His findings on the onset of previtellogenesis in females in the field from late September onwards (Tab. 7.15) indicate — very similarly

Tab. 7.15 Development of ovaries in *C. septempunctata* in the field (Paris region — Bonnemaison, 1964)

| Sample | Females possesssing ovarioles with | | Length of ovarioles | | |
|---|---|---|---|---|---|
| | germarium only | one oocyte | average | min. | max. |
| 21. 8. | 100 | 0 | 575 | 510 | 620 |
| 11. 9. | 100 | 0 | 610 | 530 | 660 |
| 30. 9. | 96 | 4 | 600 | 480 | 672 |
| 11. 10. | 16 | 84 | 730 | 640 | 870 |
| 19. 10. | 30 | 70 | 660 | 465 | 800 |
| 27. 11. | 25 | 75 | 590 | 460 | 750 |

to ourresults — some kind of lifting of the diapause inhibition in autumn. In contrast to Czechoslovakia, however, the apparently higher temperature of the Parisian region enabled the first stage of maturation to proceed in the field and to be identified directly by the dissection of field samples, whereas in central Europe the potential for maturation can only be revealed by laboratory breeding (Tab. 7.06).

When summarized, these fragmentary results indicate clearly enough that diapause in *C. septempunctata* usually lasts about 3 — 4 months (both in Czechoslovakia and in the region of Paris) and that "chilling" is not necessary for its completion. This is also confirmed by findings of Iperti (in Hodek 1966 p. 105) with populations from the Cote d'Azur. The high autumnal temperature together with the resurgence of aphids after summer drought may, under exceptional conditions (e. g. peninsula of Giens), enable the establishment of a 2nd generation. We may, however, expect important variations of this scheme in a species with such ecological plasticity as is a feature of *C. septempunctata.*

#### 7.4213 Diapause in males

As explained earlier (part 7.2) diapause in coccinellids is not connected with inactivity of the tissue of the testicular follicles. From knowledge of other insects, one might however expect inhibitory changes in the accessory glands and in the copulatory aptitude of males. When Bonnemaison (1964) reared emerged males of the fourth selected generation under 12 hr light and 14 ° or 18 °C for 15 or 25 days, he could not even find these symptoms of diapause. After transfer to favourable conditions where females were being reared (16 hr, 20 °C), the males copulated after 2 — 7 days, and the females laid apparently viable eggs (not specified) after 3 — 23 days.

### 7.422 Chilocorus spp.

*C. rubidus.* — A species of eastern Siberia is strictly univoltine. The dormancy lasts between late August to mid September and late April, while diapause (which is marked by a trough in oxygen consumption ends as early as late December to early January. The rest of dormancy is spent in mere quiescence ("spyachka" — sleep) during which it is possible to obtain oviposition by transferring the beetles to a higher temperature. Pantyukhov (1968b) found that neither "chilling", nor any particular photoperiod was necessary for the termination of diapause, but the passage of 3.5 — 4 months under any conditions (25°, 20°, 5—8° and field conditions were tested) was alone sufficient.

*C. renipustulatus.* — In contrast to *C. rubidus* diapause in the majority of *C. renipustulatus* individuals can be prevented in both the populations that have been studied. Thus only 12 % of beetles from Maïkop (southern Ukraine) entered diapause in spite of long days, whereas in the population from Leningrad the figure was 38 %.

The critical photoperiod is insensitive to temperature (within the range 20—25 °C) and is 2 hr longer in the beetles from Leningrad than in the Maïkop strain (Pantyukhov 1968) (Fig. 7.03).

*C. bipustulatus.* — Of *Chilocorus spp.* most attention has been paid to *C. bipustulatus*, whose distributional area comprises a large part of the Palaearctic region. Whereas in Central Asia (Tashkent) this typical multivoltine species has a simple type of facultative diapause which can be prevented or terminated by long-day conditions (Zaslavskiĭ and Bogdanova 1965, Zaslavskiĭ 1970), the annual cycle of the northern populations (Leningrad) is regulated in a more complex way.

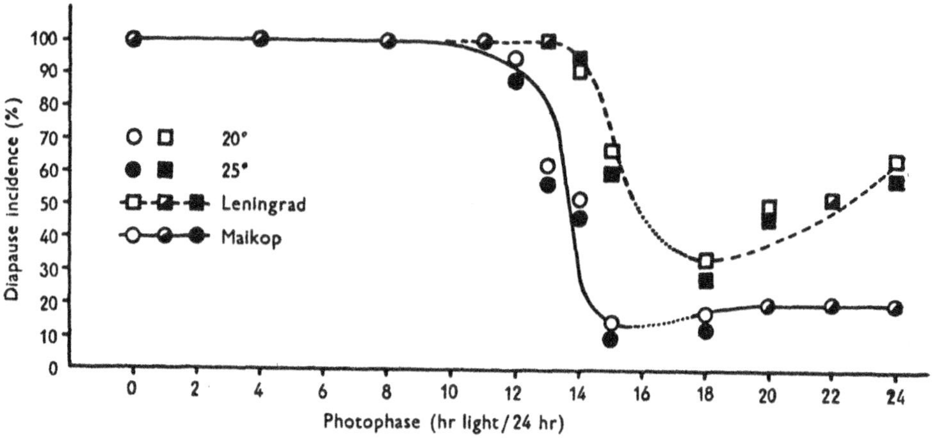


Fig. 7.03 The effect of photoperiod on the incidence of diapause in two strains of *Chilocorus renipustulatus* (Pantyukhov 1968a).

In *C. bipustulatus* from **Central Asia** (40 — 41° northern latitude) diapause is induced by a photophase shorter than 15 hr, and at 20 °C 100 % beetles enter diapause. At 24 °C, however, diapause is completely prevented in all females in spite of the short day (9 or 11hr). Under temperatures near 20° spontaneous termination of diapause does not occur providing that the beetles are maintained at diapause conditions. On the other hand, either an increase in temperature or a prolongation of photophase beyond the above-mentioned thresholds terminates diapause very quickly — oviposition takes place within a few days. It is equally easy to terminate diapause 7 — 10 days after its induction as it is after 3 months. By contrast, transfer of the ovipositing beetles from long day to diapause conditions gradually suppresses oviposition, while a return to long days restores it. Alternatively, diapause may be terminated by "chilling". After an exposure to + 8 °C for 30 — 40 days the beetles mature even in short days. Thus, sensitivity to photoperiod could be nullified by an exposure to cold.

The completion of adult diapause in *Pyrrhocoris apterus* (*Heteroptera*) closely resembles that of *C. bipustulatus* (Hodek 1968).

*C. geminus* — In the central Asian populations (Tashkent) of this closely related species (40 — 41 ° northern latitude) induction and termination of diapause is very similar to that in *C. bipustulatus*. The only difference is a higher temperature threshold — as much as 26 °C is needed to prevent diapause in short-days (Zaslavskiĭ and Bogdanova 1965).

*C. bipustulatus* from **Leningrad** (60° northern latitude). — The beetles of the northern population do not mature and do not lay eggs, if the larvae, pupae and adults are reared continuously in long days (20 hr) (Zaslavskiĭ 1970). Short-day conditions (9 hr) for at least 3 weeks before the transfer to long day are a prerequisite for normal reproductive activity. The short day treatment is equally effective when given to the larvae. The alternative way to induce oviposition is "chilling" for about a month at +7 °C followed by a transfer to long days or otherwise to "chilling" for at least 3 months before transfer even to short days. If the beetles are constantly left in long days diapause ends spontaneously after about 2 — 4 months. However, the beetles gradually cease ovipositing in continuous long days after about 1.5 — 2 months of egg laying and have to be reactivated from this "secondary" diapause by a new passage of about 25 days in short days.

The critical photoperiod for the beetles from Leningrad is 2 hr longer (17 hr) than for the central Asian population (Fig. 7.04). It does not change with temperature within the range of 20 — 27 °C.

The results of Zaslavskiĭ (1970) are of extreme importance, as they confirm for *Coccinellidae* that a number of alternative pathways lead to the triggering of the neuro-endocrine system and its control of oviposition. The failure to oviposit in long days would normally place the diapause of the Leningrad strain of *C. bipustulatus* into the category of obligatory diapause.

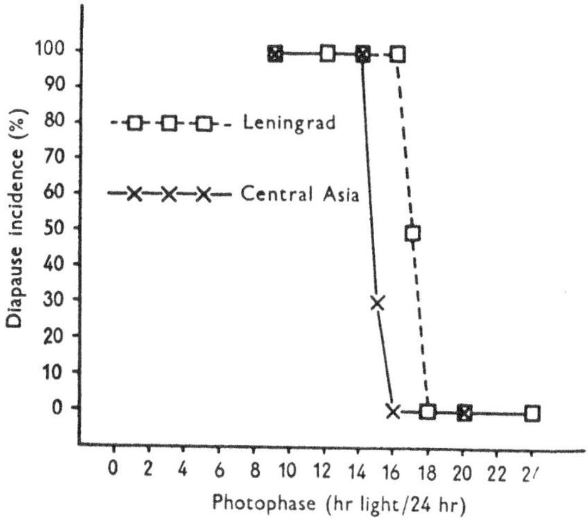


Fig. 7.04 The effect of photoperiod on the incidence of diapause in two strains of *Chilocorus bipustulatus* (Zaslavskiĭ 1970).

*C. bipustulatus* in **Israel** (32—33° northern latitude). — The populations from the eastern Mediterranean area (Israel) differ from those of central Asia in their threshold both for photoperiod and temperature (Tadmor and Applebaum 1971). A photophase of 14 hr (per 24 hr photoperiod) prevents diapause, whereas a shorter photophase (10 hr) tends to induce it (Tab. 7.16). Thus, consistent with the findings of Danilevskiĭ (ed.) (1968), the critical photoperiod is about 2 hr shorter in Israel than in the central — Asian strain. The effect of the short photophase is modified by temperature — at 22° it is completely nullified so that all females lay eggs. A cumulative sensitivity to diapause induction is exhibited both in the preimaginal stages and in the adults.

There is contradiction between early and more recent reports about the summer occurrence and efficiency of *C. bipustulatus* in Israel. Early reports concluded that

Tab. 7.16 Effect of photoperiod and temperature on oviposition in *Chilocorus bipustulatus* in Israel (Tadmor and Applebaum 1971)

| Photophase [hr/24 hr] and temperature [°C] | | | | Ovipositing females [%] | Mean pre-oviposition period [days] | Number of replicates |
|---|---|---|---|---|---|---|
| larva and pupa | | adult | | | | |
| 24 | 28° | 10 | 8°/20°[1]) | 46 | 21 | 11 |
| 24 | 28° | 10 | 18° | 60 | 17 | 10 |
| 10 | 18° | 10 | 18° | 1 | 21 | 11 |
| 24 | 28° | 10 | 22° | 100 | 10 | 7 |
| 24 | 28° | 14 | 18° | 93 | 14 | 14 |

[1]) 8° during scotophase, 20° during photophase.

*C. bipustulatus* was common only in spring and extremely scarce in summer, and was therefore an inefficient predator of scale insects (Hecht 1936, Bodenheimer 1951). The high mortality in summer is supposed to have been caused by the 'chamsin' — a dry hot wind from the desert. The more recent findings show, however, that *C. bipustulatus* is rather abundant in the summer (some authors even report the population peak in early summer) and plays an important role in retarding the build-up of scales in this period (Nadel and Biron 1964, Avidov and Rosen 1965, Rosen and Gerson 1965, Kehat 1968b, Ben-Dov and Rosen 1969). In the summer, however, the females produce many fewer eggs and in this way determine the decrease in adult numbers later on. This has led to speculation that the difference in recent years has perhaps resulted from the improved physical conditions brought about by the introduction of irrigation (Hodek 1967) as the same difference also exists between older (Bodenheimer 1951) and more recent reports (Plaut 1965) on the abundance of *Stethorus punctillum*. Kehat et al. (1971 in print) attempted to analyse the factors which inhibit the reproductive activity of *C. bipustulatus* in late summer and winter. In citrus groves they found a considerable decline in numbers of females with well-developed ovaries during July, and October — December. Transferred to the laboratory (28 °C) and provided with coccids, the non-reproductive females matured at both these times of year. The failure by these authors to induce diapause at 16 combinations of light conditions was apparently due to too high a temperature of 28 ° and 20 ° (cf. Tadmor and Applebaum 1971). Very high temperature (35 °C) additionally, decreases oviposition and survival.

### 7.423 Coccinella novemnotata

The bivoltine *C. novemnotata* (McMullen 1967a, b) undergoes diapause twice during the annual cycle. The adults of the spring generation pass the hot, dry summer months in diapause and lay eggs in the early autumn. The adults of the autumn generation pass the winter in diapause and reproduce in the early spring. In the spring, the teneral adults are subjected to a photophase which increases from 17 hr 30' to 18 hr and in the autumn decreases from about 14 hr to 13 hr. In the field, diapause in this species is induced both by "long" and "short" days. This specific need is fulfilled by photoperiodic reaction which differs from the commonly described reactions of "long-day" or "short-day" insects, but has been verified experimentally. While in some insects diapause is prevented by long photoperiods and in others by short, in *C. novemnotata* diapause is induced both by short days of 10 hr, 12 hr and (even though less) 14 hr, and by long days of 18 hr (Fig. 7.05). As well as photoperiod, temperature and the amount of prey are also involved. Particularly low temperature contributes to the induction of diapause in the autumn generation. The intermediate photoperiod of

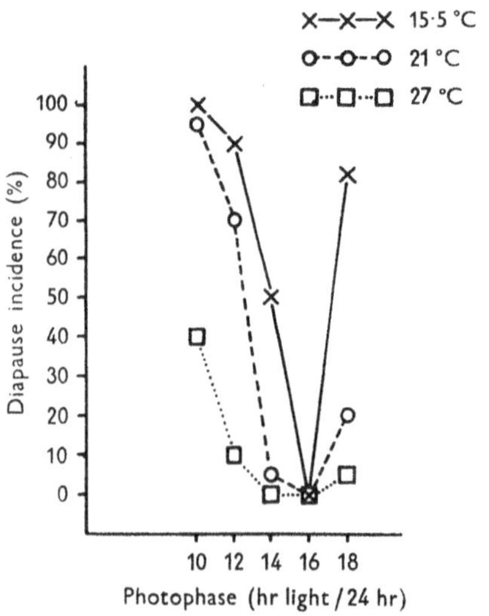


Fig. 7.05 Modification of the photoperiodic reaction by temperature in *Coccinella novemnotata* (McMullen 1967b).

16 hr consistently determines non-diapause development in 90 — 100 % of coccinellids, even when feeding rates are reduced to one-fourth and the whole temperature range of 15.5 — 32 °C is used.

Sensitive stage. — Experiments made with *C. novemnotata* (McMullen 1967b) are the only studies on *Coccinellidae* which are thorough with respect to the determination of the stage sensitive to dia ›ause induction (Fig. 7.06). The author's conclusion that it is the young adult between the ages of 1 and 7 days which is sensitive is, however, rather questionable because the percentage of diapause was judged by the dissection of females at only 14 days of age (at 21 °C), i. e. only 4 days after transfer of adults 10 days old and 7 days after transfer of adults 7 days old. In such a short time the processes controlling the maturation or regression of ovaries could not be complete (see 7.5). In order to exclude what seems indeed very probable, i. e. that adults aged more than 7 days are also responsive, it would be necessary to dissect females sufficiently far in time (at least 14 days) from the date of transfer.

The above-mentioned transfer experiments evidently did not exclude the possibility that, as in other insects with imaginal diapause (*Pyrrhocoris apterus*, *Leptinotarsa decemlineata* — Hodek 1971a), even the pre-imaginal stages are sensitive to the stimuli controlling the onset of diapause. This is known to be so in *Chilocorus bipustulatus* (Tab. 7.16) (Tadmor and Applebaum 1971). In *C. novemnotata* (Fig. 7.06) the sensitivity of preimaginal stages is indicated by retarded ovogenesis in females (marked with 2)

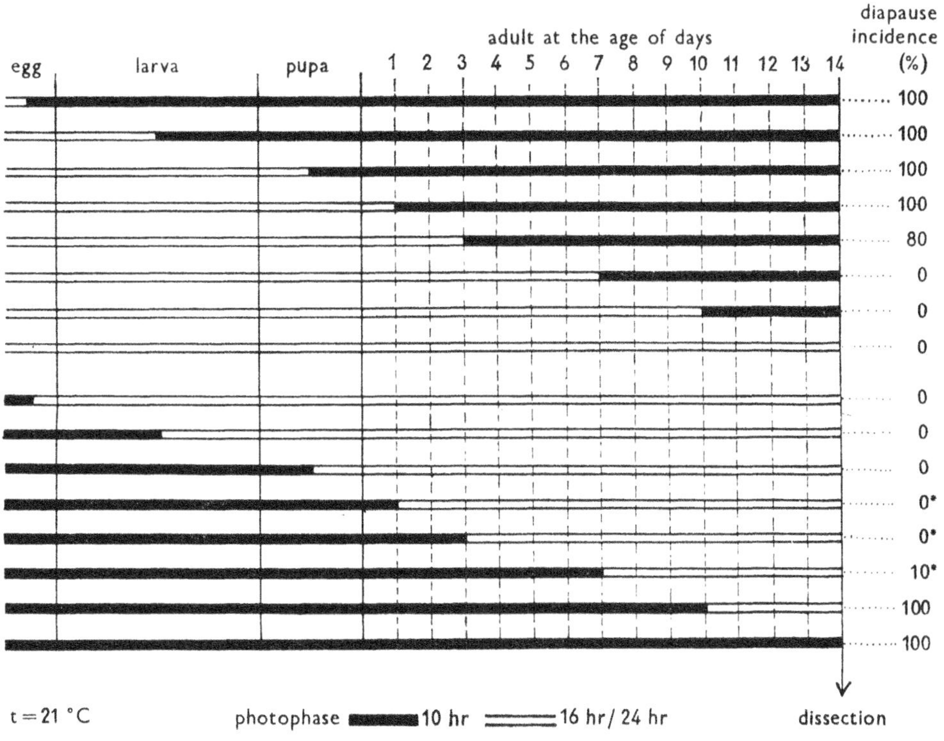


Fig. 7.06 The effect of change in photoperiod on the incidence of diapause in *Coccinella novemnotata*. Females were dissected at the age of 14 days. Under the treatments marked by* , ovogenesis in non-diapausing individuals was retarded (compared with the controls).

which were transferred to a non-diapause photoperiod as teneral or young adults.

The sensitivity of earlier stages may be masked when later stages are influenced by the reverse stimuli; the effect may thus be reversed. The sensitivity of the preimaginal stages may be distinguished when the adults are kept under neutral conditions. We do not usually search for a specific sensitive stage, but rather for the stage of development in which we can still succeed in completely preventing oviposition (i. e. to reverse the former opposite effect) by a transfer from non-diapause to diapause-inducing conditions. A suitable criterion for the sensitivity of larvae may be the length of the pre-oviposition period in females kept as larvae for different periods of time under diapause conditions (Hodek 1971a).

## 7.5 Physiological mechanisms of adult diapause

This chapter is included, although this book is written by an ecologist for ecologists. In the study of ecological causal relations which govern the onset and maintenance of diapause it appears to be helpful to have at least a general idea of the physiological pathways which regulate these phenomena.

The function of the neuro-endocrine system has not been studied very long, and so there is comparatively little knowledge about it and the findings from individual species are often contradictory. This brief survey will, therefore, suffer from oversimplification and will very soon become dated. The reader is referred to recent reviews (Engelmann 1968, De Wilde 1970) and reviews sure to appear in the future in publications such as Advances in Physiology, Annual Review of Entomology, Annual Review of Physiology, proceedings of various conferences, etc.

The following sequence of events seems to be relatively certain and common in insects. The information concerning the ambient physical or biotic conditions (token stimuli or seasonal tokens) or of endogenous information arrives at the brain. This information affects the neurosecretory cells of the pars intercerebralis of the protocerebrum which produces the brain hormone or activation hormone (AH) of which the chemical nature is still unknown. The corpora cardiaca have the function of a release organ for this neurosecretory material which apparently activates the corpora allata. (It is also thought to regulate protein metabolism.) The hormone (or hormones) liberated by the corpora allata (the juvenile hormone — JH) may have several functions. It promotes the incorporation of proteinaceous and lipid yolk into the growing oocytes, i. e. vitellogenesis in oocytes, probably via the ovarian follicular cells. (The JH is also supposed simultaneously to control various phases of metabolism, as, for example, the protein metabolism associated with egg maturation and fat metabolism.) This whole sequence of events accounts for a certain delay in the reproductive reaction of adults to a change in conditions. Because of that delay, the young females start laying eggs even when during the pre-oviposition period the conditions have changed from favourable to inhibitory. The oviposition then only ceases later on. The process initiated by the earlier stimulating conditions continues, and apparently cannot be affected by ambient "tokens" in later phases. Not until the JH disappears from the haemolymph, are vitellogenesis and ovulation arrested.

The above scheme does not apply to some insects or group of insects. In *Phasmida* and *Lepidoptera*, the extirpation of the corpora allata apparently does not prevent egg maturation and oviposition. In *Leucophaea* (*Blattaria*), in *Rhodnius* (*Heteroptera*) and in *Pieris* (*Lepidoptera*), neurosecretion from the pars intercerebralis does not control the function of the corpora allata (Engelmann 1968, De Wilde 1970).

The evidence for yet more chains of events is, however, still lacking or controversial. Above all, it is not at all certain whether the failure of ovaries to ripen during diapause is caused only by lack of JH or additionally by the presence of an inhibitory factor. Further there is the question of whether the connection between the brain and the corpora allata is only humoral, through the medium of corpora cardiaca and the haemolymph, or also nervous.

It can be assumed that the above mentioned organs also have a feed-back mechanism, e. g. that the corpora allata are somehow influenced by the maturation of ovaries, etc. No such feed-back, however, has as yet been proved (with the exception of viviparous *Blattaria* — Engelman 1968).

Another range of questions, very little studied till now, concerns the mechanism by which the whole system is affected when essential food is taken in. Some results indicate a possible direct connection between the receptors which detect food intake and the corpora allata, without the mediation of the brain. It is certain that it is not only sufficiency or lack of necessary material for the vitellogenesis which is involved. It seems that mating stimuli also activate the corpora allata.

# 8 ENEMIES OF COCCINELLIDAE

For a long time, information on predators, parasites and pathogens attacking coccinellids accumulated more or less haphazardly, mostly in a form of faunistic communications. In fact, a study of biology was undertaken only for the parasite *Perilitus coccinellae*. Only during the last 15 years have the natural enemies of coccinellids become a subject of systematic study on a wider scale. Because of this it is now possible to present preliminary ideas about the effectiveness of the enemies of coccinellids, which in some cases appears great and considerably endangers the beneficial activity of these important entomophagous insects.

Because detailed studies are really only just beginning, and because their regional distribution is rather disjointed (Italy — Domenichini, south-eastern France and north-western Africa — Iperti, the surroundings of Leningrad (USSR) — Semyanov and Lipa, central Europe — Klausnitzer and Fukuoka region of Japan — Maeta) a survey of the accumulated knowledge cannot yet provide a complete picture of the enemies of coccinellids. Many more systematic studies in more regions are needed in this field.

## 8.1 Ants

Several species of ants attend honeydew producing *Homoptera*. Aphids are often associated with *Lasius* and *Formica spp.*, coccids with *Iridomyrmex humilis* and *Oecophylla spp.* The observations and experiments of many authors that the attending ants are hostile to enemies (including *Coccinellidae*) of attended aphids or coccids have been summarized by Nixon (1951) and particularly by Way (1963)[1]). Nixon supposed that the ants do not protect the *Homoptera* actively as a source of food, but merely incidentally, since they attack swift or erratic moving organisms.

Way (1963) considers this view to be an oversimplification; ants may attack other insects at various instances. (1) Predatory ants attacking most insects, which they discover within their foraging territories, protect the attended *Homoptera* incidentally from natural enemies. This is in concert with Nixon's concept. (2) The ants, however, attack other insects also if they intrude on the nest or on food sources which the ant is monopolizing. They are then aggressive towards intruders which, away from the food source, would be tolerated. Thus, *Lasius niger* workers were not hostile to coccinellid adults which they met away from attended aphids. When food is scarce, the aggressiveness of ants at the food source and hence their protection of *Homoptera* is enhanced.

[1]) The reader is referred to this review for the quotations of original papers.

Thus, the protection given by ants to *Homoptera* is related to their value as a food source. Moreover, the attendant workers of predatory ant species are too young to behave as predators and yet protect actively their brood and *Homoptera* from intruders.

An exceptional case is that of the myrmecophilous *Coccinella divaricata* which is associated with *Formica spp.* This coccinellid is not attacked although ants do react to its movement by tapping it with their antennae (Pontin 1960). This author supposed that *Platynaspis luteorubra* is also myrmecophilous; he found one larva of this species in underground galleries of *Lasius niger.* The shape of the larva (Pl. XXXIV), untypical of other coccinellid species resembles the shape of other myrmecophiles, e. g. of *Lycaenids* and *Microdon* ( *Syrphidae*).

## 8.2 Predators of Coccinellidae and coccinellid defense mechanism

There have been numerous reports of predation by both arthropods and vertebrates on *Coccinellidae.* Older observations are listed in the survey by Schilder and Schilder (1928).

Coccinellids which are disturbed fall into thanatosis (a stupor, during which the legs and antennae are pressed closely into depressions on the abdomen) and exude a yellowish fluid from the femorotibial articulations ("reflex bleeding"). This phenomenon is also known in other insects (Heikertinger 1932, Rothschild 1961). The bitter taste and quinolene-like smell (Rothschild 1961) of the secretion has been regarded as protecting the coccinellids against their vertebrate and insect predators. This view has been strongly opposed by Heikertinger (1932) who succeeded in feeding coccinellids to several vertebrate and insect predators. Frazer and Rothschild (1960) and Lane and Rothschild (1960), however, found experimentally that coccinellids were unacceptable to some predators (6 species of mammals, 6 species of birds, several lizards and terrapins and one toad). Gas chromatography showed (Rothschild 1961) that the secretion of coccinellids is similar to that of certain insects generally rejected by predators and even to the exudation of nettles (*Urtica dioica*).

Relevant to the discussion as to whether defence mechanisms do or do not exist in coccinellids (i. e. whether the special taste and smell have protective value) is the theory that coccinellids show aposematic Müllerian mimicry. This theory has been rejected by Heikertinger (1932), but stressed by Rothschild (1961).

Several authors express a strong belief that some of the predators have a great impact on population change in coccinellids and in the case of spiders and birds as predators this is very likely so. However, no tangible evidence has as yet been published. This is a dangerous shortcoming, which limits the value both of the utilization of *Coccinellidae* and of all evaluations and prognoses of their impact.

## 8.3 Parasites

The recorded insect parasites of coccinellids belong to 3 genera of *Diptera* (*Phalacrotophora, Degeeria, Hyalomyodes*) and 6 genera of *Hymenoptera* (*Perilitus, Centistes, Metastenus, Pachyneuron, Tetrastichus, Homalotylus*). In principle, they attack only larvae, pupae and adults of coccinellids, and there exist only two records of egg parasites,

*Ooencyrtus johnsoni* from North America (Peck 1963) and *Tetrastichus ovulorum*, parasitizing eggs of various *Epilachna spp.* (Schmidt 1954). Although 7 species of *Ooencyrtus* are known in central Europe, Klausnitzer (1969b) failed to rear a single parasite from 5000 coccinellid eggs collected in the field. He supposes that the adaptation of egg parasites to coccinellids may have been impeded by the circumstance that coccinellid larvae soon after hatching eat yet unhatched eggs of the same batch. It is indeed known that the egg parasites emerge from parasitized eggs mostly after the larvae have hatched from nonparasitized eggs.

## 8.31 Diptera

### *Phalacrotophora spp.* (*Phoridae*)

Two species are known of this gregarious endoparasite of larvae and pupae of European coccinellids. The adults may be determined according to Colyer (1952):

1 (2) Posterior metatarsae very dark and considerably broadened, especially in females. ............................................................ *fasciata* Fallen
2 (1) Posterior metatarsae yellowish white and of normal shape. ...... *berolinensis* Schmitz.

What follows refers to *P. fasciata*, which is nearly always the species mentioned.

This species is reported by many authors as parasitizing a number of coccinellid species in the tribes *Chilocorini*, *Hippodamiini*, *Coccinellini* and *Psylloborini;* these reports are listed by Klausnitzer (1969b). A single pupa of *Coccinella septempunctata* may contain as many as 7 parasites, a pupa of *Adonia variegata* as many as 4 (Iperti 1964), that of *Adalia bipunctata* an average of 2, and that of *Anatis ocellata* an average of 8.4 (Klausnitzer 1969b). Maeta (1969b) found averages of 4 and 5 (range 2—7) parasites per host in *Harmonia axyridis* over two years in Japan (Nagano region). Clausen (1940) reports 8—14 parasites per single specimen of the genus *Coccinella.*

**Life cycle.** — In central Europe the development of a 2nd generation is possible (Klausnitzer 1969b). If *C. septempunctata* is used as a host, the total development time of the parasite from egg to adult emergence amounts to 30 days in May and to about 25 days in June in southeastern France. In pupae of *A. variegata*, development time is shortened by 5—6 days (Iperti 1964).

If the adults are reared on an artificial diet based on sugar, agar, honey and water, they mature sexually in 15 days. Each ovary contains only one egg (Iperti 1964). The eggs are laid on prepupae or freshly moulted pupae between the legs. Embryonic development is already advanced by the time the egg is laid, so that the larva hatches in a few hours and immediately bores into the host at the base of its leg.

**Percentage parasitization.** — In southeastern France, Iperti (1964) found by dissection of pupae of *C. septempunctata* that 9.1 % were parasitized in the region of Antibes in June 1961, whereas round Giens the figure was 6 % in May 1961 and 2.5 % in June 1962. In contrast, *A. variegata* was much more heavily parasitized: 26.6 % in July 1962 round Digne. In central Europe, Klausnitzer (1969b) reared 90 *P. berolinensis* from 1111 pupae (several parasites may emerge from one pupa) of *A. bipunctata*, and of 212 pupae of *A. ocellata* 14 (= 6.6 %) were parasitized. Clausen (1940) reports parasitization as high as 48 % of coccinellids of the genus *Adalia* by this dipteran. In Japan (Nagano region) the parasitization of *Harmonia axyridis* ranged between 7 and 18 % (Maeta 1969b).

*Degeeria luctuosa (Tachinidae)*

This endomonoparasite of adult *Coleoptera* has only been recorded from 5 species of coccinellids. Banks (1956) and Walker (1962) report two isolated cases from *Adalia decempunctata* in England. In south-eastern France Iperti (1964 and unpubl.) recorded this parasite mainly on *Adalia bipunctata* and *Synharmonia conglobata*. *Propylaea quatuordecimpunctata* had a low parasitization (0.75 — 3.5 %) in summer. Hibernating *Synharmonia conglobata* were rather more attacked in the region of Bautzen, G. D. R.: dissection of 209 specimens, collected in November 1965, revealed larvae of *Degeeria* in 8 males and 21 females (13.9 %). When breeding 226 hibernating coccinellids collected in March and April 1966, Klausnitzer (1969a) found only 14 (6.2 %) puparia of *Degeeria*. Klausnitzer (1969b) only obtained 3 puparia from 567 coccinellids of 15 species collected near the end of hibernation near Dresden and Tharandt: 2 from *Exochomus quadripustulatus* and 1 from *Myrrha octodecimguttata*.

A single larva of the parasite develops in the body cavity of the coccinellid. In the third and last larval stage, the parasite consumes all fundamental organs of the coccinellid; it finally escapes through the dorso-abdominal wall of the beetle and pupates in the soil. The adult already emerges after a week. Adult parasites emerged from both parasitized specimens of *Adalia decempunctata* found in England, whereas emergence failed to occur from puparia obtained from *Propylaea quatuordecimpunctata* in France (Iperti 1964). Domenichini (1953) describes the larval stages.

*Hyalomyodes triangulifera* (Tachinidae)

This parasite was reared by Howard and Landis (1936) and Smith (1960) from adults of *Coleomegilla maculata*. Smith (1960) records this parasite also from *Coccinella trifasciata perplexa*.

*Pericheta unicolor (Tachinidae)*

was found in May and June on *Subcoccinella vigintiquatuorpunctata* (Iperti unpubl.).

## 8.32 Hymenoptera

*Perilitus coccinellae (Braconidae, Euphorinae)*

The subfamily *Euphorinae* includes a group of endoparasites of adult *Coleoptera* of which those belonging to the genus *Perilitus* parasitize *Coccinellidae* and *Curculionidae*. Domenichini (1956) records two species which are exclusive parasites of coccinellids in the subfamily *Coccinellinae: Perilitus stuardoi*, parasitizing *Scymnini* and *Coccinellini* in South America; and the cosmopolitan *P. coccinellae*, parasitizing coccinellids in the tribes *Coelophorini*, *Hippodamiini* and *Coccinellini*.

Both are solitary endoparasites (endomonoparasites) of adult coccinellids. *P. coccinellae* is by far the best studied of all parasites. Not only have a fair number of studies been devoted to its biology, from classical works by Ogloblin (1913) and Balduf (1926) to recent studies from France (Iperti 1964), central Europe (Klausnitzer 1969b) and from Japan (Maeta 1969a), but also the changes caused by the parasite larva in the body of the coccinellid have been studied in detail (Ogloblin 1924, Sluss 1968).

**Host range.** — The list presented in table 8.01 shows a very wide range of hosts within the three coccinellid tribes mentioned above.

However, the table also shows that the parasite has a distinct preference for certain coccinellid species. *Coleomegilla maculata* is most frequently attacked in the USA (Balduf 1926, Smith 1960), *Coccinella septempunctata* in Europe (Walker 1962, Iperti

Tab. 8.01 Hosts of *Perilitus coccinellae*

| Host species | Region | References |
|---|---|---|
| *Adalia bipunctata* (?) | Europe | Telenga 1948 |
| *Adonia variegata* | Europe, China | Ogloblin 1913, Liu 1944 (2), 1950, Telenga 1948, Walker 1961, Iperti 1964 |
| *Coccinella divaricata (distincta)* | Europe | Sem'yanov and Lipa 1967 (5) |
| *Coccinella novemnotata* | USA, Canada | Riley 1888 (1), Folsom 1909 (1), Hudon 1959 (3) |
| *Coccinella quinquepunctata* | Europe | Walker 1961, Sem'yanov and Lipa 1967 (5), Klausnitzer 1969b |
| *Coccinella repanda* | Formosa | Sonan 1939 (2) |
| *Coccinella sanguinea* | USA | Balduf 1926 |
| *Coccinella septempunctata* | Europe, China | Ogloblin 1913, Goriainov 1917 (1), Tullgren 1916 (1), Sonan 1939 (2), Liu 1944 (2), Bryden and Bishop 1945, Telenga 1948, Cavalloro 1949, Liu 1950, Walker 1961, Iperti 1964, Sem'yanov and Lipa 1967, Klausnitzer 1969b |
| *Coccinella septempunctata bruckii* | Japan | Koide 1961 (4), Maeta 1969a |
| *Coccinella trifasciata perplexa* | Canada | B. C. Smith 1960b |
| *Coccinella undecimpunctata* | Europe | Bryden and Bishop 1945, Walker 1961 |
| *Coccinella undecimpunctata aegyptiaca* | Egypt | Ibrahim 1955b |
| *Coleomegilla maculata* (=*Ceratomegilla fuscilabris*) | USA, Canada | Riley 1888 (1), Weed and Hart 1889 (1), Folsom 1909 (1), Cushman 1922 (1), Balduf 1926, Hudon 1959 (3), B. C. Smith 1960b, |
| *Coelophora inaequalis* | Hawaii | Timberlake 1918 (1) |
| *Eocaria muiri* | Japan | Koide 1961 (4), Maeta 1969a |
| *Harmonia arcuata* | Formosa | Sonan 1939 (2) |
| *Harmonia axyridis* | China, Japan | Liu 1950, Koide 1961 (4), Maeta 1969a |
| *Harmonia quadripunctata* | Europe | Iperti 1964 |
| *Hippodamia convergens* | USA | Timberlake 1916 (1), Cushman 1922 (1) Balduf 1926, Sluss 1968, |
| *Hippodamia parenthesis* | USA | Balduf 1926 |
| *Hippodamia tredecimpunctata* | USA, USSR | Cutright 1924 (1), Balduf 1926, Sem'yanov and Lipa 1967 |
| *Lemnia biplagiata* | Formosa | Sonan 1939 (2) |
| *Macronaemia hauseri* | China | Liu 1944 (2) |
| *Menochilus sexmaculatus* | Formosa, Japan | Sonan 1939, (2), Maeta 1969a |
| *Micraspis discolor* | Formosa | Sonan 1939 (2) |
| *Olla abdominalis* | USA, Hawaii | Timberlake 1918 (1) |
| *Propylaea japonica* | Japan | Koide 1961 (4), Maeta 1969a |
| *Propylaea qutuordecimpunctata* | Europe | Ogloblin 1913, Walker 1961, Iperti 1964 |
| *Semiadalia undecimnotata* | Europe | Iperti 1964 |
| *Synharmonia conglobata* | Europe | Iperti 1964 |
| !! *Epilachninae* | never parasitized | Balduf 1926, Liu 1950 |

(1) (2) (3) (4) (5) References marked in this way are not included in the list of references in this book; they are quoted by the following authors: (1) Balduf 1926, (2) Liu 1950, (3) B. C. Smith 1960b, (4) Maeta 1969a, (5) Klausnitzer 1969b.

1964, Klausnitzer 1969b) and *Coccinella septempunctata bruckii* in Japan (Maeta 1969a). On the other hand, some coccinellid species (within the three tribes mentioned) are not parasitized by *Perilitus*, or at least the parasite does not develop in them. *Adalia bipunctata* and *A. decempunctata* apparently rank among such species, although *P. coccinellae* females readily attack the adults in experiments (Bryden and Bishop 1945,

Walker 1962, Klausnitzer 1969b). It is also very difficult, although not impossible, to obtain emergence of an adult parasite from the coccon following parasitization of *Propylaea quatuordecimpunctata* (Iperti 1964). According to Klausnitzer (1969b) the parasitization of *Anatis ocellata* is impeded by the size of this coccinellid.

**Life cykle.** — *Perilitus coccinellae* has thelytokous parthenogenesis (i. e. unfertilized eggs hatch, but give rise only to females). The female inserts her eggs into coccinellid adults. Only very rarely has it been demonstrated (in *Coleomegilla maculata* — Cushman 1922, Smith 1960) that parasite adults can result if eggs are laid in coccinellid larvae. The parasite has also been reported to lay eggs in pupae of *Hippodamia convergens* (Sluss 1968) and of *Coccinella septempunctata bruckii* (Maeta 1969a).

The female especially attacks moving coccinellids, pursues them in its typical posture, i. e. with the abdomen flexed to the front between the legs and the ovipositor extending slightly forward of the head. After several minutes of antennal palpating and probing with the ovipositor, an attempt to oviposit follows involving a powerful forward thrust of the ovipositor. About one-half of the attempts result in actual egg deposition (Sluss 1968). The female inserts its ovipositor in the gap between the elytra and abdominal sternites, or according to Iperti (1964), into the intersegmental seam, most often between the head and the thorax, or again according to Sluss (1968), in the coxae or in the abdomen at a ratio of about 1 : 2. The rather petiolate egg (Fig. 8.01) deposited first increases in size for about 3 days, during which period its length increases four times. The egg hatches in about 5 days, releasing the first-instar larva and the trophic cells (see chapter Response of the host to parasitism, p. 204).

In a single coccinellid, many first-instar larvae can be found (superparasitism), as many as 47 in *Coccinella septempunctata bruckii* (Maeta 1969a), but always just one single second-instar larva is eventually left. The grasping mandibles, with which first--instar larvae (but not the larvae of other instars) are equipped, apparently serve to destroy the surplus larvae (Ogloblin 1924).

The larva passes through four instars, and, as in any insect, the length of larval development depends on temperature. In south-eastern France (Iperti 1964) this takes about 12—14 days in summer, and 20—21 days in spring. Under 21° ± 3 °C in the laboratory the egg hatches in about 5 days and larval development in *Hippodamia convergens* lasts 8—9 days (Sluss 1968). The morphology and anatomy of the larva has been described in detail by Ogloblin (1913, 1924).

The adult larva escapes from the coccinellid through the membrane between the 5th and 6th (Cushman 1922) or between the 6th and 7th abdominal tergites (Sluss 1968, Maeta 1969a). The beetle does not move about half-an-hour before the escape of the parasite larvae; its body and appendages tremble and jerk at times (Balduf 1926). The emerged larva spins a cocoon on the side of the abdomen of the host. The behaviour of spinning has been described in detail by Balduf (1926), who also specified the shape and structure of the cocoon. The caudal end is broadly rounded, whereas the cephalic end is pointed, hard and thick. The wall of the cocoon has two layers: (1) the outer, loose, meshwork and loops of silk which are directly attached to (2) the inner, solid and continuous layer. The development of the pupa takes 7—9 days in southeastern France (Iperti 1964) and 5—9 days in Japan (Fukuoka) (Maeta 1969a).

After completing its development inside the pupa and after escaping from it, the adult parasite bites through the cocoon at the cephalic end by means of elongated rectangular mandibles with terminal and post-terminal teeth. It is an exacting activity which takes about 24 hr and has been described by Balduf (1926) who writes admiringly: "... the labor is so protracted and energetic that one marvels at the endurance of the insect".

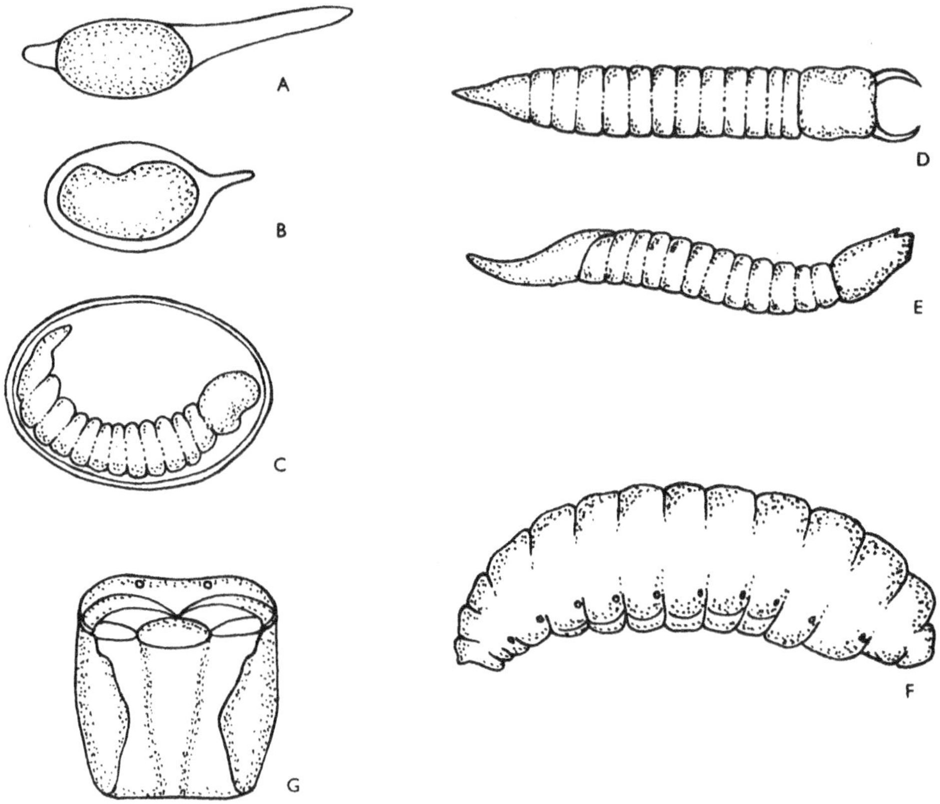


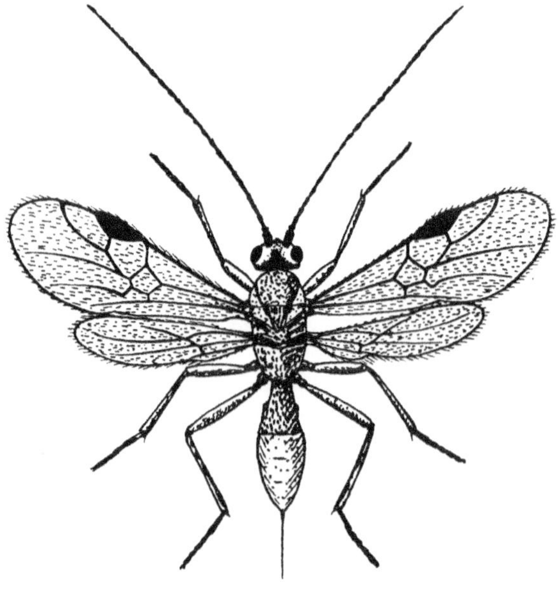

Fig. 8.01 I. Developmental stages of *Perilitus coccinellae*. A — egg newly deposited in host; B — egg aged four days; C — egg shortly before hatching; D — lst-instar larva. dorsal view; E — lateral view; F — mature larva: G — head capsule of the lst instar larva (the approximate relation of lengths A or G : B : C : D or E : F = 1 : 1.5 : 2 : 6 : 13).

II. Female of *Perilitus coccinellae* (del. J. Zelený).

Of the imagos, virtually only females have been recorded. In the literature there are only a few references to males. One of the references cannot be checked and the other specimen turned to be the male of another insect (Balduf 1926). When it leaves the cocoon, the female already contains ripe eggs and is able to parasitize immediately (Ogloblin 1924, Iperti 1964). The adults may live actively in the laboratory for about 5 days, if fed on honey solution and attacking coccinellids, but as long as 18 days when isolated in the dark although also fed on honey (Balduf 1926). Iperti (1964) recorded a maximum longevity of 21 days at 14 °C and 85 % relative humidity. Maeta (1969a) obtained a longevity of 2—6 days for starved adult parasites and 2—8 days for parasites fed on honey solution, apparently at normal laboratory temperatures.

**Voltinism.** — *Perilitus coccinellae* is a multivoltine species. Klausnitzer (1969b) reports that a second generation possibly develops in central Europe, where, according to Ogloblin (1924), there ought to be three generations. Iperti (1964) considers there are probably 3—4 generations in southeastern France. 4—5 generations are suggested for Illinois, USA, by Balduf (1926), and also for Italy by Goidanich (1943) and Cavalloro (1949). In Japan (Fukuoka), Maeta (1969a) reports the development of 3 generations of the parasite in *Coccinella septempunctata bruckii*, while the first generation of coccinellids is parasitized by the 1st and 2nd parasite generations.

The **percentage parasitization** fluctuates considerably according to regions, seasons of the year and the host. It is of doubtful value to compare data from the literature, as it is not always clear whether percentage parasitization has been ascertained by breeding — so that it is the percentage of emerged parasites (as cocoons or adults) which has been recorded — or whether the parasitization has been revealed by dissection. Moreover, some authors (Walker 1961) do not mention the total number of coccinellid individuals studied.

Balduf (1926) reports — on the basis of dissections — a range in parasitization of 1.3 — 36.0 % in *Coleomegilla maculata*, 4.0 — 21.1 % in *Hippodamia convergens* and 9.5 — 12.8 % in *H. parenthesis* (Table 8.03). According to Sluss (1968) the degree of parasitization of *H. convergens* rarely exceeds 10 % in California. In Italy, Goidanich (1943) ascertained a parasitization rate as high as 25 % for *Adonia variegata*, but usually a figure nearer 10—15 %. In Ontario, Canada, *Coccinella trifasciata perplexa* was parasitized at a rate of approximately 7 %, and *Coleomegilla maculata lengi* at 16 % (Smith 1960). Iperti's (1964) dissections of *Adonia variegata* in southeastern France in 1960—61 gave 5—25 % parasitization which is similar to the figure Goidanich gave for Italy.

Tab. 8.02 Parasitization of *Harmonia axyridis* and *Coccinella septempunctata bruckii* by *Perilitus coccinellae* in Japan (Maeta 1969a).

| Host | | Date of sampling | | Number dissected | | Percentage parasitized | | Locality[1]) |
|---|---|---|---|---|---|---|---|---|
| Species | Gener. | pupae | adults | ♀♀ | ♂♂ | ♀♀ | ♂♂ | |
| *H. a.* | 1 | June | — | 64 | 65 | 10.9 | 4.6 | Kurume |
| *H. a.* | 0 | — | November | 106 | 58 | 0.9 | 1.7 | Ina |
| *C. s.* | 1 | May | — | 133 | 132 | 30.1 | 21.2 | Kurume |
| *C. s.* | 1 | June | — | 29 | 28 | 3.4 | 7.1 | Ina |
| *C. s.* | 1 | — | June | 33 | 37 | 18.2 | 3.0 | Ina |
| *C. s.* | 1 | — | August — September | 96 | 86 | 37.5 | 25.6 | Kurume |
| *C. s.* | 0 | — | November, January, March | 35 | 31 | 11.1 | 3.2 | Ina |

[1]) Kurume, Fukuoka or Ina, Nagano.

Tab. 8.03 Parasitization of *Coleomegilla maculata*, *Hippodamia convergens* and *H. parenthesis*, by *P. coccinellae* in Urbana, Illinois, USA (Balduf 1926)

| Host species | Date of sampling | Number dissected | Percent parasitized | Locality |
|---|---|---|---|---|
| *C. m.* | March — April | 400 | 14.8 | winter quarter in a woods margin |
| *C. m.* | April | 51 | 36.0 | winter quarter along a fence in town |
| *C. m.* | April | 104 | 23.0 | active near winter site |
| *C. m.* | August — September | 157 | 1.3 | sewage disposal plant |
| *C. m.* | September | 349 | 2.5 | corn; West Lafayette, Ohio |
| *H. c.* | June | 32 | 18.8 | various plants |
| *H. c.* | July — August | 19 | 21.1 | various plants |
| *H. c.* | September | 100 | 4.0 | various plants |
| *H. p.* | June | 86 | 12.8 | various plants |
| *H. p.* | July | 111 | 10.8 | various plants |
| *H. p.* | August | 95 | 11.6 | various plants |
| *H. p.* | September | 21 | 9.5 | various plants |

Iperti found a higher parasitization rate in *Coccinella septempunctata* (10—27 %) and lower rates in *Semiadalia undecimnotata* and *Propylaea quatuordecimpunctata* (never > 8 %). Walker's (1962) result of 20 % parasitization of *Coccinella septempunctata* for England (Berkshire) is very similar to Iperti's figure for this species. Klausnitzer (1969b) reared 4 specimens of *Perilitus* from 1500 coccinellids collected in spring 1967 — 3 from *Coccinella septempunctata* and 1 from *Coccinella quinquepunctata*. In Japan, parasitization by *P. coccinellae* ranged between 1 and 11 % in *Harmonia axyridis*, and between 3 and 38 % in *Coccinella septempunctata bruckii* (Maeta 1969a. Tab. 8.02).

Balduf (1926) estimated that, in the area of Urbana in 1925, *Perilitus coccinellae* decreased the efficiency of coccinellids by 1/8. Iperti (1964) regards *Perilitus coccinellae* as the main parasite of migrating coccinellids (e. g. *Coccinella septempunctata* and *Semiadalia undecimnotata*). The relatively high potential fecundity (100 eggs), parthenogenesis and polyvoltinism of this parasite suggest that it could potentially be still much more effective. Its efficiency is apparently decreased mainly by superparasitism and by its preference for migrating coccinellid species which already enter diapause in summer (see chapter 7), and simultaneously cause a developmental arrest of the 1st instar parasite larvae.

**Response of the host to parasitism.** — Sometimes superparasitism or parasitization of a small host results in the premature death of the host. In the course of normal parasitism, the developing larva prevents the ovaries from developing (Balduf 1926). If females with ripe ovaries are parasitized, it takes some time before the developing larva of *P. coccinellae* causes ovarial regression. Parasitized females of *Coleomegilla maculata* lay fertile eggs up to 10 days before the emergence of *P. coccinellae* (Smith 1960).

On the other hand, the parasitization had no effect on mating behaviour of males until after the parasite emerged, and viable sperms were transferred to the female beetles at each mating. This fact does not in itself prove that testes had not been affected, because viable sperm can be stored for a long time. However, no changes were indeed found in the tissue of the testes of parasitized males.

The feeding of parasitized *Hippodamia convergens* was significantly less than that of non-parasitized beetles, while there was no significant effect on respiration rate. Although the gut, muscle, nerve and other tissues were examined, only the fat body showed structural changes as a result of parasitism (Sluss 1968).

Opaque globules of various sizes, which serve as nourishment for the parasite larva, are conspicuous when parasitized coccinellids are dissected. Such globules have been described in *Coccinella septempunctata* by Ogloblin (1924) and in *H. convergens* by Sluss (1968).

When a parasite larva hatches from the egg, clumps of unseparated trophic cells are released; each clump consists of 30 to 100 cells. These cells constitute one of the embryotic membranes (Ogloblin 1924, Sluss 1968). In the literature they are designated as "trophamnion" cells, although it is not certain if they in fact originate from the amnionic membrane. Next day they divide into individual, spherical opaque cells, about 50μ in diameter. In the course of larval development they keep growing, so that 2 days before emergence of the larva from the host body they already measure 370μ by 440μ. The surface of the trophic cells is covered by many microvillae. The profuse mitochondria suggest that these cells are very active metabolically. Their main function is to absorb and store nutrients from the host beetles (Sluss 1968). At first there are about 500 trophic cells (Ogloblin 1924). The strongly mandibulate first-instar larva apparently feeds directly on the fat body. The larvae of later instars eat the trophic cells (Ogloblin 1924, Sluss 1968). Most of the trophic cells present are ingested by the parasite larva during its development. The remaining few disintegrate, so that none are apparent 1 day after the parasite has emerged (Sluss 1968).

As long as the trophic cells are still small, the haemolymph distributes them into the whole body, even to the legs. They can be found even in the drop of haemolymph produced by 'reflex bleeding' at the articulation between femur and tibia (see chapter 8.2). 'Reflex bleeding' can thus be used to prove the presence of a *P. coccinellae* larva without dissecting the host (Ogloblin 1924).

After emergence of the parasite, the host coccinellid is considerably weakened and usually remains attached both to the cocoon of the parasite and to the substrate. Most observations suggest that the host coccinellid dies within the next few days (Balduf 1926: after 2–5 days, Smith 1960: within 10 days, Sluss 1968: 3–4 days). Bryden and Bishop (1945) and Walker (1962) detached parasitized individuals of *C. septempunctata* from the cocoons of *Perilitus* and fed them. The coccinellids recovered the use of their legs and appeared healthy. However other workers who repeated these experiments found that the beetles were nevertheless so weakened by parasitization that they died within 1–2 weeks in any case (Iperti 1964, Klausnitzer 1969b). Thus death is not the result of the coccinellid's imprisonment by the cocoon fibres and starvation as Balduf (1926) had thought. Total recovery of a parasitized host has only been described by Timberlake (1918), who records the case of the large coccinellid *Olla abdominalis*. Five out of nine individuals escaped the wrapping up in connection with the cocoon: they recovered, resumed feeding and were subsequently reparasitized, one individual giving issue to a second mature larva of *Perilitus*! It is probable that the degree to which the host is damaged depends on its size.

### *Homalotylus spp.* (*Chalcidoidea, Encyrtidae*)

This genus contains several species of endoparasites of the larvae and pupae of coccinellids.

According to Hoffer (1963), Graham (1969) and Klausnitzer (1969b), the nomenclature of two species, *H. flaminius* and *H. eytelweini*, is confused. The data on *H. flaminius* in the literature may partly refer to *H. eytelweini*. According to Hoffer (1963) Graham

(1969) and Klausnitzer (1969b) it is possible to distinguish the two species by the following characters:

1 (2) Hind tarsae white, radial vein short and curving away from the front edge of the wing. ........................................................ *flaminius* Dalm.
2 (1) Hind tarsae black, radial vein longer and not curving away from the front edge of the wing (i. e. almost parallel with it). ........................ *eytelweini* Ratzb.

Klausnitzer (1969b) thinks that *H. flaminius* is a solitary parasite of *Scymnini*, whilst *H. eytelweini* is a gregarious parasite of *Coccinellini*, *Chilocorini*, *Hippodamiini* and *Psylloborini*.

*H. platynaspidis*, which as yet has been reared solely as a solitary parasite from *Platynaspis luteorubra* (Hoffer 1963, Klausnitzer 1969b), is apparently monophagous.

**Life cycle.** — Eggs are laid in young (Rubtsov 1954) or even mature (Telenga 1948) coccinellid larvae, mostly while the host is attached to the substrate at ecdysis (Iperti 1964). After several days the parasitized larva again attaches itself to the substrate as at ecdysis, but instead of casting a skin it swells and the cuticule becomes hard and dark. The adult parasite leaves the host through a small round opening. When polyparasitism is involved, individual parasite larvae pupate in chambers separated from each other by remnants of dry host tissue and each adult makes its own opening through which it leaves the host. In larvae of the genus *Chilocorus* there usually develop 1—3 parasites (Telenga 1948, Domenichini 1956), about 3 parasites develop in *Propylaea quatuordecimpunctata* larvae, and 4—6 in older larvae of *C. septempunctata* (Iperti 1964). The adults are already sexually mature at emergence, and longevity in the laboratory is 6—11 days (Iperti 1964). In the open the adults feed on honeydew of aphids and coccids (Rubtsov 1954).

*Homalotylus* hibernates as a pre-nymph inside the host. In south-eastern France the parasites took as long as 97—145 days to emerge from several parasitized larvae of *Coccinella septempunctata* collected in May at an altitude of 600—1000 m (Iperti 1964).

The length of development in non-hibernating generations fluctuates with temperature during the year. Development thus takes between 25 and 45 days for *C. septempunctata* and is shorter for *Propylaea quatuordecimpunctata* (Iperti 1964). Rubtsov (1954) reports about one month. Telenga (1948) from the Ukraine SSR gives 17—19 days in summer.

**Voltinism.** — *Homalotylus* is polyvoltine. For south-eastern France, 5—6 generations per year are reported by Iperti (1964), 3 generations for the Black Sea coast of USSR by Rubtsov (1954) and 5 generations for the region of Krasnodar (Ukraine SSR) by Telenga (1948).

**The percentage parasitization.** — Parasitization by *Homalotylus* may be as high as 90 — 95 % in *Chilocorus bipustulatus* (at the Black Sea coast — Rubtsov 1954, the same as in North Africa — Smirnoff 1957, ref. in Yinon 1969b) and as high as 100 % in *Rodolia cardinalis* in India (Subramanyam 1950, ref. in Yinon 1969b). *Homalotylus* is therefore generally considered as a very significant mortality factor which limits the entomophagous efficiency of coccinellids. Other coccinellids, where *Homalotylus* has been specifically stressed as an important agent, are *C. septempunctata* and *Propylaea quatuordecimpunctata* in south-eastern France — Iperti 1964, and *Chilocorus bipustulatus* in the coastal plain of Israel — Yinon 1969b.

**Hyperparasites.** — Iperti (1964) reared adults of a proctotrupid of the genus *Lygocerus* from *C. septempunctata* larvae which had been parasitized by *Homalotylus*. Yinon (1969) reared two species of hyperparasites (an encyrtid *Achrysopophagus aegyptiacus* and a pteromalid *Pachyneuron siculum* from parasitized *Chilocorus bipustulatus* larvae by *Homalotylus*. Domenichini (1956) reared a pteromalid *Pachyneuron chilocori* from parasitized *Chilocorus bipustulatus* (hyperparasitization 86 %).

*Tetrastichus spp.* (*Chalcidoidea, Eulophidae*)

According to Klausnitzer (1969b), 4 species of this genus parasitize *Coccinellidae*. It is very difficult to identify the species of the genus *Tetrastichus*, so that it is possible that not all data attributed to *Tetrastichus coccinellae* really concern this species.

Tab. 8.04 Survey of the *Tetrastichus* species which parasitize *Coccinellidae* (after Klausnitzer 1969b).

| *Tetrastichus* | Parasitized stage of the host | Food specialization |
|---|---|---|
| *coccinellae* Kurdj. | larvae, pupae | *Chilocorini, Coccinellini* |
| *epilachnae* (Giard.) | larvae, pupae | *Epilachninae* |
| *neglectus* Domenichini | larvae, pupae | *Chilocorini, Coccinellini, Scymnini* |
| *sempronius* Erdoes | larvae | mainly *Chrysopa spp.*, but also *Chilocorus bipustulatus* |

*Tetrastichus* is a gregarious parasite. As many as 25 *T. coccinellae* can develop in one *Coccinella septempunctata*, but more often only 10—15 parasites are found (Iperti 1964, Klausnitzer 1969b). Fewer parasites develop in *Adalia bipunctata*. For 103 parasitized pupae Klausnitzer (1969b) ascertained an average of 10.2 (minimum 3, maximum 14) *T. neglectus* per pupa. In coccinellids of the genus *Chilocorus*, Telenga (1948) usually found 6—8 parasites but sometimes even more, as many as 18.

**Life cycle.** — The female lays eggs in 3rd and 4th instar larvae; the 4th instar larvae manage to pupate (Telenga 1948). According to Iperti (1964), eggs are also laid in pupae. In the host's body the developing larvae are not in separate chambers (contrast *Homalotylus*). All the adult parasites which emerge from the host do so through the same single opening. Parasites of one sex only emerge from any one host pupa (Klausnitzer 1969b). The females are mature at emergence; several minutes later they copulate and so can immediately (Iperti 1964) or the next day (Telenga 1948) parasitize another host. In the laboratory at 20—25 °C and 70 % relative humidity, the longevity of the adult fluctuated between 5 and 20 days, at 14 °C it may reach 25 days (Iperti 1964).

Depending on temperature the whole development of the parasite lasts 20—32 days; Telenga (1948) reports as little as 12—16 days for the Krasnodar region (Ukraine SSR), while Iperti (1964) reports a minimum value of about 20 days for south-eastern France in July.

**Voltinism.** — Owing to the very fast development of *Tetrastichus* the annual number of generations is high. Both Telenga (1948) and Iperti (1964) report 7 generations for the Ukraine and south-eastern France. In the Ukraine, however, diapausing larvae (= prepupae) could be found from the 4th generation onwards.

**Percentage parasitization.** — There is general agreement that parasites of the genus *Tetrastichus*, similarly to *Homalotylus*, considerably limit the population increase of aphidophagous and coccidophagous coccinellids. The maximum parasitization that Iperti (1964) recorded in *Adalia bipunctata* reached 60 % in July 1962. According to Telenga (1948), *Tetrastichus coccinellae* is even more significant than *Homalotylus flaminius* in the Krasnodar region. Rubtsov (1954) complains of the ruinous effect of both these parasites on coccinellids of the genus *Chilocorus* on the Black Sea coast. The coccinellids of the 1st (= spring) generation are parasitized at 10 %, by the 2nd

(= summer) generation the parasitization has already risen to 50—70 %, and by the 3rd generation often 90 % or more are parasitized. As a result, only 1—10 % of the beetles hibernate, thus permitting an intense reproduction in the following spring by the coccids which are their prey.

### 8.33 Acarina

Apart from *Acarina* showing phoresis (e. g. *Pediculoides ventricosus*), coccinellids are attacked also by parasitic *Acarina* of the genus *Podapolipus* (*Podapolipodidae*). The parasite develops on the inner side of elytra. 4—5 females fix themselves, lose their appendages, feed on haemolymph and lay several hundred eggs. The hatched nymphs fill the space between the tergites and the elytra. Mostly *Adalia bipunctata* is attacked (up to 40%), but also *Coccinella septempunctata* and *Synharmonia conglobata*. *Semiadalia undecimnotata* does not suffer from the attacks of this parasite. Although the parasitization of coccinellids by *Podapolipus* is not lethal, it causes an important fall in vitality of coccinellids.

### 8.34 Nematoda

*Parasitylenchus coccinellinae* (*Tylenchida, Allantonematidae*)

This nematode has been found as an endoparasite of the adults of multivoltine coccinellids in south-eastern France (Iperti 1964) and described as a new species (Iperti and van Waerebeke 1968). In dissections of coccinellids, nematodes were found from May to October with the maximum in late summer and early autumn. Either only adult female nematodes were found which had not yet oviposited, or ovipositing females together with female larvae and males. From 1 to 140 adult females could be found in one host; the number of larvae and young adults could amount to about 10.000 individuals.

**Life cycle.** — The young migrating fertilized female penetrates into the body cavity of the coccinellid. How this penetration occurs is not yet certain — there are three possible ways: (1) through the tracheae, (2) during coccinellid copulation through the sexual organs (where *P. coccinellinae* is abundant) or (3) through the softer cuticle between the sclerites (in a young infesting female the stylet is particularly well developed). The aerial environment of coccinellids is disadvantageous for the propagation of nematodes. Infection may occur in aggregations, when rain or dew can create the damp environment favourable for parasites among the coccinellids.

Within 4 days after penetration into the beetle, the body of the nematode female is almost entirely filled with ovaries, and after a short time the uterus is filled with free larvae and developing eggs containing coiled larvae. The larvae undergo their first moult while still in the uterus, then they moult twice more in the body cavity of the host. The third moult gives rise to males which are already adult, whereas the female gonads are still ovarial primordia of 5—9 cells. At this stage the larval females are fertilized. These young females evidently moult for the last (fourth) time immediately after penetrating into the host.

At the onset of the hibernation of the coccinellids, the development of *P. coccinellinae* eggs ceases and the female nematode becomes an ovisac.

**Percentage parasitization.** — Multivoltine species of coccinellids are most often attacked, especially *Propylaea quatuordecimpunctata* which is highly parasitized (up to

70 %), *Synharmonia conglobata* rather less (20 %), and least *Adalia bipunctata* and *Adonia variegata* (less than 10 %). In monovoltine species, *P. coccinellinae* has been found only incidentally in *Semiadalia undecimnotata* (Iperti and van Waerebeke 1968).

Although the nematodes are not quickly lethal to their host, they will reduce and may even completely stop the maturation of ova. They certainly also consume the reserves of the host.

*Howardula sp.* (*Tylenchida, Allantonematidae*)

In England, Hariri (1965) found adults and larvae of nematodes in 9 individuals of *Adalia bipunctata;* the nematodes were determined by Dr. G. O. Poinar as probably a new species of *Howardula.* The gonads of coccinellids were little affected by the parasite, but the fat bodies were much smaller than those of unparasitized beetles.

*Mermis sp.* (*Mermitidae*)

This endomonoparasite was found in south-eastern France by Iperti (1964) in four aphidophagous coccinellids during both diapause and activity. Hibernating *Adonia variegata* (2.5—4.2 %) and the aestivating *Semiadalia undecimnotata* (1.2 %) were attacked most often while the nematode was found only occasionally in active *Propylaea quatuordecimpunctata* and *Coccinella septempunctata.*

## 8.4 Pathogens

### 8.41 Protozoa

*Gregarinidae* (*Sporozoa*)

Laudého et al. (1969) have recently provided a literature review surveying the occurrence of these protozoa in coccinellids. A number of species, whose synonomy has not been reliably clarified, are involved. They destroy intestinal cells of coccinellid larvae and adults. Digestive troubles increase in the host with the number of gregarines which can, within a month, rise from only 10 individuals to several thousands in the central intestine of each infected coccinellid. A great number of gametocysts are then formed, and these may cause death of the beetle by blocking the intestine near the rectal pouch. In adults of *Pharoscymnus anchorago* which survive the danger of blockage, longevity is then reduced to one-half and fecundity to one-third. Laudého et al. found a high infection rate (more than 50 %) in two coccidophagous coccinellids inhabiting palm groves of Mauretania; these were the indigenous *P. anchorago*, and *Chilocorus bipustulatus* introduced from Iran. Two other introduced species showed different sensitivities. *Chilocorus stigma* was very sensitive to infestation by gregarines, whereas these protozoa were never found in *C. distigma.*

In areas with a milder climate, infections by gregarines are less massive. Iperti (1964) located them in 5 aphidophagous coccinellids in south-eastern France and, for example, in *Propylaea quatuordecimpunctata* infection was only around 10 % infestation during the whole year.

### 8.42 Fungi

*Beauveria* group *bassiana*

The coccinellids hibernating in aggregations are very often infected and killed by the pathogenous fungi of the genus *Beauveria*, group *bassiana*. The body cavity of the attacked coccinellids is then filled with a homogeneous cheese-like mass where the individual coccinellid organs cannot be discerned.

In *Semiadalia undecimnotata* the occurrence of mycosis is related to the type of hibernation shelter and its altitude (Iperti in Hodek 1966, p. 189). Below 1000 m the rate of mortality which was registered at three different sites for 4 years amounted to 19.7%. At three other sites elevated above 1000 m the mortality rate was only 1.5 % at the same period. The incidence of mycosis was much lower in rock cracks or in piles of stones (1.3 and 8.6%) than when the coccinellids hibernated among plants (34 %). To reduce the high mycosis, Iperti (in Hodek 1966, p. 189) placed artificial traps simulating rock cracks (Photo 4?,50) on the mountain tops lacking rocks. The coccinellids preferred the artificial "hibernation quarters" over the original sites between plants and their mortality was substantially reduced.

Hibernating adults of *Coccinella septempunctata* are also often attacked by *B. bassiana;* Iperti (pers. comm. 1971) reports a mean 50 % infection in south-eastern France.

*Laboulbenia*

Iperti (unpubl.) found a non-pathogenous fungus of the genus *Laboulbenia*, very bizarre in form, on the body surface of adult coccinellids. In south-eastern France, *Adalia bipunctata* and *A. decempunctata* are attacked in spring, *Pharoscymnus semiglobosus* in Morocco all over the year.

## 8.5 Comparison of parasitization and disease

Iperti (1964 and unpubl.) made a detailed study of the parasitization of 5 coccinellid species (*Adonia variegata, Semiadalia undecimnotata, Coccinella septempunctata, Adalia bipunctata, Propylaea quatuordecimpunctata*) in south-eastern France (Alpes-Maritimes and Basses-Alpes). He observed parasitization in all the three developmental stages and for some coccinellid species in both active and diapausing adults.

The abundant material (7617 specimens) which he acquired over 4 years (1961—64) makes possible an interesting comparison of different coccinellid species and stages, though at certain points such comparison relies on limited numbers — e. g. the data for *P. quatuordecimpunctata* pupae obtained from only 22 specimens must be regarded as preliminary.

**Total parasitization.** — The above mentioned 5 coccinellid species can be divided into 3 groups on the basis of the parasitization of larvae and pupae (fig. 8.02). (1) *A. variegata* and especially *S. undecimnotata* show low parasitization of these stages, and exclusively by the phorid *Phalacrotophora fasciata.* (2) In one coccinellid species, *A. bipunctata*, chiefly the pupae were strongly parasitized by the eulophid *Tetrastichus coccinellae.* (3) In *P. quatuordecimpunctata* and *C. septempunctata* it was the larvae which were strongly parasitized and the main parasite was the encyrtid *Homalotylus flaminius.* Parasitization of pupae was lower than that of larvae and included all three preimaginal parasite species.

From the above, it can be seen that the pre-imaginal stages are parasitized only by insects. By contrast, the adults are also heavily attacked not only by *Nematoda,*

*P. quatuordecimpunctata* is often strongly infected by a new species of the genus *Parasitylenchoides* and *A. variegata* rather less by *Nematoda* belonging to the genus *Mermis*, but in addition by pathogens (protozoans and fungi). *Gregarina* occurred in active adults of all observed species, but were no more than rare in *A. bipunctata* and *S. undecimnotata*.

In adults the only generalization which can be made is that a rather intense parasitization by the braconid *Perilitus coccinellae* occurs in hibernating individuals of all three species studied. In addition, hibernating *S. undecimnotata* were infected heavily also by fungi of the genus *Beauveria*. By contrast with active adults (see below) *Gregarina* have not been found in hibernating coccinellids.

When comparing parasitization of active coccinellids, we find a unique situation in *P. quatuordecimpunctata*. The contribution of a massive nematode (*Parasitylenchus*) infection raises the overall level of infestation by all agencies to 35 % which is much higher than is found in any other coccinellid species.

Next most infested are *C. septempunctata* and *A. variegata* with 20 % and 12 %. Active adults of *A. bipunctata* and *S. undecimnotata* may be considered by comparison as rather "healthy", as total infestation is only 6 %. Of the insect parasites, the braconid *Perilitus coccinellae* is strikingly abundant in active *C. septempunctata*, whereas in other species it forms only a negligible percentage (1—2 %) and in *A. bipunctata* it is completely missing. The other insect parasite of adults, the tachinid *Degeeria luctuosa* is completely absent from *C. septempunctata* and only parasitizes the other four species to a small extent (0.3—2.1 %).

If we compare all the five coccinellid species according to the total sum of parasitization and disease of the nonhibernating stages, *S. undecimnotata* would appear to be the 'healthiest' species. If we regard this species as the standard for 'health', *P. quatuordecimpunctata* stands out as by far the most sickly by a factor of 10. Intermediate are *C. septempunctata* (a factor of 5) and *A. variegata* (a factor of 3).

All the above figures used for comparison are the averages. In the course of every season, the percentage of parasitization increases. Thus, about 70 % of coccinellids die instead of entering dormancy (Iperti unpubl.).

**Differences between individual years.** — Percentage of infection by individual parasites and pathogens often varies from year to year (Iperti 1964 and unpubl.). Exceptions are perhaps seen in the rather constant attack of *A. bipunctata* pupae by *Tetrastichus coccinellae*, of larval and pupal *A. variegata* by *Phalacrotophora fasciata* and of *P. quatuordecimpunctata* larvae by *Homalotylus flaminius*. By contrast, parasitization of *C. septempunctata* larvae by *T. coccinellae* which had risen considerably in May and June 1961, was very low in the same period of 1962, and in 1963 *Tetrastichus* did not occur at all before July. Infection of *C. septempunctata* by *Gregarina* also differed from season to season: in 1961 they were not found in a single sample, in 1962 they were found only in May, and in 1963 the infection by gregarines showed a gradual rise from June to August.

The comparisons which have been presented here are probably limited to the one region (south-eastern France) where such studies have been carried out. Unfortunately, similar data are not yet available from other regions, although they would be very useful, especially from the point of view of employing coccinellids in biological and integrated control. The task of accumulating such data is one of the projects of international cooperation in aphid predator studies within the framework of I. B. P. (see chapter 9). The only other extensive data comes from two regions of the USSR and from Israel (see 8.32) but unfortunately concerns only one genus of coccinellid (*Chilocorus*) and is limited to insect parasites only.

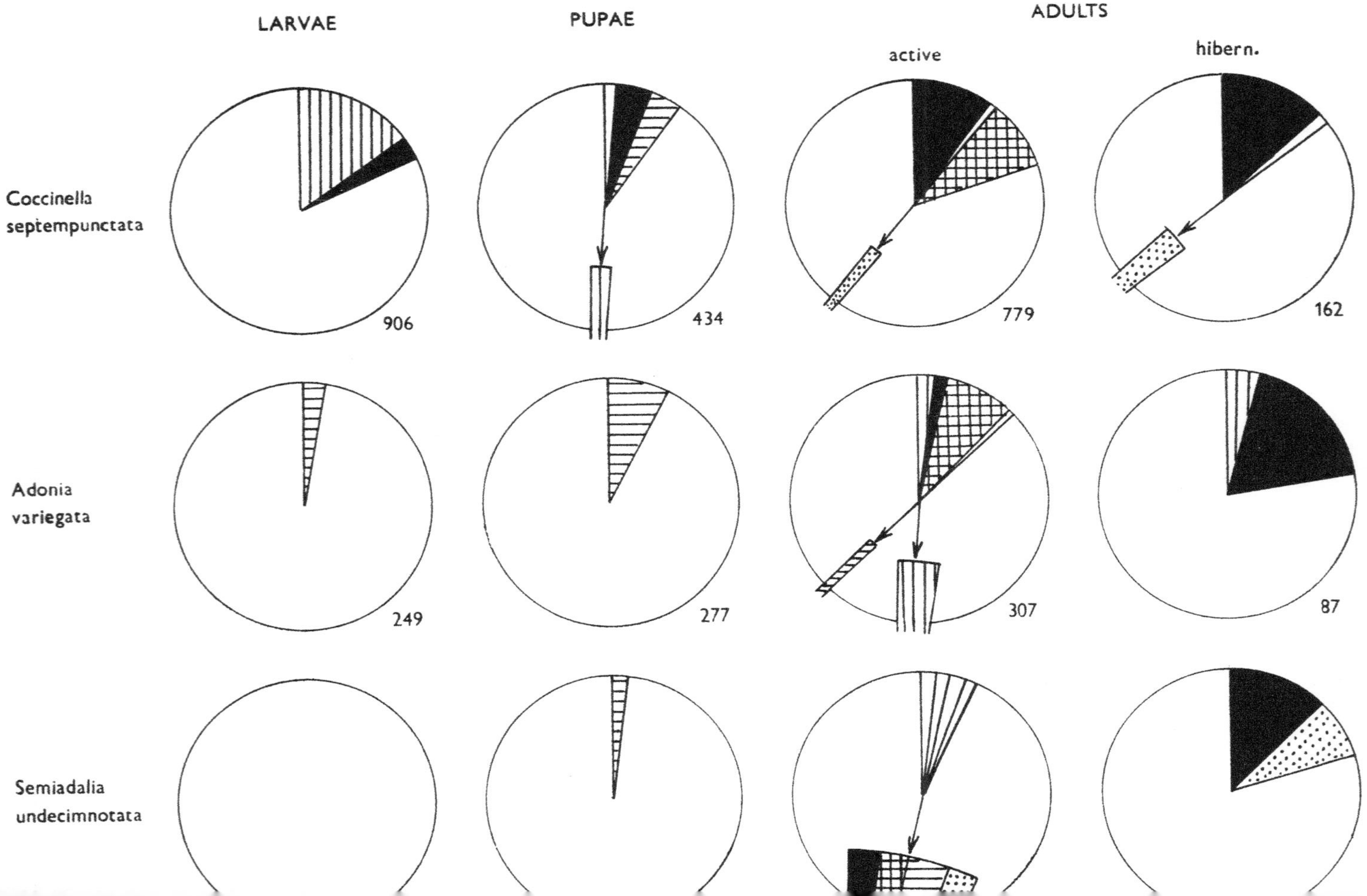
LARVAE
PUPAE
ADULTS
active
hibern.
Coccinella septempunctata
906
434
779
162
Adonia variegata
249
277
307
87
Semiadalia undecimnotata

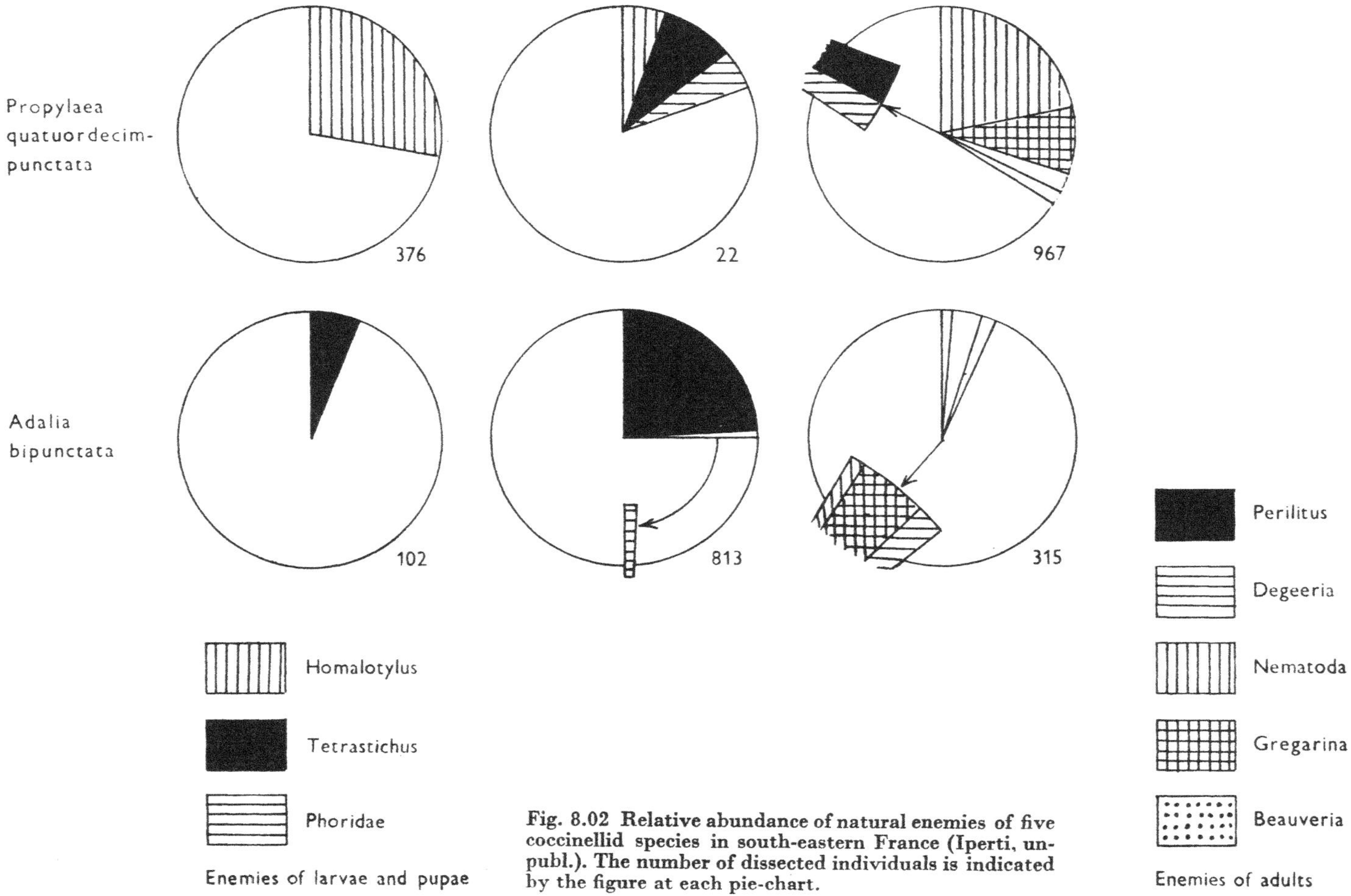


**Fig. 8.02 Relative abundance of natural enemies of five coccinellid species in south-eastern France (Iperti, unpubl.). The number of dissected individuals is indicated by the figure at each pie-chart.**

Tab. 8.05 Parasitization of coccinellid predators of the sorghum aphid in N. Transvaal (van Rensburg unpubl.)[1])

| Coccinellid species | Parasitization | |
|---|---|---|
| | of larvae by *Homalotylus flaminius* [%] | of adults by *Perilitus coccinellae* [%] |
| *Cheilomenes propinqua* | 19.8 | 24.0 |
| *Cheilomenes lunata* | 81.3 | 10.4 |
| *Lioadalia flavomaculata* | 69.5 | 36.2 |
| *Exochomus concavus* | 68.6 | 6.6 |
| *Scymnus moreletti*[2]) | 54.9 | |

[1]) Neither the eggs nor the pupae of any of the coccinellid species studied were attacked by parasites.
[2]) *S. moreletti* larvae were parasitized by an unidentified *Homalotylus sp.*

# 9 EFFECTIVENESS AND UTILIZATION

The use of predaceous coccinellids for the control of pests differs with climatic area (the contrast between the temperate zone and the tropics and subtropics) and with the type of prey (aphids, coccids, mites). A further important consideration is whether an indigenous or imported exotic prey is involved.

As is true in other examples of biological control, comparatively the easiest way is, of course, to import a missing coccinellid against a prey which has been unintentionally introduced into a new area without its predator. This was the basis of the almost miraculous success with *Rodolia cardinalis* (see 9.21). This success was further assured by the fact that the rate of increase ratio between predator and prey was in favour of the coccinellid. In general it can be said that this ratio is much more favourable for the multivoltine coccidophagous coccinellids in the tropics and subtropics than it is for the usually univoltine (or only partially bivoltine) aphidophagous coccinellids in temperate areas. This is evidently the reason why coccids are particularly successful subjects for biological control (DeBach 1964) (see 9.21).

In temperate zones there is a major difference in strategy for utilizing aphidophagous coccinellids between the permanent habitats (forests, orchards) and the fields, especially annual crops. The coccinellids which breed in forests also hibernate there (see 7.324); it would therefore be comparatively easier than in annual crops to use supporting procedures to augment their numbers in the long term. No attempts in this direction have been reported, for the simple reason that aphids in forests are generally considered as economically neutral within that habitat. Orchards somewhat resemble forests in that at least a part of the population of some coccinellid species (e. g. *Stethorus punctillum*, *Adalia bipunctata*) both breed and hibernate there (see 7.323). However, the majority of coccinellids which are of economic significance in orchards move to other habitats even in the course of the breeding season (see 5.32), and they also hibernate elsewhere. Thus the coccinellid fauna in an orchard is, to a large extent, re-formed every year and is influenced by conditions in neighbouring habitats.

This applies to annual field crops to a much greater extent. As a rule, they are secondary host plants of pest aphids. Whether the coccinellids will or will not have an impact on the aphid populations is often determined within a short period of time, and the deciding factors are: the previous development of aphids on their primary hosts, the proximity to the crop of the hibernation sites of coccinellids (see 5.33), the meteorological conditions in the critical period (see 9.1), etc.

The most frequent problem which applies generally in all habitats is to provide coccinellids on the crop in sufficient numbers during the critical period to have an impact on the pest population. This can be achieved in a variety of ways:

(1) by the release of sufficient numbers of coccinellids
   (a) collected elsewhere, e. g. in hibernation quarters;
   (b) cultured in the laboratory;
(2) by attracting coccinellids from other habitats or by retaining them on the crop
   (a) by providing alternative food;
   (b) by disturbing the adjacent habitats or strata;
(3) by the general support of coccinellids in the longer term and over larger areas. Here may be suggested
   (a) in the period after the population crash of aphids the provision of some alternative food to enable the coccinellids to lay down sufficient reserves before hibernation (see 6.111 and 7.311),
   (b) the planting of suitable hibernation sites, protection of hibernation sites (especially against burning), or the erection of artificial shelters to prevent mycosis (see 8.421) and mammalian predation.

In spite of the considerable study devoted to coccinellids, there are still **gaps in our knowledge** which must be filled before coccinellids can be utilised or supported rationally:

(1) the mobility of coccinellid adults in terms of distance and direction of migration;
(2) the qualities of a habitat that attracts the immigrant coccinellid adults;
(3) the influence of meteorological factors, mainly temperature, on the searching efficiency and feeding rate of the predator in comparison with the rate of reproduction of the prey (see 9.1);
(4) the population density of the prey
   (a) necessary to retain the predator;
   (b) to enable reproduction;
   (c) to assure a high survival of first instar larvae which are least efficient in searching for prey;
(5) the composition of artificial essential food to enable economic mass production;
(6) the composition of cheap alternative foods
   (a) for attracting and maintaining coccinellids in the habitat during the critical period;
   (b) for the nutrition of coccinellids before hibernation;
(7) methods of treating coccinellids to reduce their tendency to fly and thus disperse at the end of hibernation. If they are collected and transported to crops, dispersal will nullify attempts at mass release.

Coccinellids are, of course, only one segment — even though a significant one — in a whole complex of factors utilizable in pest management. A variety of other known predators (*Anthocoridae* and other *Heteropteran* families, *Syrphidae*, *Itonididae*, *Chrysopidae*, *Hemerobiidae*) belong to this complex. There is some evidence that some other aphid predators are being overlooked because of their nocturnal feeding activities. The *Dermaptera*, for example, also apparently prey on aphids (Schlinger et al. 1959, Skuhravý 1960, Smith 1966, Asgari 1966, Way and Banks 1968). Furthermore, it seems probable that some *Carabidae* (Skuhravý 1959) visit plants at night and prey on aphids, chiefly those which have fallen from the plants. Some *Cantharidae* (Smith 1966, Way and Banks 1968) and spiders are other overlooked aphid predators. Other important factors are parasites and pathogens (Hagen and van den Bosch 1968), also the other tools of integrated control, e. g. modification of cultivation schemes, premature defoliation of host plants, breeding plants resistant to aphids, etc.

In addition, the prey of coccinellids is hardly ever the only pest in a crop. Coccinellids must, therefore, be incorporated into the complex of the **integrated control** measures for a given crop. "Integrated pest control is a pest population management system that utilizes all suitable techniques in a compatible manner to reduce pest populations and maintain them at levels below those causing economic injury" (Smith and Reynolds 1966). This approach is being recognized as essential for the sound future of agricultural production. Other approaches may give dramatic results on a short-term basis, but in the long run we must learn how best to apply basic ecological principles to the inherently artificial crop environment. However, these basic principles are expressed differently in every ecosystem and therefore no general scheme can be designed — integrated control procedures must be specific for each crop and for each climatic area. Each agroecosystem must be analyzed for the most serious "key pests", their competitors and natural enemies and the main alternative food supplies of the pest.

## 9.1 Evaluation of the impact

There still exists only little concrete, numerical evidence of the impact of coccinellids (Smith 1966, Hagen and van den Bosch 1968, van Emden et al. 1969, Hodek et al. 1971). This, in fact, also applies to other entomophagous insects. More reliable data on the efficiency of coccinellids (and other natural enemies acting synergistically) are highly desirable, particularly for two reasons:
(1) we must know to what extent and at what period we can rely upon their action;
(2) we should be able to calculate the cost efficiency ratio of supporting procedures (e. g. mass production, import, spraying with artificial food, etc.). To do this, we must be able to evaluate the effect of such treatments.

The control potential of coccinellids can be examined from several angles. The significance of coccinellids (and natural enemies in general) in the full population dynamics of aphids, i. e. in the long-term regulation of aphid numbers, is not yet sufficiently understood. This is because, with the exception of a few aphids (see 9.11), most evidence relates to short-term population dynamics restricted to a small area (R. F. Smith 1966). But even such studies enable us to recognize which factors determine the impact of natural enemies at individual phases of the aphid cycle (see 9.12). Very often the impact is judged by hindsight from a co-incidence of events observed from the sampling data of both prey and natural enemy populations; this method is not very likely to yield reliable conclusions. There is a need for much more extensive use of the more efficient experimental methods (see 9.13). In most cases it appears that coccinellids on their own are unable to check pest infestations effectively and must be helped in various ways (see 9.2 and 9.3).

### 9.11 Role of coccinellids in the regulation of aphid numbers

Our ideas on the efficiency of aphidophagous insects have broadly remained unchanged for 40 years, since Blattný (1925) hypothesized that the alternation of years with an overpopulation by aphids and years with lower aphid population is caused by the action of natural enemies. In a year with many aphids the enemies have suitable conditions for rapid multiplication, and thus the next year the large number of enemies can keep the aphid population at a low level. Because numbers of prey are thus reduced, the population increase of aphidophagous insects in the second year is inhibited and thus the third year will again be favourable for overpopulation by aphids (Fig. 9.01 b). Similar assumptions are still generally accepted, though they are now much more sophisticated (Müller 1966, Behrendt 1969). However, the hypothesis has still not been proved unequivocally.

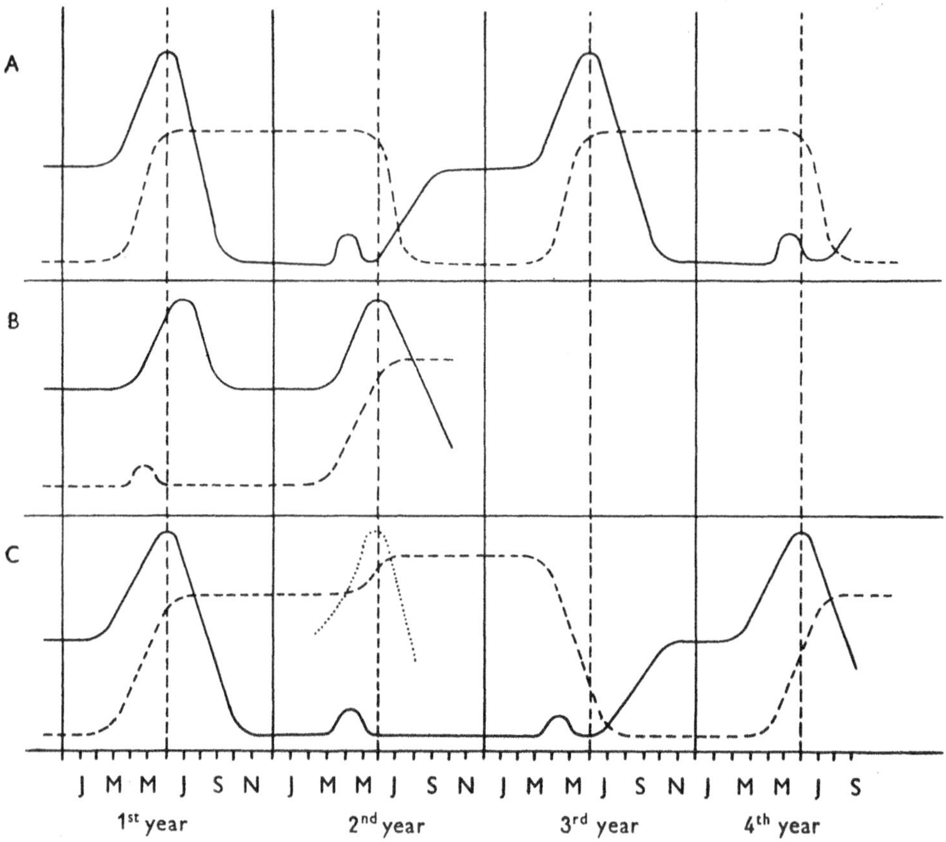


Fig. 9.01 Scheme of predator/prey relationships between coccinellids (— — —) and aphids (———) (Müller 1966).

A: the normal case with an alternation of aphid and coccinellid peaks:

B: coldweather delays the aphid peak; the coccinellid population remains low. and therefore the aphid is again abundant in the following year;

C: during low population levels of the main host aphid the coccinellids can prey on another aphid (. . . . . . . ) and thus show high population levels in two successive years; this results in a low population of aphids.

Müller (1966) includes two modifications of Blattný's hypothesis. Cool weather delays the aphid peak and the population of coccinellids therefore remains low. This enables a repeated outbreak of aphids in the following year and therefore aphid overpopulation can occur in two consecutive years (Fig. 9.01b). The opposite situation holds when, during low population levels of one prey species, the coccinellids can build up a high population by preying on another species. Thus they may prevent the outbreak of the original species of prey for a second year (Fig. 9.01c).

This scheme is relevant to the fluctuations of *Aphis fabae* in a mild climate. Way and Banks (1958) in England, Palyi (1960) in the "Black-soil" region of the USSR and Behrendt (1969) in the German Democratic Republic, are all of the opinion that, in the long-term population dynamics of this aphid, the most important period for the impact of natural enemies follows the aphid peak on crops. In this period, the activity of natural enemies determines the number of overwintering aphid eggs on primary host-plants (*Euonymus*). A correlation has deen found between a low number of eggs and a subsequent low summer occurrence of the aphid on secondary host-plants (sugar-beet etc.). The summer level of aphios is also affected (apart from other factors) directly by the number of natural enemies, mainly overwintered coccinellids (Hodek et al., in Hodek 1966, p. 331, Hodek et al. 1966). The number of the latter is dependent on their survival during hibernation (esp. with respect to mycoses end late frosts). The spring impact of coccinellids is modified by the effect of meteorological conditions on their feeding activity and by the distance between the hibernation sites.

In both these ideas concerning the critical phase for the action of coccinellis, it has been decided already at the beginning of the vegetation season whether or not coccinellids (and other natural enemies) will have an economic impact.

An entirely different pattern of fluctuation in aphid populations holds in California (Hagen and van den Bosch 1968) for alfalfa aphids (*Acyrthosiphon pisum* and *Therioaphis trifolii*), where there can be two or three population peaks each year. The principal factor responsible for this contrast with mid-Europe is the climate which, in the warmer lowland California, not only speeds aphid development but favours anholocycly. There is a close synchronization between natural enemies and aphids, and less drastic population oscillations occur. All coccinellid species do not have equal impact. The *Hippodamia* species, which only have a facultative adult dormancy, are more sensitive to the occurrence of aphid; they reproduce if there are sufficient aphids and if temperatures are high enough. Most *Coccinella spp.* do not react to summer aphid abundance; they go into diapause anyway. Microclimatic changes caused by harvesting the alfalfa have important effects on both natural enemies and aphids (see 9.232).

In middle lowland California, coccinellids are extremely important in biological control of alfalfa aphids, particularly in spring-time. This knowledge was used effectively in the development of an integrated control of alfalfa aphids, and was largely responsible for chemical treatments in spring-time being abandoned.

## 9.12 Factors determining the impact of coccinellids

### 9.121 Theoretical evaluation of coccinellids as predators

Efficiency of a natural enemy as a controlling agent is determined by a variety of characteristics (Bombosch 1963, DeBach 1964, van Emden in Hodek 1966 p. 227, van Emden et al. 1969, Hodek et al. 1966, Hodek 1967). Some of these characteristics appear to indicate that *Coccinellidae* might be effective predators.

The searching capacity of coccinellids in general is quite high (see 6.31), although the adults of some species tend to leave the habitat if the prey density falls below a critical level (see 9.124). The random searching behaviour of coccinellids is an advantage e. g. in the critical period of the arrival of aphids on the secondary host-plants; at the beginning of an infestation the coccinellids will find isolated aphids, to which other predators which orientate to the smell of aphid colonies would not yet be attracted. Coccinellids occupy all the habitats and niches of their prey (see 5.3) and the similarity in taxes (phototaxis, geotaxis) of coccinellids and their prey cause them to occur in the same places (see 6.31).

There is controversy about the relative controlling efficiency of specific and general predators. Whilst specific natural enemies have provided the most successful examples of control using introduced species, a specific predator is more endangered by prey scarcity than is a more general teeder. Furthermore, it the better utilization of indigenous predators is considered, it seems probable that best results will come from the manipulation of the complex of non-specific or relatively non-specific species of which coccinellids are often outstanding. Coccinellids have as a valuable characteristic a degree of host specificity connected with dependence on certain aphid species for breeding (see 6.11), which is associated with considerable polyphagy in terms of alternative food for the adults (see 6.111 and 6.12). This is a very advantageous characteristic, as far as such ephemeral prey as aphids are concerned. Coccinellids are very voracious (particularly the fourth instar larvae) and their voracity or at least the number of prey killed increases with an increase in prey density (see 6.22). They are at the same time able to adjust their rate of development and adult weight to the abundance of food (see 6.23). Coccinellids can survive periods of scarcity of food by such adjustment and by feeding on alternative food (see 6.111) or by cannibalism (see 6.13).

Coccinellids are amenable to culture (see 9.2), although with some difficulties, particularly in the case of the aphidophagous species. If satisfactory artificial foods can be developed, then the demands on space and labour, which at present limit the amount of culturing possible, will be much reduced. The fact that both larvae and adults of coccinellids are entomophagous is of paramount importance for their impact. It enables them to attack immediately on arrival; some other predators, e. g. *Syrphidae*, are incapable of this.

As regards the rate of increase of coccinellids, fecundity is generally fairly high, and both developmental rate and number of generations (as compared to these characteristics of their prey) are favourable in the case of coccidophagous species (particularly in warmer regions), but unfavourable in aphidophagous ones.

#### 9.122 The phases of prey population, with respect to their sensitivity or insensitivity to coccinellid impact

This section is a simplified generalization of a series of events, which of course vary with different prey, different habitats and climatic conditions. In the relationship between coccinellids and coccids in the subtropics (or tropics), such phases are not discrete. The section is, however, vitally relevant to the relationship between coccinellids and aphids in temperate regions. To facilitate the argument, Smith's (in Hodek 1966, p. 285) division of short term population dynamics of aphids is used with slight modifications.

(A) Initiation phase. — Often it is during this period, when the aphid population establishes itself in a new environment and begins to increase, that its future fate is determined. It has been observed with *Aphis fabae* on secondary host-plants (Way et al. 1954, Hodek et al. 1962, 1966) that a check of the population can occur in this phase. The coincidence of predator and prey in both time and space is the most critical element at the very beginning of this phase (Smith in Hodek 1966 p. 285), and therefore the vicinity of hibernation quarters of *Coccinellidae* to the crops is of special importance (Hodek 1964, Skuhravý and Novák 1966).

As aphidophagous coccinellids are very mobile, a certain aphid population level will be necessary to keep the predator on the crop in this phase. This crucial factor has largely been neglected and has been evaluated only for *Scymnus minimus* (Kuenen 1947), for *Stethorus punctillum* (Putman 1955), for *Hippodamia convergens* and *Olla abdominalis* (Sluss and Hagen in Hodek 1966, p. 243), and for coccinellids on sugar-beet fields (Skuhravý and Novák 1966). On carrots (Dunn 1965), the abundance of coccinellids does not seem to be correlated with the population density of *Cavariella aegopodii*.

Skuhravý and Novák (1966) found that. at the beginning of the immigration to sugar-beet fields, the coccinellids flew to all fields, regardlesss of the extent of aphid infestation of these fields. The coccinellids only remained, however, on fields with a sufficient aphid density — 100 or more aphids per 100 plants were sufficient to maintain a non-declining population of coccinellids.

(B) First increase. — If the aphid population is not checked at the very beginning, its increase is then affected mainly by the balance between the rate of reproduction of the aphids and the voracity of the aphidophagous insects (van Emden in Hodek 1966, p. 227); both are governed by weather conditions (Hodek et al. 1965). Dunn (1952) found that only at constant temperatures above 51—52 °F (= 11 °C) can one *Coccinella septempunctata* adult consume more than the number of nymphs produced by one *Acyrthosiphum pisi* mother. Some field confirmation of a similar differential effect of temperature on the reproductive capacity of *Aphis fabae* and on the preying efficiency of *Coccinella septempunctata* adults was obtained in field cage experiments on sugar-beet (Hodek et al. 1965). In the warmer season of 1964 the impact of coccinellids was more than twice as great as in 1963. In 1963 the coccinellids failed to prevent overpopulation by aphids in all cages where the ratio of aphids to beetles was 90: 1 or greater; in 1964, however, the coccinellids destroyed the population of aphids completely within 10 days, even in cages with as many as 200 aphids per beetle (Fig. 9.02). Sometimes it happens that a catastrophic weather event checks the increase of aphids.

(C) Peak phase. — When the population increase of aphids passes a certain limit, further activity of natural enemies makes no contribution to economic impact on that crop, even if the enemies invade in large numbers from neighbouring habitats (Szalay —Marzsó 1957, Way and Banks 1958, Hodek et al. 1966). At worst, the aphidophagous insects may in fact even benefit the aphids if they remove sufficient individuals to decrease intraspecific competition (Way, in Hodek 1966, p. 283). At best, the impact of natural enemies shortens the time during which aphids remain at peak density (Sluss and Hagen, in Hodek 1966, p. 243). From the point of view of integrated control, however, this phase is very significant. Chemical treatment should not be carried out as

(1) it can no longer prevent yield loss;
(2) the outbreak will anyway cease spontaneously;
(3) the natural enemies are then usually present in masses and should not be destroyed (see 9.231), but preserved for later phases of the season and for next year (*Aphis fabae* — Hodek et al. 1966).

Similar conclusions were reached by Sluss and Hagen (in Hodek 1966, p. 243) for *Chromaphis juglandicola*.

(D) Crash. — The crash of the prey outbreak is brought about by the complex action of several factors: meteorological conditions (particularly high temperature), changes in the physiological condition of the host plant, mycosis, action of natural enemies. With regard to the last mentioned, the importance of syrphids and parasites increases in this and the previous phase (C), unless the parasites were heavily hyperparasitized earlier. If the numbers of aphids are reduced to very low levels, the coccinellids (and other predators) do not remain in the habitat, and the aphid populations are allowed to resurge later in the season (Sluss and Hagen, in Hodek 1966, p. 243).

(E) Scarcity or absence of aphids. — If after the crash the aphids on a given crop still maintain small population fluctuations, the coccinellids may stay there until the onset of dormancy and produce a further generation, then accumulating enough reserves for overwintering. During the last two phases (D and E) the survival ability of the predator is of importance, and this depends on the relation of the population density of prey present to the minimum required by the predator. The first larval instar is limiting in such a situation (see 6.31). If all food has disappeared, cannibalism becomes relatively advantageous in that it makes survival possible for at least a part of the population. However, the feeding of old larvae on soft pupae or newly emerged adults has no adaptive advantage (see 6.13).

Very often the aphids disappear completely from the habitat and the coccinellids search for them in other habitats (see 5.32). Usually they find the prey in humid places and on weeds. Where they do not find their essential prey, extensive polyphagy with respect to alternative food (including even different types of phytophagy, see 6.111 and 6.12) increases survival. The last two phases are very critical for the population dynamics of coccinellids and attention should be paid to these phases when searching for ways of utilizing coccinellids fully. During phase (E) coccinellids may be conserved by the provision of alternative, even artificial, food (see 9.233).

## 9.13 Methods for studying the effectiveness of Coccinellidae[1]

The scarcity of real evidence on the impact of natural enemies (Hagen and van den Bosch 1968) arises from the fact that most conclusions have been based on correlations of data obtained by indirect methods (9.131). Wherever possible, direct (9.133) or semi-direct methods (9.132) should be used in preference, in conjunction with long-term ecosystemic studies. A survey of all such methods has been given in detail (Hodek et al. 1972) and is therefore only brief here.

### 9.131 Indirect methods

At the beginning of this century the collapse of high aphid populations was often attributed to predation by coccinellids, as these were observed in large numbers on the crop at the aphid peak. Unfortunately, observations or even counts of aphids and their natural enemies may merely demonstrate a coincidence of events and they do not prove causal relationships. However, such relationships can be established from a combination of more refined sampling procedures and a detailed knowledge of the biology of the insects concerned. For example, the losses of aphids by emigration can be determined from the proportion of alate nymphs in successive samples. Losses from disease and parasites can also be accurately assessed from sample data.

In order to interpret the population dynamics of prey and predator from sampling data realistically, auxiliary studies on essential biological, ethological and ecological characteristics are required for the predator: essential and alternative prey (see 6.113), prey-searching behaviour (see 6.31), voracity (see 6.2), induction of diapause affecting voltinism (see 7.42), other components of the rate of increase, such as fecundity, rate of development (see 4.12) and longevity (see 4.2), and the impact of natural enemies of the predator (see 8).

The weak point of indirect methods persists through even accurate sampling: causal relationships are being deduced by hindsight from a coincidence of events. The many factors (extremes of weather conditions such as heavy rain or very high temperature,

[1]) A comprehensive account of all sampling methods is given by Southwood (1966). Many methods specifically developed for studying the effectiveness of natural enemies are discussed by DeBach (1964).

physiological state of the plant, fungal attacks, emigration of alatae etc.) affecting aphids interrelate in a very complex manner, leaving an intuitive hypothesis as the only course open to the investigator. Sampling can, however, estimate the ratio prey/natural enemies and if we know the critical ratio, it will help us in assessing the feasibility of releasing coccinellids at a particular time, mostly in phase (A) (see 9.122).

Two types of sampling are most used. (1) Numbers on plants may be counted directly in the field, or after transfer of a sample to the laboratory. In both cases some mobile predators may escape. (2) Sweeping on herbaceous plants or beating off trees and shrubs are simple quick methods for large areas but being affected by physical conditions and by the height of plants (Dumas et al. 1964) they give differential results even for related species. Interception traps (mainly special sticky traps) which catch insects more or less at random are important for detecting predator mobility (Ewert and Chiang, in Hodek 1966, p. 195).

The size of populations may be expressed as numbers, biomass (weight) (Szalay-Marzsó 1958), or energy content. Graphic methods facilitate understanding of prey/predator relationships. Abundance may be simply plotted arithmetically against a time scale. When comparing several curves, the time scale goes from front to rear and the abscissa indicates the origins of separate curves (Fig. 9.02). Correlations can be expressed directly as a function by a plot of number of prey against number of predators without any time scale. A log scale can be used for a geometrically increasing population (Hughes 1963) (Fig. 9.03). A useful illustration technique has been used by van Emden (in Hodek 1966, p. 227). In two horizontal axes he plotted the number of prey against the number of predators with % of prey eaten on the vertical axis (Fig. 9.04). The horizontal plot was joined up in sequence of time.

A very promising indirect approach is to estimate predator efficiency on the basis of simplified models (Bombosch 1963, Hughes 1963, van Emden in Hodek 1966, p. 227).

### 9.132 Semi-direct methods

Some methods combine observation or sampling with experimentation. van Emden's (1963) method of inspecting marked aphid colonies of the fairly sessile *Brevicoryne brassicae* was developed to measure mortality factors directly in the field; thus the correlation of prevailing weather conditions and the appearance and abundance of predators with changes in aphid numbers is not assumed vaguely a posteriori, as it inevitably is with indirect methods. Similarly, the influence of predators on marked colonies of the more or less sessile *Adelges piceae* was followed by Karafiat and Franz (1956) and Franz (1958).

An important obstacle in evaluating the role of natural enemies from counts of prey and predator is the changing physiological state of the host-plant. This variable has been standardized by sampling predators on trap plants infested in the laboratory with a constant number of prey (Müller 1966).

### 9.133 Direct, i. e. experimental methods

Logically, the most convincing test of predator impact would be a comparison of similar host populations with and without natural enemies.

The idea itself is obvious, but actual experiments are technically difficult. Natural enemies can be killed (insecticides), removed (hand) or excluded (cages). All of these

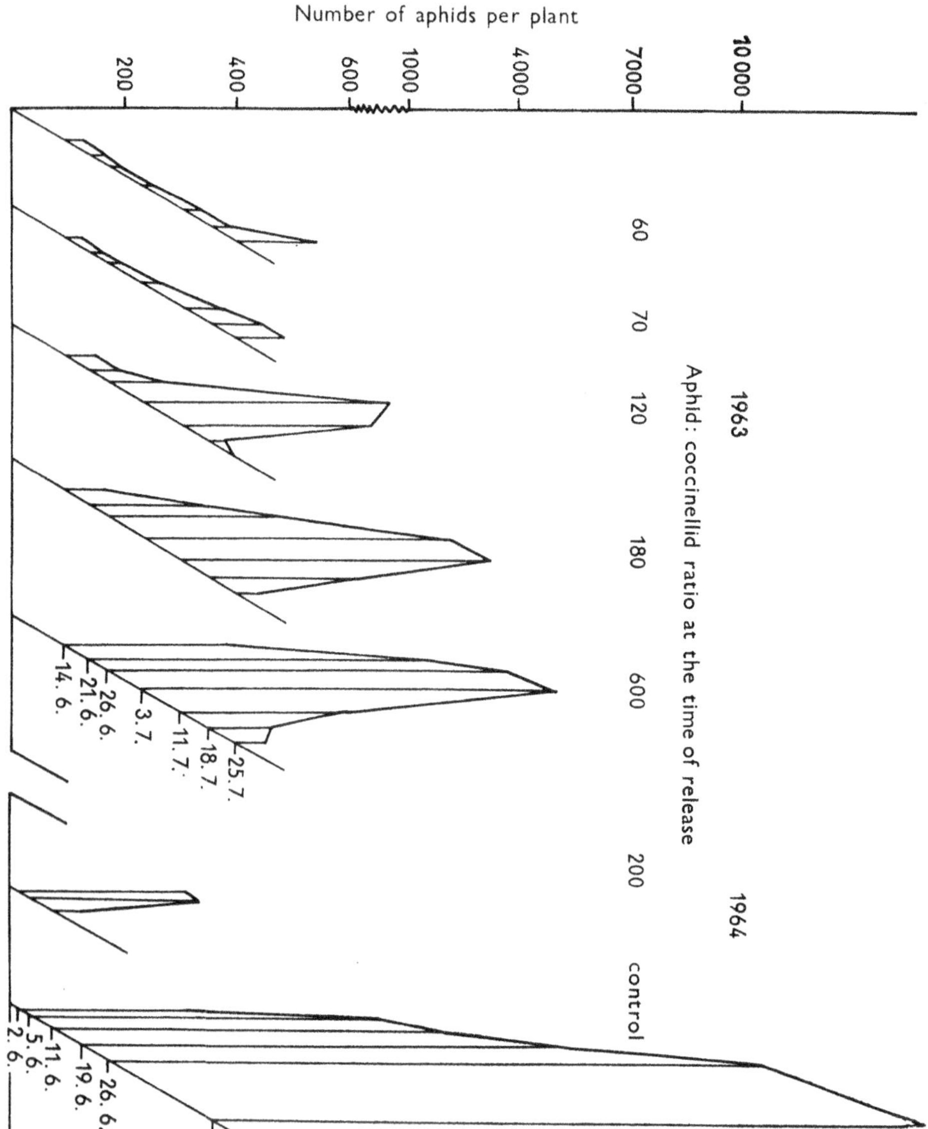
Number of aphids per plant
10 000
7000
4000
1000
600
400
200
1963
1964
Aphid: coccinellid ratio at the time of release
60
70
120
180
600
200
control
14. 6.
21. 6.
26. 6.
3. 7.
11. 7.
18. 7.
25. 7.
2. 6.
5. 6.
11. 6.
19. 6.
26. 6.
19. 7.

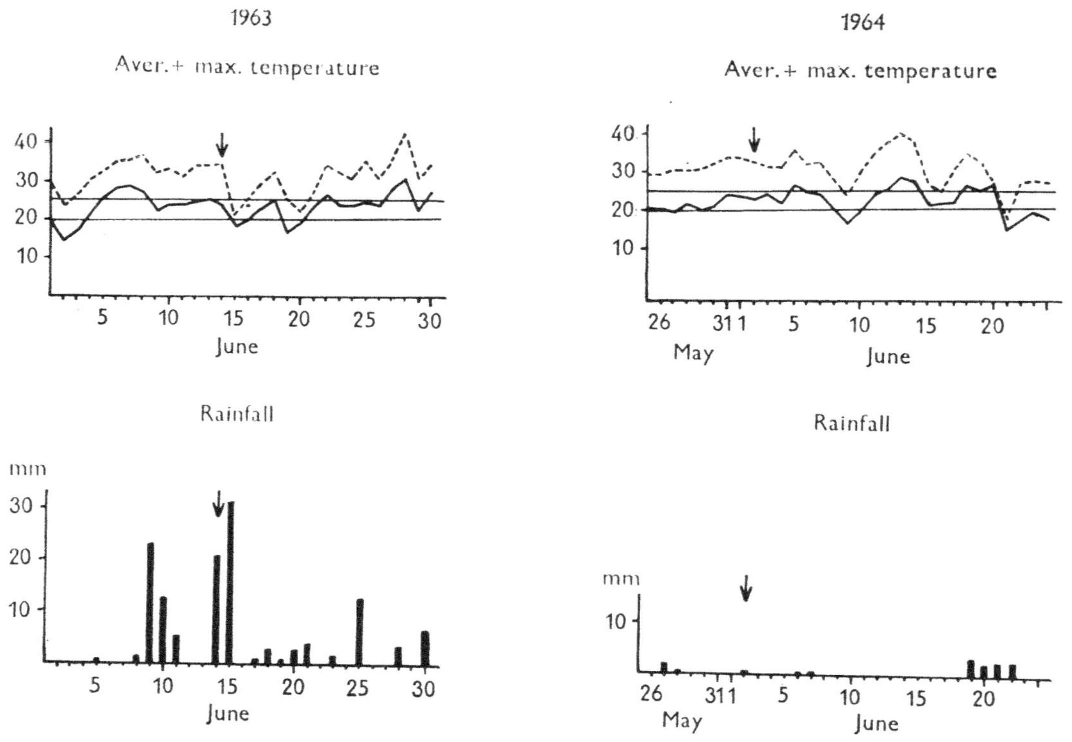


**Fig. 9.02** Impact of *Coccinella septempunctata* on *Aphis fabae* on sugar-beet inside nylon cages (Hodek et al. 1965). The arrow shows the date of coccinellid release.

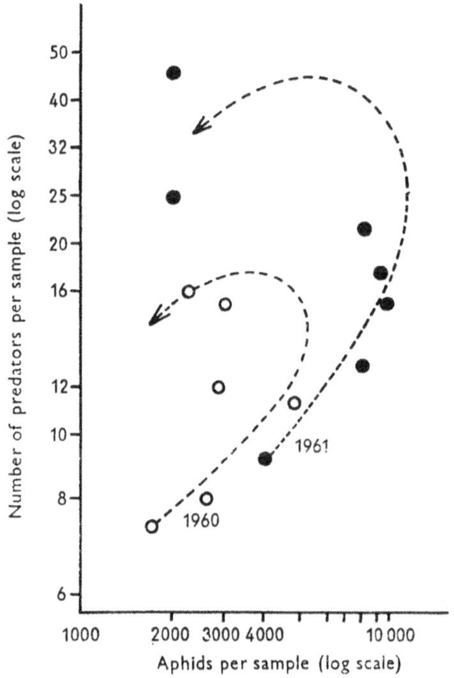


Fig. 9.03 Time trends in the relation between number of aphids and number of predator larvae showing the lag caused (Hughes 1963).

Fig. 9.04 The efficiency of predator action in a field infestation of *Brevicoryne brassicae* (calculated from van Emden 1964, 1965) (van Emden in Hodek 1966, p. 227).

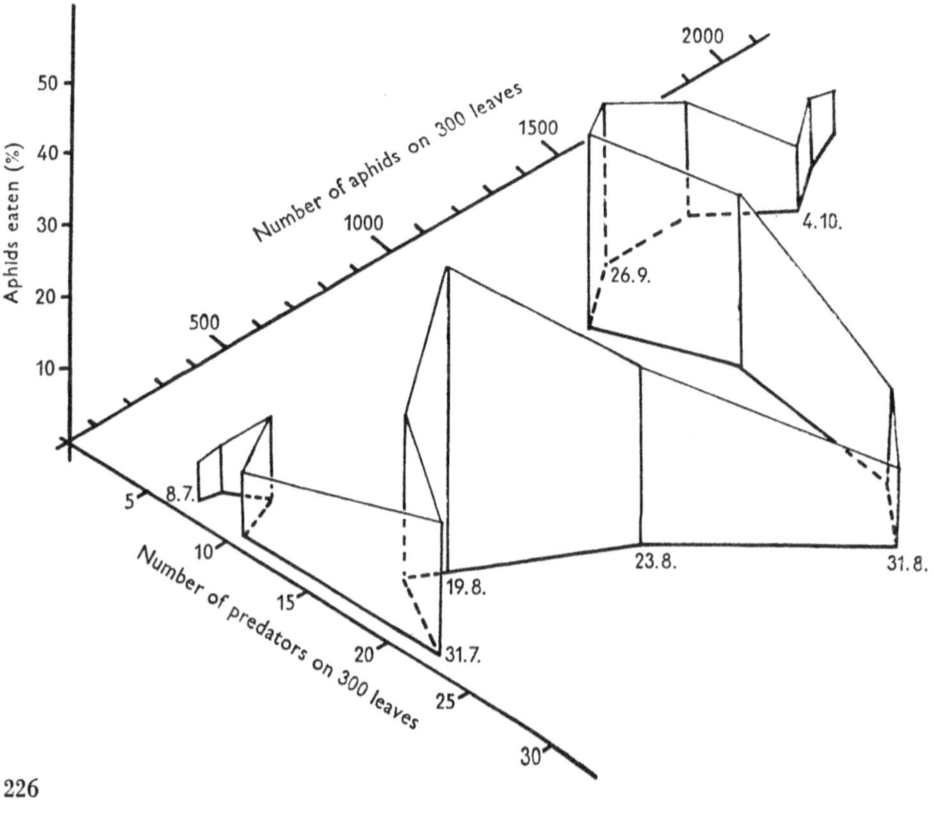

techniques have disadvantages. Chemical exclusion (DeBach 1946, 1964) needs a selective insecticide in absolute terms, i. e. it kills the predator but leaves the prey unaffected. Unfortunately, although DDT has been widely used for the purpose and Sevin has also been found satisfactory (Meier in Hodek 1966, p. 323), insecticides are rarely, if ever, absolutely selective. Also, insecticides must be sufficiently residual to kill immigrating predators. Both these limitations result in a certain underestimation of the role of natural enemies.

Hand picking of natural enemies can be only recommended for calibration of other methods (Fleschner 1958, Matsumoto and Nishida 1966), because it is very tedious. There is also a risk of damaging the host-plant and disturbing the prey.

While both hand removal and insecticides require constant repetition, cages ensure permanent exclusion. They have, however, two important disadvantages: the microclimate inside the cage is changed and the natural emigration of the prey is prevented. Both disadvantages can partly be overcome by adding control cages in the form of a "double cage" (Sparks et al. 1966) or slatted wooden framework (Way and Banks 1968). Both allow some movement of natural enemies and emigrants, and reproduce many of the climatic conditions of cloth-covered cages.

Numbers of natural enemies can also be varied experimentally by reducing them with ants ("biological check" of DeBach et al. 1949) or increasing them by adding artificial food (Schiefelbein and Chiang 1966, Smith in Hodek 1966, p. 19).

A known initial number of prey and natural enemies may be confined together in a simplified artificial system. Natural physical conditions can be retained (within the limitations discussed above) by using outdoor cages and thus the effect of differences between individual years can be studied (Hodek et al. 1965, see Fig. 9.02).

The effectivenes of natural enemies can be studied in a still more simplified system under controlled laboratory conditions. Only the essential "skeleton" of the prey/natural enemy relationship then remains. No long-term laboratory systems with coccinellids have yet been studied.

## 9.2 Use in biological control

The term biological control has two basic different meanings (DeBach 1964). (1) From the ecological viewpoint, biological control is defined as the action of parasites, predators, or pathogens in maintaining another organism's population density at a lower average than would occur in their absence. Man's manipulation of natural enemies is not implicit in this definition. (2) The same term is, however, more frequently used for "the study and utilization of parasites, predators and pathogens for the regulation of host population densities", i. e. exactly man's activity.

In the previous section (9.1) the biological control function of coccinellids in sense (1) has been discussed as well as the methods by which this activity can be studied. In the old mixed and scattered farming, which affected the ecosystem relatively little, such a spontaneous impact of natural enemies (including *Coccinellidae*) on pests was quite probable. Great acreages of monoculture, although creating ideal conditions for pests, deteriorate the environment for natural enemies, and greatly limit the possibility of their preventing the increase of the pest population above the economic level on their own accord. Because of the danger of the wide use of pesticides, the only sensible pest management is to gradually build a complex system of integrated controls (see 9 Introduction). An important constituent of these measures is the use of natural enemies, i. e. biological control in the sense (2) comprising three types of activity: importation (see 9.21), augmentation and conservation (see 9.22) (DeBach 1964).

## 9.21 Importation of exotic species

The classical and still most frequent approach to biological control is to import the natural enemy to an area where it is lacking. From the economic point of view it is an extremely efficient procedure: if after a successful establishment the natural enemy reduces the target host population, activity then continues without any further effort from man.

**Coccidophagous species.** — Most often the missing biotic agent is imported to follow the host which has been unintentionally introduced into a new area. This was the case with the introduction of *Rodolia cardinalis* which has still remained the most spectacular success story of biological control (DeBach 1964). In 1887, the infant citrus industry in California was threatened with destruction by massive infestations of *Icerya purchasi*. In December 1888 and January 1889 the first 129 specimens of *R. cardinalis* were sent from Australia. After propagation in the laboratory, 10,555 specimens were released in June 1889 and established successfully. In one year, the production of oranges increased about three times, and the cost of the project was a mere $ 1500! *R. cardinalis* was later introduced to other countries [DeBach (1964) listed 29 countries] and produced complete (or at least substantial) control of *I. purchasi* everywhere. It also succeeded in controlling the other species of this genus.

Even though the *Rodolia*/*Icerya* relationship is more favourable for control than may be encountered with some other coccidophagous coccinellids (e. g. *Cryptolaemus montrouzieri* must be periodically replenished — see 9.222), coccids in the tropics and subtropics seem on the whole to be very suitable subjects for biological control by coccinellids. In the table of "cases of biological control of pest insects by imported entomophagous insects" (DeBach 1964), 11 coccid species in addition to *Icerya spp.* are listed which were controlled by the coccinellids *Chilocorus cacti*, *C. distigma*, *C. nigritus*, *C. politus*, *Cryptognatha nodiceps*, *Cryptolaemus montrouzieri*, *Rhizobius ventralis*, *Rodolia pumilla* and *Telsimia nitida*.

Also the introduction of the coccinellid *Lindorus lophantae* to Italy, North Africa and to the Black Sea coast of the USSR resulted in successful control of coccids, particularly of the subfamily *Diaspidinae* (Yakhontov 1960). The European *Chilocorus bipustulatus* seems only tenuously established in California (Huffaker and Doutt 1965). *Orthezia insignis* attacking ornamentals in Kenya, has been substantially controlled by *Hyperaspis jocosa*, imported from Hawaii (DeBach 1964).

**Aphidophagous species.** — The 12 successful examples of coccid control by coccinellids contrast strikingly with only two examples available for aphids — *Aphis sacchari* on sugar cane in Hawaii (Pemberton 1948) and *Therioaphis maculata* on alfalfa in California — and even in both examples predators did no more than contribute to control. Moreover, the coccinellids preying on *T. maculata* are native species. The earliest known introduction of an insect for biological control is the importation of *Coccinella undecimpunctata* into New Zealand in 1874 for the control of aphids (Dumbleton 1936, Read 1965). The species is now generally distributed throughout the country, but its impact has been curtailed by the inadvertent introduction of its parasite (Dumbleton 1936).

*Aphidecta obliterata* was imported from Germany to North Carolina against *Adelges piceae*, the cause of considerable damage to *Abies spp.* The aphid was accidentally introduced into North America in about 1900 and was discovered in N. Carolina in 1957. The coccinellid was released in 1960 and 1963. After four years of survival, it has been reported as a permanent addition to the predator complex (Amman 1966).

The Palaearctic *Coccinella septempunctata* is reared for repeated mass releases (see 9.222) against potato aphids in Maine (U.S.A.).

Sometimes instances of inadvertent establishment are reported without a known introduction, or instances of belated establishment many years after seemingly unsuccessful importation. Inadvertent establishment is reported by Putman (1955): the native species in Ontario, *Stethorus punctum*, has become replaced by the Palaearctic *S. punctillum*. It would certainly be extremely valuable for the theory of introductions to study such a process as it occurs; it is a pity that this was neglected. *S. punctillum* began to invade North America both from east and west, in both cases probably after accidental importation.

An example of belated establishment concerns *Cryptolaemus montrouzieri*, which is very well established round Bangalore (Puttarudriah et al. 1952), although it had hitherto remained unnoticed. The only reference to importation is in the list of introductions in Washington, D. C.: "imported into India in 1898, not established".

**Recent attempts at introduction.** — Attempts to control the coccid *Parlatoria blanchardi* infesting the date palm in the region of Adrar in Mauretania (Iperti and Laudého 1968, Iperti and Brun 1969, Iperti and Laudého 1969), have commenced with the importation of four species of coccinellids: *Pharoscymnus ovoideus* (from Iran), *Chilocorus bipustulatus*[1]) (from Iran), *C. stigma* (from California) and *C. distigma* (from Senegal). These coccinellids are propagated in the Valbonne quarantine laboratory in the Station de Lutte Biologique at Antibes. They are reared on *Pseudaulacaspis pentagona* on potatoes and on three coccids on a gourd (pastèque): *Quadraspidiotus perniciosus*, *Chrysomphalus ficus*, and *C. dictyospermi*, and they are sent regularly to Adrar. After release, *C. bipustulatus* (from Iran) reproduces successfully and the infestation by *P. blanchardi* decreases. The dry, hot period of the year, however, is unfavourable for the permanent establishment of *C. bipustulatus*.

In Central Asia, Savoïskaya (1970b) is attempting to introduce four coccinellid species to the Alma-Ata area. Two of the species come from the Far East (*Aiolocaria mirabilis* and *Harmonia axyridis*) and two come from southern Kazakhstan (*Calvia punctata* and *Exochomus undulatus*). *C. punctata* and *H. axyridis* are aphidophagous, *E. undulatus* is coccidophagous (but can also be fed on aphids), and *A. mirabilis* is being introduced against the chrysomelid *Melasoma populi*, a pest of poplars. Both the Far Eastern species have already been released (*H. axyridis* for 3 years), but it would be premature to say whether or not they have successfully established.

**General considerations.** — Although a long series of successful introductions of various insect species has been made, we still do not understand the conditions which determine the success of establishment of a species in a new area. DeBach (1964) emphasizes that the only valid generalization from "more than 220 cases involving about 110 species of pests being controlled to a greater or less degree by natural enemies" is: "Over a period of time, the number of successes attained will be proportional to the amount of research and importation work carried out". All other hypotheses which have been put forward based on the accumulated evidence are contradicted by exceptions, so that any easy advice may deter potentially successful projects. Wilson (1971) also itresses that there can be no advance certainty of a successful outcome of any particular smportation. The importation of natural enemies is thus still doomed to the method of trial and error. This, however, does not imply that researchers should merely rely on luck. The probability of success is much greater, if natural enemy importations are based on deductions from main ecological principles. Thus it is, for instance, advisable

[1]) There is an essential difference in the beetles from the Near East and Europe in their biological characteristics. Up to now it is not clear, whether just biotypes or ssp. are involved (Iperti and Brun 1969).

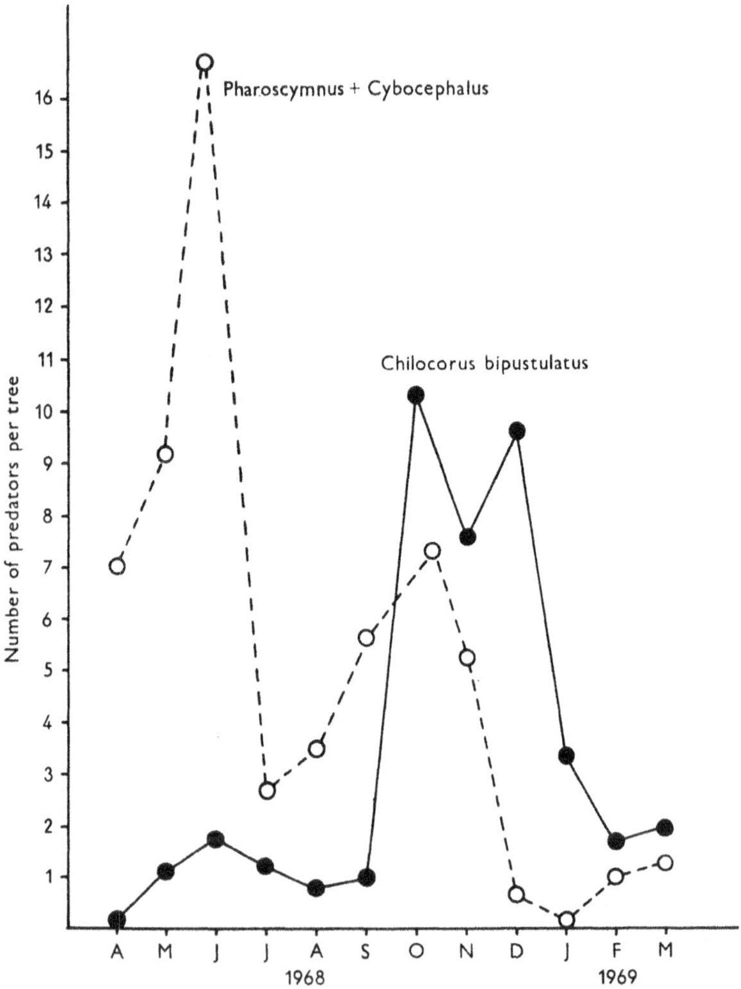


Fig. 905 Occurrence of the introduced *Chilocorus bipustulatus* var. *iranensis* and of endemic predators (*Pharoscymnus anchorago* and *Cybocephalus sp.)* on date palms in Mauretania (Laudého et al. 1970). (Larvae and adults are summed.)

to import insects from areas which are ecologically similar to the region where they are to be introduced. Also genetic studies may help in making decisions about importations. The genetic structure of the source population may differ in marginal and central areas of the species' distribution (Remington 1968). Ecologically marginal populations are small inbred units with high homozygosity and are closely adapted to a narrow range of conditions, while the ecologically central populations are numerous outcrossed units with high heterozygosity and tend to be adapted to a wide range of environments. As these assumptions by Remington seem inconsistent with his own conclusions (esp. his table), the criticism by Lucas (1969) should also be consulted. The best chance of establishment appears to be the importation of a large sample of the ecologically central population.

## 9.22 Conservation and augmentation

This involves two phases — first, the maintenance of existing natural enemies by avoiding harmful practices, such as use of inappropriate pesticides, and secondly the augmentation of natural enemy populations either directly by releasing them in the environment or indirectly by making the environment more favourable for them.

### 9.221 Selective breeding

This method has largely been neglected as regards coccinellids. Yakhontov (1957) has succeeded in raising the viability of coccinellids by inducing heterosis through intraspecific cross-breeding of individuals from different climatic regions. This type of breeding does not appear to have large-scale application, as the heterotic effect disappears in the $F_2$-generation.

### 9.222 Colonization in the open

More often *Coccinellidae* have been used for periodic colonizations. In several cases the coccinellids for release have been produced by laboratory mass rearing.

*Cryptolaemus montrouzieri* was introduced into California in 1892 by Koebele against *Pseudococcus spp.* (mealybugs) infesting citrus. Permanent establishment was achieved only along the coast of southern California; in other areas, *C. montrouzieri* could not survive the winter. Even in the narrow zone of permanent establishment, the coccinellid failed to reduce the mealybug populations in time to prevent damage. The pest can be, however, kept under control by periodic re-colonizations of insectary-cultured *C. montrouzieri* released together with the encyrtid *Leptomastix dactylopii*. The coccinellid is cultured on mealybugs on potato sprouts. About 20 adults per tree are released, and each tree receives 1—5 releases each year. The production of 30 million beetles in 1960 cost 17,000 $. The annual cost for the co-operating growers in whose groves *Cryptolaemus* and *Leptomastix* are periodically released is about 10 $ per acre (Fischer 1963, DeBach 1964, p. 441).

Having proved successful in California, *C. montrouzieri* was subsequently imported into several subtropical regions. It has also been used with success on the Black Sea coast of the USSR (Telenga 1948, Rubtsov 1954, Yakhontov 1960).

The Palaearctic *Coccinella septempunctata* has been imported to the USA against aphids on potatoes with the aim of periodic re-colonizations, and is propagated in the laboratory at Maine (Shands et al. 1966, 1970). The recently described method for producing eggs seems to be sufficiently low in cost to be economic and to enable large-scale colonizations. One person can tend and collect the daily eggs from 2800 beetles in 112 cages. If each of ca. 1800 females deposited 25 eggs per day, one technician would then collect 315,000 eggs per week. So far 64 % of the possible production has been achieved, the major limiting factor being diapause. The mass-production technique involves two ingenious improvements. Gelatin-coated, A-fluted, single-phase corrugated cardboard serves as the oviposition substrate, and the eggs can be readily removed by dissolving the gelatin in lukewarm water. The separated eggs are sprayed in water in the field. One piece of corrugated cardboard is laid onto another with the flutes touching, and thus forming a multicelled sandwich. The sandwich appears to attract gravid females and shelters the eggs from cannibalism. Cannibalism is also decreased by a 5 cm layer of fluffy excelsior in the oviposition cages which provides increased room for the movement of non-ovipositing beetles.

In most cases of periodic re-colonization, field-collected coccinellids have been used. Between 1908 and 1914 huge numbers of *Hippodamia convergens* adults were collected from winter aggregations in the Sierra Nevada mountains of California and distributed to cantaloupe growers in the Imperial Valley for the control of melon aphid, *Aphis gossypii* (Carness 1912a, b). Such transfers, however, did not bring about control because the beetles dispersed rapidly. Scarcely any of the marked beetles could be found after three days in the area where they had been released (Hagen 1962, DeBach 1964). Recently, Hagen (in DeBach 1964, p. 444) has observed that feeding synthetic diets to *H. convergens* adults collected from aggregations somehow reduces this tendency to disperse.

Cooke (1963) made 4 massive releases (40,000 to 150,000 of beetles per release) of *H. convergens* against *Acyrthosiphon pisum* on alfalfa fields in the Blue Mountain area, and obtained some reduction in aphid population. However, the reduction was uneconomic in relation to the cost of the releases. Within 4—5 days the coccinellids left the fields with equal rapidity, whether or not they had previously been fed (for a couple of days on a formula containing honey and protein hydrolysates). Iperti (pers. comm. 1963) and Savoïskaya (in Hodek 1966, p. 318) reduced the dispersion of coccinellids after release by the diametrically opposite approach, i. e. by starving the coccinellids.

In the Georgian SSR, Bogdanova (1956) has achieved control of *Chloropulvinaria floccifera* by transferring *Hyperaspis campestris* from tree habitats to tea plantations, where it has then established. In south-eastern Kazakhstan, the release of field-collected *Adalia bipunctata* and *Coccinella septempunctata* (Savoïskaya, in Hodek 1966, p. 318) has reduced aphid infestation of orchards.

Coccinellids have even been released successfully against their alternative food. Yakhontov (1960) reported economic impact of the coccinellids *Brumus octosignatus* and *Semiadalia undecimnotata* against their alternative prey, the weevil *Phytonomus variabilis* on alfalfa in Central Asia (Tab. 9.01). The author evaluated the control achieved as incomplete, although crop yield was increased.

Tab. 9.01 Control of alfafa weevil *Phytonomus variabilis* after release of *Coccinelidae* in Central Asia (Yakhontov 1960)

| Coccinellid species released | *Brumus octosignatus* | *Semiadalia undecimnotata* |
|---|---|---|
| Number of weevil larvae per coccinellid | 20—40 | 20—80 |
| Increase of dry alfalfa yield [kg/hectare] compared with control plots | by 60—270 | by 160—660 |

Some authors make recommendations on colonization by coccinellids without having attempted it by themselves. Dosse (1957) advises the release of young larvae of *Stethorus gilvifrons* at the beginning of the population increase of tetranychid mites on apple trees in the Lebanon, and Sundby (1968) in Norway similarly recommends the release of *C. septempunctata* against aphids in the fields early in the season.

### 9.223 Use in glass-houses

A special case of augmentative manipulation of coccinellids is their use in the closed environment of the glass-house. *Cryptolaemus montrouzieri* has been released with success in glass-houses for the control of mealybugs on flower crops (DeBach 1964, p. 443). In one case, where *C. montrouzieri* was not effective because of low temperature, *Exochomus flavipes* was liberated along with *Anagyrus kivuensis* and *Chrysopa plorabunda*, and this gave effective control of *Pseudococcus citri* (Doutt 1951).

Gurney and Hussey (1970) tested 4 species of coccinellids with the aim of using them for biological control of aphids on cucumbers and chrysanthemums in glasshouses, and arrived at somewhat discouraging conclusions. They apparently also found it too laborious to be economical to rear coccinellids on an aphid diet. It is possible, however, that other species may be more suitable than the species tested. These included *Coleomegilla maculata* which is preferentially pollinivorous (see 6.12), and *Coccinella septempunctata*, which had been collected in the hibernation sites in the Himalayas while in diapause, and therefore was possibly not in the most suitable life-phase.

It is likely that coccinellid species which are small in size and also have an adequate range of optimum temperature will be better suited to breed in glass-houses. The promising results of preliminary tests with *Adonia variegata* in France (Jourdheuil, pers. comm.) seem to support this suggestion.

### 9.224 Sensible use of pesticides

The application of insecticides has often caused outbreaks of new or old pests previously kept in control by introduced natural enemies. An example was the resurgence of *Icerya purchasi* following destruction of *Rodolia cardinalis* by DDT and malathion (DeBach 1947, Bartlett and Legace 1960).

Natural enemies can to a large extent be conserved by supervised use of chemical control. The ecologically damaging zeal for exterminating pests must be completely abandoned. Pesticides should never be applied as a routine preventative measure; if their application is inevitable, selective compounds should be used at an appropriate time. All aspects of this subject have been discussed by Bartlett (in DeBach 1964, p. p. 489) and Newsom (1967). Bartlett has also tabulated pesticides in respect of their toxicity to natural enemies. In this list, the pesticides least toxic to coccinellids are: butylpheroxyisopropyl chloroethyl sulphite, captan, chlorfenson, chlorobenzilate, bis (4-chlorophenoxy) methane, 1,1-dichloro-2,2-bis (4-ethylphenyl) ethane, 2,4-dichlorophenyl benzenesulphonate, dicofol, dinocap, ferbam, gamma-BHC, glyodin, ryania, sabadilla, schradan, tetradifon, trichlorphon, zineb.[1]) Comparative studies on the sensitivity of coccinellids to pesticides have been carried out by many authors; the most recent are: Atallah and Newsom 1966b, Bartlett 1965, Campbell and Hutchins 1964, Cherry and Pless 1969, Hamilton and Kieckhefer 1968, 1969, Hukusima 1964, Kehat and Swirski 1964, Laster and Brazzel 1968, Lindgren and Ridgway 1967, Meier (in Hodek 1966, p. 323), Nagui 1969, Nohara et al. 1965, Takeda et al. 1965, Wiackowski and Dronka 1968, Wiackowski and Nowacka 1968, Yun and Ruppel 1964, Zelený (in Hodek 1966, p. 337), Zelený 1969.

In general, contact insecticides are highly toxic to coccinellids whereas systemic insecticides are highly selective or are made selective if they are applied via the soil. There is considerable variation in the resistance of individual coccinellid species. Usually larvae are more sensitive than pupae or adults. Some insecticidal compounds may be added to Bartlett's list of pesticides with low toxicity to coccinellids: thiometon (Zelený), demeton and menazon (Wiackowski and Nowacka), and metox (Wiackowski and Dronka). *Coleomegilla maculata* is resistant to DDT; the mechanism was studied by

[1]) In order to facilitate reference to the chemical compounds quoted, the standard "chemical names" (Martin 1968) have been used.

Atallah and Nettles (1966) and ascribed at least in part to the ability of this coccinellid to metabolize DDT to DDE (1,1 — dichloro — 2,2- bis/p-chlorophenyl/ ethylene).

Coccinellids can also be preserved by suitable timing of chemical applications. Hodek et al. (in Hodek 1966, p. 331) advise that aphicides are best applied to sugar-beet during the two weeks following the end of the main immigration of *A. fabae* from *Euonymus*. Savoïskaya (in Hodek 1966, p. 317) gives recommendations for the timing of the first two sprays in apple orchards in Alma-Ata. In general, chemical treatments should be avoided during the peak phase of the aphid infestation (see 9.122).

In apple orchards in Japan, Hukusima (1968, 1969) combined applications of selective fungicides and acaricides with the release of the coccinellids *Harmonia axyridis* and *Propylaea japonica* and the spider *Misumena tricuspidata*. The predators appeared to survive. If, however, aphicides must be applied at times when coccinellids should be preserved, intrinsically specific chemicals as nicotin, schradan, menazon or pirimicarb should be used (Way, pers. comm. 1971). Alternatively, specificity can be conferred on an otherwise non-specific chemical if it is applied in granular formulation on the foliage or is placed in the soil.

### 9.225 Modification of cultural practices

There have been many suggestions for promoting the effectiveness of coccinellids by cultural means. Destruction of the ground cover drives the beetles into orchard trees and has been accomplished by cutting the grass in south-eastern Kazakhstan orchards (Savoïskaya, in Hodek 1966, p. 317) and by cultivating the ground under walnut trees in California (Hagen and Sluss, in Hodek 1966, p. 245). Careful timing can establish a good beetle/aphid ratio in the trees. Adjacent parts of a crop may serve as refuges for coccinellids and enable them to re-invade strips treated with insecticide (Dyadechko 1953, Sem'yanov 1965c) or crops harvested in strips (Schlinger and Dietrick 1960). An example given by the latter authors shows the adverse effects of mowing the entire field of alfalfa at one time (regular farming) in contrast to mowing in alternate strips (strip-farming). Although the aphid occurred at similar densities under both systems early in the season, it rose to densities causing considerable damage in the regular-farmed field, but numbers became negligible in the strip-farmed field (Tab. 9.02). In addition to these numerical differences, several species of parasites which were effective in the strip-farmed field never appeared in the regular-farmed one.

Tab. 9.02 The effect of strip-farming and regular farming on the average number of natura enemies of *Therioaphis maculata* (Schlinger and Dietrick 1960)

| Group of natural enemies | Regular farming[1]) (thousands of individuals/acre) | Strip farming[1]) (thousands of individuals/acre] |
|---|---|---|
| Coccinellid adults | 46 | 205 |
| Coccinellid larvae | 11 | 232 |
| Chrysopid larvae | 195 | 206 |
| Hymenopterous parasites | 70 | 287 |
| „Big-eyed" *Heteroptera* | 199 | 401 |
| An aphidophagous spider | 105 | 1094 |
| Totals | 626 | 2435 |

[1]) June 1959, Brawley, California.

Burning old grass early in spring, which is practised in central Europe, is very detrimental to coccinellids. The burning is done on headlands, dykes and other habitats in arable land which are used by coccinellids as dormancy sites; many coccinellids are thus destroyed.

Tamaki and Halfhill (1968) have used defoliants to suppress *Myzus persicae* on peach trees. The aphids concentrated on the remaining foliage and were then easy prey for predators migrating into the orchards in autumn. The number of overwintering aphid eggs was thus reduced.

#### 9.226 Providing alternative food

A whole range of natural enemies suffer from a lack of pollen and nectar in pure monocultures. Pollen and nectar are particularly required by the adults of hymenopterous parasites and adult syrphids, but this food is also important for coccinellids when essential prey is missing (see 6.111). The artificial provision of certain flowering plants, which has been recommended for other natural enemies, would be valuable for coccinellids also.

It has been shown that populations of predators, including *Coccinellidae*, can also be increased by treating the crops with alternative artificial food (Schiefelbein and Chiang 1966, Smith 1965a, b). This could be used to attract the coccinellids to the required habitats especially at the time when the pest begins to occur; perhaps also after the end of the infestation in order to maintain a high number of coccinellids, mainly for the next season.

#### 9.227 Providing shelter

That adjacent shelter used by coccinellids for hibernation can produce a higher and usually earlier occurrence of the predator on the crop has been demonstrated for sugar-beet (Bombosch 1965), potatoes (Fenjves 1945, Galecka, in Hodek 1966, p. 255) and brussels sprouts (van Emden 1965a, b, c). In Czechoslovakia, it has been recommended (Hodek et al. 1962) that patches and strips of woodland should be planted to compensate for the amalgamation of small fields into large crop acreages.

In south-eastern France, Iperti (in Hodek 1966, p. 189) has used artificial traps which simulate rock cracks to effect a considerable reduction in the high *Beauveria* mortality which takes place in *Semiadalia undecimnotata* sheltering amid vegetation for hibernation. (Photo 50). These traps also facilitate the collection and transport of hibernating coccinellids for use in biological control. On the Black Sea coast in the Caucasus, large moderately heated cages placed around the trees have been used for overwintering *Rodolia cardinalis* and *Cryptolaemus montrouzieri*. The predators from such cages show a greater viability than those bred in the laboratory and the cost of the spring release is reduced (Sysoev 1958).

Nohara (1962) recommends that citrus branches be wrapped in cloth to provide artificial hibernation quarters for *Chilocorus kuwanae*. With the same aim — enabling the predators to overwinter — Tamaki and Weeks (1968) put black painted aluminium bands (lined with thick paper and tarred burlap) on the main branches of peach trees.

Such banding on trees may also benefit the predator in another way. Smith and Armitage (1931) found that bands on the trunks of citrus trees benefited *C. montrouzieri* by attracting both the prey (*Pseudococcus gahani*) and the coccinellid, thereby increasing the intensity of predation.

## 9.3 Conclusions

There is abundant evidence that coccinellids are often a major cause of mortality of aphids, coccids and spider mites. In regions where coccids are serious pests particular coccinellid species with a high degree of prey specificity may be paramount in the biological control of a pest (e. g. *Rodolia*) and there may be further opportunities for spectacular biological control through introductions of such coccinellids.

Coccinellids that prey on aphids are relatively less specific, and this characteristic, together with the aphids' ephemeral occurrence on particular host-plants, makes it likely that coccinellids will provide their contribution as a major component of the overall biological control complex in integrated control practices. Despite the very considerable amount of work already done on aphidophagous coccinellids much still needs to be done before they can be used to best advantage. In particular much more behavioural and ecological work is required of the type needed for population models in which the quantitative effects of the different natural enemies are separately determined. Such knowledge could then be used to manipulate those coccinellid species which play or could play a key role in depressing aphid numbers.

# References

Aleksidze G. N. 1970: Adalia bipunctata. Zashch. Rast., 1970: 12 (in Russian).
Allee W. C., Emerson A. E., Schmidt K. P., Park O., Park Th. 1949: The principles of animal ecology. 837 pp., Saunders, Philadelphia.
Allen D. C., Knight F. B., Foltz J. L. 1970: Invertebrate predators of the jack-pine budworm, Choristoneura pinus, in Michigan. Ann. ent. Soc. Am., 63: 59—64.
Amman G. D. 1966: Aphidecta obliterata (Coleoptera: Coccinellidae), an introduced predator of the balsam woolly aphid, Chermes piceae (Homoptera: Chermidae), established in North Carolina. J. econ. Ent., 59: 506—508.
Andrewartha H. G. 1952: Diapause in relation to the ecology of insects. Biol. Rev., 27: 50—107.
Andrewartha H. G., Birch L. C. 1954: The distribution and abundance of animals. 782 pp., Univ. Chicago Press.
Asgari A. 1966: Untersuchungen über die im Raum Stuttgart Hohenheim als wichtigste Prädatoren der grünen Apfelblattlaus (Aphidula pomi Deg.) auftretenden Arthropoden. Z. angew. Zool., 53: 35—93.
Atallah Y. H., Killebrew R. 1967: Ecological and nutritional studies on Coleomegilla maculata (Coleoptera: Coccinellidae). IV. Amino acid requirements of the adults determined by the use of $C^{14}$—labeled acetate. Ann. ent. Soc. Am., 60: 186—188.
Atallah Y. H., Nettles W. C. jr. 1966: DDT-metabolism and excretion in Coleomegilla maculata DeGeer. J. econ. Ent., 59: 560—564.
Atallah X. H., Newsom L. D. 1966a: Ecological and nutritional studies on Coleomegilla maculata DeGeer (Coleoptera: Coccinellidae). I. The development of an artificial diet and a laboratory rearing technique. J. econ. Ent., 59: 1173—1179.
Atallah Y. H., Newsom L. D. 1966b: Ecological and nutritional studies on Coleomegilla maculata DeGeer (Coleoptera: Coccinellidae). III. The effect of DDT, Toxaphene, and Endrin on the reproductive and survival potentials. J. econ. Ent., 59: 1181—1187.
Atwal A. S., Sethi S. L. 1963: Biochemical basis for the food preference of a predator beetle. Curr. Sci., 32: 511—512.
Avidov Z., Rosen D. 1965: Studies towards an integrated control programme of citrus pests in Israel. 12. Int. Congr. Ent. London, 1964, pp. 572—573.
De Bach P. 1946: An insecticidal check method for measuring the efficacy of entomophagous insects. J. econ. Ent., 39: 647—658.
De Bach P. 1947: Cottony-cushion scale, vedalia and DDT in central California. Calif. Citrogr., 32: 406—407.
De Bach P. (Ed.) 1964: Biological control of insect pests and weeds. 844 pp., Chapman & Hall Ltd., London.
De Bach P., Dietrick E. J., Fleschner C. A. 1949: A new technique for evaluating the efficiency of entomophagous insects in the field. J. econ. Ent., 42: 546.
Balachowsky A. 1930: L'extension de la cochenille australienne (Icerya purchasi Mask.) en France et de son prédateur Novius cardinalis Muls. Annls Epiphyt., 16: 1—24.
Balduf W. V. 1926: The bionomics of Dinocampus coccinellae Schrank. Ann. ent. Soc. Am., 19: 465—489.
Balduf W. V. 1935: The bionomics of entomophagous Coleoptera. (13. Coccinellidae — lady beetles). 220 pp., John S. Swift Co. Inc., Chicago, New York.
Banks C. J. 1954a: Random and non-random distributions of Coccinellidae. J. Soc. Br. Ent., 4:211—215.

Banks C. J. 1954b: The searching behaviour of coccinellid larvae. Br. J. Anim. Behav., 2: 37—38.
Banks C. J. 1955: An ecological study of Coccinellidae associated with Aphis fabae Scop. on Vicia faba. Bull. ent. Res., 46: 561—587.
Banks C. J. 1956: A second record of a tachinid (Dipt.) parasite bred from one of the Coccinellinae (Col., Coccinellidae). Entomologist's mon. Mag.. 92: 188.
Banks C. J. 1957: The behaviour of individual Coccinellid larvae on plants. Br. J. Anim. Behav., 5: 12—24.
Bänsch R. 1964: Vergleichende Untersuchungen zur Biologie und zum Beutefangverhalten aphidovorer Coccinelliden, Chrysopiden und Syrphiden. Zool. Jb. Syst., 91: 271—340.
Bartlett B. R. 1965: The repellent effects of some pesticides to hymenopterous parasites and coccinellid predators. J. econ. Ent., 58: 294—296.
Bartlett B. R., Legace C. F. 1960: Interference with the biological control of cottony-cushion scale by insecticides and attempts to re-establish a favourable natural balance. J. econ. Ent., 53: 1055—1058.
Beck S. D. 1968: Insect photoperiodism. 288 pp., Academic Press, New York — London.
Behrendt K. 1969: Über langjährige Massenwechselbeobachtungen an der Schwarzen Bohnenblattlaus, Aphis fabae Scopoli (Homoptera: Aphididae). TagBer., 10, Wanderversamm. dtsch. Ent.
Ben-Dov Y., Rosen D. 1969: Efficacy of natural enemies of the California red scale on citrus in Israel. J. econ. Ent., 62: 1057—1060.
Benham B. R., Muggleton J. 1970: Studies on the ecology of Coccinella undecimpunctata Linn. (Col. Coccinellidae). Entomologist, 153—170.
Benkevich V. I. 1958: Biology of Coccinella septempunctata. Uchen. Zap. Orekh. Zuev. pedag. Inst., 11: 127—133 (in Russian).
Benner, Franz, Ansorge 1905: Discussion. Z. Ent., 30.
Berker J. 1958: Die natürlichen Feinde der Tetranychiden (Stethorus punctillum). Z. angew. Ent., 43: 115—172.
Bielawski R. 1959: Biedronki -- Coccinellidae. Klucze Oznacz. Owad. Polski, 19, 76, 92 pp., Warszawa (in Polish).
Bielawski R. 1961: Die in einem Krautpflanzenverein und in einer Kieferschonung in Warszawa /Bielany auftretenden Coccinellidae (Coleoptera). Fragm. faun. (Warszawa), 8: 485—525.
Bielawski R., Klausnitzer B. 1970: Redescription von Chilocorus malasiae Cr. samt Beschreibung der Larvae (Col.. Coccinellidae). Bull. Acad. pol. Sci., Cl. 2, Sér. Sci. biol., 18: 343—345.
Binaghi G. 1941: Gli stadi preimaginali del Pullus auritus Thunb. e dello Scymnus rufipes Fabr. Memorie Soc. ent. ital., 20: 148—161.
Binaghi G. 1941: Larve e pupe di Chilocorini. Memorie Soc. ent. ital., 20: 19—36.
Blackman R. L. 1965: Studies on specificity in Coccinellidae. Ann. appl. Biol., 56: 336—338.
Blackman R. L. 1967a: The effects of different aphid foods on Adalia bipunctata L. and Coccinella 7-punctata L. Ann. appl. Biol., 59: 207—219.
Blackman R. L. 1967b: Selection of aphid prey by Adalia bipunctata L. and Coccinella 7-punctata L. Ann. appl. Biol., 59: 331—338.
Blattný C. 1925: Het voorspellen van het massaal optreden van schadelijke insekten. Tijdschr. PlZiekt., 31: 139—144.
Blattný C., Osvald C. V. 1949: Contribution à la connaisance des signes géneriques et de la biologie du Stethorus punctillum Weise. Věst. Čs. spol. zool., 13: 30—40 (in Czech, Engl. summ.).
Blunck H. 1914: Die Entwicklung des Dytiscus marginalis vom Ei bis zur Imago. Z. wiss. Zool., 111: 76—151.
Bodenheimer F. S. 1943: Studies on the life-history and ecology of Coccinellidae. I. The life-history of Coccinella 7-punctata L. in four different zoogeographical regions. Bull. Soc. Fouad I. Ent., 27: 1—28.
Bodenheimer F. S. 1951: Citrus entomology in the Middle East. 663 pp., Dr. W. Junk, The Hague.
Bodenheimer F. S., Neumark S. 1955: The Israel pine Matsucoccus (Matsucoccus josephi n. sp.). 122 pp., Kiryath Sepher Ltd, Jerusalem.
Bogdanova N. L. 1956: Hyperaspis campestris Herbst (Coleoptera, Coccinellidae) as destroyer of Chloropulvinaria floccifera Westw. (Homoptera, Coccidoidea). Ent. Obzor., 35: 311—322 (in Russian).
Bombosch S. 1963: Untersuchungen zur Vermehrung von Aphis fabae Scop. in Samenrübenbeständen unter besonderer Berücksichtigung der Schwebfliegen (Diptera, Syrphidae). Z. angew. Ent., 52: 105—141.
Bombosch S. 1965: Untersuchungen über die Disposition und Abundanz von Blattläusen und deren natürlichen Feinden. 12. Int. Congr. Ent., London 1964, pp. 578—580.

Bonnemaison L. 1964: Observations écologiques sur la Coccinelle à 7 points (Coccinella septempunctata L.) dans la région parisienne (Col.). Bull. Soc. ent. Fr., 69: 64—83.
Börner C., Heinze K. 1957: Aphidina. In P. Sorauer: Handbuch der Pflanzenkrankheiten, 5/4, Berlin—Hamburg.
Van den Bosch R., Schlinger E. I., Dietrick E. J., Hall I. M. 1959: The role of imported parasites in the biological control of the spotted alfalfa aphid in southern California in 1967. (Coccinellid activity). J. econ. Ent., 52: 142—154.
Böving A. 1917: A generic synopsis of the Coccinellid larvae in the United National Museum with a description of the larva of Hyperaspis binotata Say. Proc. U. S. natn. Mus., 51: 621—650.
Brassler K. 1930: Ist Coccinella septempunctata L. wirklich nur Blattlausfresser? Z. PflKrankh. PflPath. PflSchutz, 40: 511—513.
Brown H. D. 1969: The predacious Coccinellidae associated with the wheat aphid, Schizaphis graminum (Rond.) in the Orange Free State. D. Sc. Thesis, Univ. Stellenbosch, S. Africa.
Bryden J. W., Bishop M. W. H. 1945: Coccinella 7-punctata, parasitized by Perilitus coccinellae (Hym., Braconidae) in Cambridgeshire. Entomologist's mon. Mag., 81: 51—52.
Butt F. H. 1951: Feeding habitats and mechanism of the Mexican bean beetle. Cornell Univ. Agr. Exp. Sta., Mem. 306, 32 pp., Ithaca, New York.
Camargo F. 1937: Notas taxonomicas e biologicas sobre alguns Coccinellideos do genera Neocalvia Crotch, predadores de larvas de genero Psyllobora Chevrolat (Col. Coccinellidae). Revta Ent. Rio de J., 7: 362—377.
Campbell W. V., Hutchins R. E. 1964: Toxicity of insecticides to some predaceous insects on cotton. J. econ. Ent., 45: 786—789.
Capra F. 1947: Note sui Coccinellidi (Col.) III. La larva ed il regime pollinivoro di Bulaea lichatschovi Hummel. Mem. Soc. ent. ital., Suppl. 26: 80—86.
Carnes E. K. 1912a: Collecting ladybirds (Coccinellidae) by the ton. Mon. Bull. Calif. Commn Hort., 1: 71—81.
Carnes E. K. 1912b: An explanation of the hibernating habits of Hippodamia convergens. Mon. Bull. Calif. Commn Hort., 1: 177—188.
Cavalloro R. 1949: Contributo alla biologia dei "Coccinellidi predatori" nell Umbria con particolare riguardo a "Coccinella septempunctata L.". 26 pp., Tipografia Perugina già Santucci, Perugia.
Cherry E. T., Pless C. D. 1969: Bioassay of leaves from tobacco grown on soil treated with certain systematic insecticides. J. econ. Ent., 62: 1313—1316.
Chumakova B. M. 1962: Experiments in rearing of predatory beetle Cryptolaemus montrouzieri Muls. on an artificial diet. Zesz. probl. Postep Nauk roln., 35: 195—200.
Clausen C. P. 1940: Entomophagous insects. 688 pp., McGraw-Hill, New York-London.
Clements F. E., Shelford U. E. 1939: Bio-Ecology. 425 pp., Wiley Inc., New York.
Colburn R., Asquith D. 1970: A cage used to study the finding of a host by the ladybird beetle, Stethorus punctum. J. econ. Ent., 63: 1376—1377.
Collyer E. 1964: Phytophagous mites and their predators in New Zealand orchards. N. Z. J. agric. Res., 7: 551—568.
Colyer Ch. N. 1952: Notes on the life history of the British species of Phalacrotophora Enderlein (Dipt., Phoridae). Entomologist's mon. Mag., 88: 135—139.
Cooke W. C. 1963: Ecology of the pea aphid in the Blue Mountain area of eastern Washington and Oregon. Tech. Bull. U. S. Dep. Agric., 1287, 48 pp.
Cushman R. A. 1922: The identity of Ichneumon coccinellae Schrank (Hym.). Proc. ent. Soc. Wash., 24: 241—242.
Danilevskii A. S. 1961: Photoperiodism and seasonal development in insects. 243 pp., Izd. Leningrad. Univ. (in Russian).
Danilevskii A. S. (Ed.) 1968: Photoperiodic adaptations in insects and acari. 271 pp., Izd. Leningr. Univ. (in Russian, Engl. summ.).
Dauguet P. 1949: Les Coccinellini de France. 46 pp., Paris.
Davidson I. 1942: On the speed of development in insect eggs at constant temperatures. Aust. J. exp. Biol. med. Sci., 20: 233—239.
Davidson I. 1944: On the relationship between temperature and rate of development of insects at constant temperatures. J. Anim. Ecol., 13: 26—38.
Delucchi V. 1954: Pullus impexus (Muls.) (Coleoptera, Coccinellidae) a predator of Adelges piceae (Ratz.) (Hemiptera, Adelgidae), with notes on its parasites. Bull. ent. Res., 45: 243—278.
Dempster J. P. 1960: A quantitative study of the predators on the eggs and larvae of the brown beetle, Phytodecta olivacea Forster, using the precipitin test. J. Anim. Ecol., 29: 149—167.
Dickson R. C., Laird E. F., Pesho G. R. 1955: The spotted alfalfa aphid (Therioaphis maculata).

(Predator relationship: Hemiptera, Coccinellidae, Syrphidae, Neuroptera). Hilgardia, 24: 93—118.
Dixon A. F. G. 1958: The escape responses shown by certain Aphids to the presence of the Coccinellid Adalia decempunctata (L.). Trans. R. ent. Soc. Lond., 110: 319—334.
Dixon A. F. G. 1959: An experimental study of the searching behaviour of the predatory coccinellid beetle Adalia decempunctata (L.). J. Anim. Ecol., 28: 259—281.
Dixon A. F. G. 1970: Factors limiting the effectiveness of the coccinellid beetle, as a predator of the sycamore aphid, Drepanosiphum platanoides (Schr.). J. Anim. Ecol., 39:739—751.
Dixon A. F. G., Martin-Smith M., Subramanian G. 1965: Constituents of Megoura viciae Buckton. J. chem. Soc., 1965: 1562—1564.
Dobrzhanskiĭ F. G. 1922a, b: Imaginal diapause in Coccinellidae. Mass aggregations and migrations in Coccinellidae. Izv. Otd. prikl. Ent., 2: 103—124, 229—234.
Dobzhansky T. 1924: Über geographische und individuelle Variabilität von Adalia bipunctata und Adalia decempunctata. Russk. ent. Obozr., 18: 201—211.
Dobzhansky T. 1925: Über das Massenauftreten einiger Coccinelliden im Gebirge Turkestans. Z. wiss. InsektBiol., 20: 249—256.
Dobzhansky T. 1933: Geographical variation in lady-beetles. Am. Nat., 67: 97—126.
Dobzhansky T., Sivertzew-Dobzhansky N. P. 1927: Die geographische Variabilität von Coccinella septempunctata. Biol. Zbl., 47: 556—569.
Domenichini G. 1953: Degeeria luctuosa (funebris) Meug. (Dipt. Larv.) e Perilitus deceptor Wesm. (Hym. Bracon.) parassiti di Melasoma aenea L. adulta. Boll. Zool. agr. Bachic. Milano, 19: 103—140.
Domenichini G. 1956: Contributo alla conoscenza dei parassiti e iperparassiti dei Coleoptera Coccinellidae. Boll. Zool. agr. Bachic. Milano, 22: 215—246.
Dosse G. 1967: Schadmilben des Libanons und ihre Prädatoren. Z. angew. Ent., 59: 16—48.
Douglass J. R. 1930: Hibernation of the convergent ladybeetle Hippodamia convergens Guer., on a mountain peak in New Mexico. J. econ. Ent., 23: 288.
Doutt R. L. 1951: Biological control of mealybugs infesting commercial greenhouse gardenias. J. econ. Ent., 44: 37—40.
Downes J. A. 1965: Adaptations of insects in the arctic. A. Rev. Ent., 10: 257—274.
Dumas B. A., Boyer W. P., Whitcomb W. H. 1964: Effect of various factors on surveys of predaceous insects in soyabeans. J. Kans. ent. Soc., 37: 192—201.
Dumbleton L. D. 1936: The biological control of fruit pests in New Zealand. N. Z. Jl. Sci. Technol., 18: 588—592.
Dunn J. A. 1949: The parasites and predators of potato aphids. Bull. ent. Res., 40: 97—122.
Dunn J. A. 195: The effect of temperature on the pea aphid-ladybird relationship. 2nd Rept., pp. 21—23, Natl Veg. Res. Sta., Wellesbourne.
Dunn J. A. 1960: The natural enemies of the lettuce root aphid, Pemphigus bursarius (L.). Bull. ent. Res., 51: 271—278.
Dunn J. A. 1965: Studies on the aphid, Cavariella aegopodii Scop. I. On willow and carrot. Ann. appl. Biol., 56: 429—438.
Dyadechko N. P. 1953: On the preservation of predators controlling tetranychid mites in orchards. Sad i ogorod, 1953, 2: 40 (in Russian).
Dyadechko N. P. 1954: Coccinellids of the Ukrainian SSR, 156 pp., Kiev (in Russian).
Eastop V. F.. Pope R. D. 1966: Notes on the ecology and phenology of some British Coccinellidae. Entomologist, 99: 287—289.
Eastop V F.. Pope R. D. 1969: Notes on the biology of some British Coccinellidae. Entomologist, 102: 162—164.
Eichhorn O.. Graf P. 1971: Sex-linked colour polymorphism in Aphidecta obliterata L. (Coleoptera: Coccinellidae). Z. angew. Ent., 67: 225—231.
Ellingsen I.-J. 1969a: Fecundity, aphid consumption and survival of the aphid predator Adalia bipunctata L. (Col., Coccinellidae). Norsk ent. Tidsskr., 16: 91—95.
Ellingsen I.—J. 1969b: Effect of constant and varying temperature on development, feeding, and survival of Adalia bipunctata L. (Col., Coccinellidae). Norsk ent. Tidsskr., 16: 121—125.
van Emden F. I. 1949: Larvae of British beetles. VII (Coccinellidae). Entomologist's mon. Mag., 85: 265—283.
van Emden H. F. 1963: A field technique for comparing the intensity of mortality factors acting on the cabbage aphid, Brevicoryne brassicae (L.) (Hem.: Aphididae) in different areas of a crop. Entomologia exp. appl., 6: 53—62.
van Emden H. F. 1965a: The role of uncultivated land in the biology of crop pests and beneficial insects. Scient. Hort., 17: 121—136.

van Emden H. F. 1965b: The importance of adjacent uncultivated land in relation to crop pest insects. 12. Int. Congr. Ent., London 1964, 577 pp.
van Emden H. F. 1965c: The effect of uncultivated land on the distribution of cabbage aphid (Brevicoryne brassicae) on an adjacent crop. J. appl. Ecol., 2: 171—196
van Emden H. F., Eastop V. F., Hughes R. D., Way M. J. 1969: The ecology of Myzus persicae. A. Rev. Ent., 14: 197—270.
Engelmann F. 1968: Endocrine control of reproduction in insects. A. Rev. Ent., 13: 1—26.
Ermolenko S. F. 1963: Histological and histochemical study of the fat body in connection with maturation of gonades in Cryptolaemus montrouzieri Muls. (Coleoptera, Coccinellidae). Ent. Obozr., 42: 56—76.
Essig E. O. 1912: The walnut plant louse (Chromaphis juglandicola (Kalt.) Walker. Mon. Bull. Calif. Commn Hort., 1: 190—194.
Evans A. C. 1936: A note on the hibernation of Micraspis sedecimpunctata L. (var. 12-punctata L.) (Coccinell.) at Rothamsted Experimental Station. Proc. R. ent. Soc. London, 11: 116—119.
Evenhuis H. H. 1960: Observations and experiments on the enemies of aphids, that are harmful for apple growing in Nova Scotia, in relation to the modified spray program of Dr A. D. Pickett. OnderzVersl. Inst. plantenziektenk. Onderz., 1960: I—20.
Evenhuis H. H. 1968: The natural control of the apple-grass aphid Rhopalosiphum insertum, with remarks on the control of apple aphids in The Netherlands in general. Neth. J. Pl. Path., 74: 106—117.
Ewert M. A., Chiang H. C. 1966: Dispersal of three species of coccinellids in corn fields. Can. Ent., 98: 999—1001.
Fabre J.—H. 1879: Souvenirs entomologiques, I. Etudes sur l'instinct et les moeurs des insectes. Paris, 5ème ed.
Fenjves P. 1945: Beiträge zur Kenntnis der Blattlaus Myzus (Myzodes) persicae Sulz., Überträgerin der Blattrollenkrankheit der Kartoffel. Mitt. schweiz. ent. Ges., 19: 489—611.
Fischer T. W. 1963: Mass culture of Cryptolaemus and Leptomastix natural enemies of citrus mealybug. Bull. Calif. agric. Exp. Stn, 797, 39 pp.
Fleschner C. A. 1950: Studies on searching capacity of the larvae of three predators of the citrus red mite (Paratetranychus citri) (Stethorus picipes, Conwentzia hageni, Chrysopa californica) Hilgardia, 20: 233—265.
Fleschner C. A. 1958: Field approach to population studies of Tetranychid mites on citrus and avocado in California. 10. Int. Congr. Ent. Montreal 1956, 2: 669—676.
Fomenko R. B., Zaslavskiĭ V. A. 1970: Genetics of gregarious behaviour in the coccinellid Chilocorus bipunctatus L. (Coleoptera, Coccinellidae). Dokl. Akad. Nauk SSSR, 192: 229—231 (in Russian).
Forbes S. A. 1883: The food relations of the Carabidae and Coccinellidae. Bull. Ill. St. Lab. nat. Hist., 1: 33—64.
Ford E. B. 1964: Ecological genetics. 335 pp., Methuen, London, J. Willey, New York.
Franz J. M. 1958: The effectiveness of predators and food in limiting gradations of Adelges (Dreyfusia) piceae (Ratz.) in Europe. 10. Int. Congr. Ent. Montreal 1956, 4: 781—787.
Frazer J. F. D.. Rothschild M. 1962: Defence mechanisms in warningly coloured moths and other insects. 11. Int. Congr. Ent. Vienna 1960, 3: 249—256.
Fulmek L. 1957: Insekten als Blattlausfeinde. Annln naturh. Mus. Wien, 61: 110—227.
Fuzeau-Braesch S. 1961: Les déterminismes de la diapause chez les insectes. Annls Biol., 37: 44—69.
Fuzeau-Braesch S. 1966: Etude de la diapause de Gryllus campestris (Orthoptera). J. Insect Physiol., 12: 449—455.
Gage J. H. 1920: The larvae of the Coccinellidae. Illinois biol. Monogr., 6: 232—294.
Gagné W. C., Martin J. L. 1968: The insect ecology of red pine plantations in central Ontario. V. The Coccinellidae (Coleoptera). Can. Ent., 100: 835—846.
Galecka B. 1966: The role of predators in the reduction of two species of potato aphids, Aphis nasturtii Kalt. and A. frangulae Kalt. Ekol. pol., 14: 245—274.
George K. S. 1957: Preliminary investigations on the biology and ecology of the parasites and predators of Brevicoryne brassicae (L.). Bull. ent. Res., 48: 619—629.
Geyer J. W. C. 1947: A study of the biology and ecology of Exochomus flavipes Thunb. (Coccinellidae, Coleoptera). Part 1. 2. J. ent. Soc. Sth Afr., 9: 219—234, 10: 64—109.
Ghabn A. 1951: Studies on the biology and control of Epilachna chrysomelina F. in Egypt. Bull. Soc. Fouad I. Ent., 35: 77—106.
Goidanich A. 1943: Due Coccinelle igrofile e pollinivore sul riso (Hippodamia 13-punctata L., Anisosticta 19-punctata L.). Risicoltura, 33: 145—156, 169—177.

Goodarzy K., Davis D. W. 1958: Natural enemies of the spotted alfalfa aphid in Utah. J. econ. Ent., 51: 612—616.
Graham M. W. R. de V. 1969: Synonymic and descriptive notes on European Encyrtidae (Hym., Chalcidoidea). Polskie Pismo ent., 39: 211—319.
Gumoś H., Wiśniewski J. 1960: Intensity of appearing of Coccinellidae in pine woods. Polskie Pismo ent., B, 19/20: 217—223.
Günther V. 1958: Ergebnisse der zoologischen Expedition des Nationalmuseums in Prag nach der Türkei. 22. Coleoptera-Coccinellidae. Sb. ent. Odd. nár. Mus. Praha, 32: 19—36.
Gurney B., Hussey N. W. 1970: Evaluation of some coccinellid species for the biological control of aphids in protected cropping. Ann. appl. Biol., 65: 451—458.
Hafez M., El-Ziady S. 1952: On the morphology of Hyperaspis vinciguerrae Capra. On the histology of the alimentary canal of Hyperaspis vinciguerrae Capra (Coleoptera: Coccinellidae). Bull. Soc. Fouad I. Ent., 36: 247—291, 293—310.
Hagen K. S. 1962: Biology and ecology of predaceous Coccinellidae. A. Rev. Ent., 7: 289—326.
Hagen K. S., van den Bosch R. 1968: Impact of pathogens, parasites, and predators on aphids. A. Rev. Ent., 13: 325—384.
Hamilton E. W., Kieckhefer R. W. 1968: Integrated control of cereal aphids: tests with malathion and parathion. Proc. N. cent. Brch Am. Ass. econ. Ent., 23: 158—160.
Hamilton E. W., Kieckhefer R. W. 1969: Toxicity of malathion and parathion to predators of the English grain aphid. J. econ. Ent., 62: 1190—1192.
Hariri G. 1965: Records of nematode parasites of Adalia bipunctata (L.) (Col., Coccinellidae). Entomologist's mon. Mag., 101: 132.
Hariri G. 1966a: Laboratory studies on the reproduction of Adalia bipunctata (Coleoptera, Coccinellidae). Entomologia exp. appl., 9: 200—204.
Hariri G. 1966b: Changes in metabolic reserves of three species of aphidophagous Coccinellidae (Coleoptera) during metamorphosis. Entomologia exp. appl., 9: 349—358.
Hariri G. 1966c: Studies on the physiology of hibernating Coccinellidae (Coleoptera): changes in the metabolic reserves and gonads. Proc. R. ent. Soc. Lond. (A), 41: 133—144.
Harpaz I. 1958: Bionomics of the 11-spotted ladybird beetle, Coccinella undecimpunctata L., in a subropical climate. 10.Int. Congr. Ent. Montreal 1956, 2: 657—659.
Havnvik J. I., Frye R. D. 1969: An evaluation of insect predators and prey species found on corn. Proc. N. cent. Brch Am. Ass. econ. Ent., 24: 18—19.
Hawkes O. A. M. 1920: Observations on the life history, biology, and genetics of the lady-bird beetle, Adalia bipunctata (Mulsant). Proc. zool. Soc. Lond., 1920: 475—490.
Hawkes O. A. M., Marriner T. F. 1927: A preliminary account of the life-history of Coccinella 11-punctata (L.). Trans. R. ent. Soc. Lond., 75: 47—52.
Heathcote G. D. 1969: Notes on some plant bugs and predators of aphids caught on sticky traps in sugar beet fields in southern England. I. I. R. B., 4: 25—29.
Hecht O. 1936: Studies on the biology of Chilocorus bipustulatus (Coccinell.) an enemy of the red scale Chrysomphalus aurantii. Bull. Soc. Fouad I. Ent., 20: 299—326.
Heikertinger F. 1932: Die Coccinelliden, ihr "Ekelblut", ihre Warntracht und ihre Feinde. Teil I. II. Biol. Zbl., 52: 65—102, 385—412.
Hodek I. 1956: The influence of Aphis sambuci L. as prey of the ladybird beetle Coccinella septempunctata L. Věst. Čs spol. zool., 20: 62—74 (in Czech, Engl. summ.).
Hodek I. 1957a: The influence of Aphis sambuci L. as food for Coccinella 7-punctata L. II. Čas. Čs. spol. ent., 54: 10—17 (in Czech, Engl. summ.).
Hodek I. 1957b: The larval food consumption of Coccinella 7-punctata L. Zool. listy, 6: 3—11 (in Czech, Engl. summ.).
Hodek I. 1958: Influence of temperature, rel. humidity and photoperiodicity on the speed of development of Coccinella septempunctata L. Čas. Čs. spol. ent., 55: 121—141 (in Czech, Engl. summ.).
Hodek I. 1959: Ecology of aphidophagous Coccinellidae. Int. Conf. Insect Path. biol. Control, Praha, 1958: 543—547.
Hodek I. 1960a: Hibernation-bionomics in Coccinellidae. Čas. Čs. spol. ent., 57: 1—20 (in Czech, Engl. summ.).
Hodek I. 1960b: The influence of various aphid species as food for the lady-birds Coccinella 7-punctata L. Ontogeny of Insects (Acta Symp. Praha), 1959: 314—316.
Hodek I. 1962a: Experimental influencing of the imaginal diapause in Coccinella septempunctata L. (Col., Coccinellidae), II. Čas. Čs. spol. ent., 59: 297—313.
Hodek I. 1962b: Essential and alternative food in insects. 11. Int. Congr. Ent. Vienna 1960, 2: 696—697.

Hodek I. 1964: Die Schonung der natürlichen Feinde als Hilfe bei der integrierten Bekämpfung der Blattläuse. TagBer. dt. Akad. LandwWiss. Berlin, 60: 37—53.
Hodek I. 1965: Several types of induction and completion of adult diapause. 12. Int. Congr. Ent., London 1964: 431—432.
Hodek I. (ed.) 1966: Ecology of aphidophagous insects. Proc. Symp. Liblice near Prague, 1965, 360 pp., Academia, Praha, Dr. W. Junk, Haag.
Hodek I. 1967: Bionomics and ecology of predaceous Coccinellidae. A. Rev. Ent., 12: 79—104.
Hodek I. 1968: Diapause in females of Pyrrhocoris apterus L. (Heteroptera). Acta ent. bohemoslov., 65: 422—435.
Hodek I. 1970: Termination of diapause in two coccinellids (Coleoptera). Acta ent. bohemoslov., 67: 218—222.
Hodek I. 1971a: Sensitivity of larvae to photoperiods controlling the adult diapause of two insects. J. Insect Physiol., 17: 205—216.
Hodek I. 1971b: Termination of adult diapause in Pyrrhocoris apterus (Heteroptera: Pyrrhocoridae) in the field. Entomologia exp. appl., 14: 212—222.
Hodek I., Čerkasov J. 1958: A study of the imaginal hibernation of Semiadalia undecimnotata Schneid. (Coccinellidae, Col.) in the open, I. Věst. Čs. spol. zool., 22: 180—192 (in Czech, Engl. summ.).
Hodek I., Čerkasov J. 1960: Prevention and artificial induction of the imaginal diapause in Coccinella 7-punctata L. Nature, 187: 345.
Hodek I., Čerkasov J. 1961: Prevention and artificial induction of imaginal diapause in Coccinella septempunctata L. (Col., Coccinellidae). Entomologia exp. appl., 4: 179—190.
Hodek I., Čerkasov J. 1963: Imaginal dormancy in Semiadalia undecimnotata Schneid. (Coccinellidae, Col.) II. Changes in water, fat and glycogen content. Věst. Čs. spol. zool., 27: 298—318.
Hodek I., Hagen K. S., van Emden H. F. 1972: Methods for studying effectiveness of natural enemies. In Aphid Technology — I. B. P. Handbook, Academic Press, London and New York 344 pp. (147—188).
Hodek I., Holman J., Starý P., Štys P., Zelený J. 1966: Natural enemies of Aphis fabae in Czechoslovakia. 126 pp., Academia, Praha (in Czech, Engl. summ.).
Hodek I., Honek A. 1970: Incidence of diapause in Aelia acuminata (L.) populations from southwest Slovakia (Heteroptera). Věst. Čs. spol. zool., 34: 170—183.
Hodek I., Honek A. 1971: Termination of adult diapause. 13. Int. Congr. Ent., Moscow, 1968, 1: 385.
Hodek I., Landa V. 1971: Anatomical and histological changes during dormancy in two Coccinellidae. Entomophaga, 16: 239—251.
Hodek I., Novák K., Skuhravý V., Holman J. 1965: The predation of Coccinella septempunctata L. on Aphis fabae Scop. on sugar beet. Acta ent. bohemoslov., 62: 241—253.
Hodek I., Starý P., Štys P. 1962: The natural enemy complex of Aphis fabae and its effectiveness in control. 11. Int. Congr. Ent., Vienna 1960, 2: 747—749.
Hodson A. C. 1937: Some aspects of the role of water in insect hibernation. Ecol. Monogr., 7: 271—315.
Hoffer A. 1963: Descriptions of new species of the family Encyrtidae from Czechoslovakia (Hym., Chalcidoidea). Acta Ent. Mus. Nat. Pragae, 35: 549—592.
Horion A. 1961: Faunistik der mitteleuropäischen Käfer. Bd. 8, Teil 2, 375 pp., Kommisionsverlag Feyel, Überlingen-Bodensee.
Howard N. F., Landis B. J. 1936: Parasites and predators of the Mexican bean beetle in the United States. Circ. U. S. Dep. Agric., 418: 1—12.
Huffaker C. B., Doutt R. L. 1965: Establishement of the coccinellid, Chilocorus bipustulatus Linnaeus, in California olive groves. Pan-Pacif. Ent., 44: 61—63.
Hughes R. D. 1963: Population dynamics of the cabbage aphid, Brevicoryne brassicae (L.) J. anim. Ecol., 32: 393—424.
Hukusima S. 1949: Microclimatic effect upon the feeding activity of the adult of Epilachna vigintioctomaculata Motschulsky in potato-fields of different furrow widths. (Activity fluctuation in insects and environmental condition, VIII.) Publ. Breeding Assoc. Potato Hokkaido, 24: 1—6.
Hukusima S. 1950: Effect of microclimate caused by the different ridging in potato-field on the feeding activity of adult of Epilachna vigintioctomaculata Motschulsky. (Activity fluctuation in insect and environmental condition, IX.) Kontyu, 18: 18—20.
Hukusima S. 1951a: On the activity of the adult of Epilachna vigintioctomaculata Motschulsky, in potato-fields furrowed in different directions. (Activity fluctuation in insects and environmental condition, V.) Oyo-Kontyu, 6: 178—183 (in Japanesse, Engl. summ.).

Hukusima S. 1951b: Intercropping and fencing cultivation of potatoes for controlling Epilachna vigintioctomaculata Motschulsky. Agriculture Hort., Tokyo, 26: 1103—1104.
Hukusima S. 1955: On the insect association in potato-fields furrowed in different directions (Studies on the insect association in the field, 3.) Jap. J. appl. Zool., 19: 155—163.
Hukusima S. 1963a: Comparative acaricidal activity on the pest-predator complex in young apple orchards. Res. Bull. Fac. Agric. Gifu Univ., 18: 76—87.
Hukusima S. 1963b: Maintenance of the balance of pest-predator complex in apple orchards with insecticide-fungicide spray treatments. Res. Bull. Fac. Agric. Gifu Univ., 18: 61—75.
Hukusima S. 1964: Evaluation of effect of acaricide-insecticide or acaricide fungicide combinations on the balance of pest-predator complex in apple orchards. Res. Bull. Fac. Agric. Gifu Univ., 19: 38—54.
Hukusima S. 1966a, b: An ecological context for the effect of pesticide stress on pest-predator complex in apple orchards. Petroleum oils: effect on balance of pest-predator complex in apple orchards. Res. Bull. Fac. Agric. Gifu Univ., 22: 60—78, 79—94.
Hukusima S. 1968: Integrating arthropod predator releases and chemical manipulations for pest control in apple orchards. Res. Bull. Fac. Agric. Gifu Univ., 26: 40—63.
Hukusima S. 1969: Ecological implications in population trends of common arthropod predators and major pests in apple orchards under different control programs. Res. Bull. Fac. Agric. Gifu Univ., 28: 64—87.
Hukusima S., Kamei M. 1970: Effects of various species of aphids as food on development, fecundity and longevity of Harmonia axyridis Pallas (Coleoptera: Coccinellidae). Res. Bull. Fac. Agric. Gifu Univ., 29: 53—66.
Hukusima S., Sakurai H. 1963; Comparison of physiological natures between predaceous Coccinellidae reared on artificial diet and aphid alone. Rept. Tokai Brch Ent. Soc. Japan, No. 15: 12—21 (in Japanese).
Hukusima S., Sukurai H. 1964: Aphid consumption by adult Coccinella septempunctata bruckii Mulsant in relation to temperature (Coleoptera: Coccinellidae). Ann. Rep. Plant Protection North Japan, No. 15: 126—128 (in Japanese).
Ibrahim M. M. 1955a, b: Studies on Coccinella undecimpunctata aegyptiaca Reiche. 1. Preliminary notes and morphology of the early stages. 2. Biology and life-history. Bull. Soc. ent. Égypte, 39: 251—274, 395—423.
Imms A. D. 1947: Insect natural history. London.
Iperti G. 1964: Les parasites des Coccinelles aphidiphages dans les Basses-Alpes et les Alpes--Maritimes. Entomophaga, 9: 153—180.
Iperti G. 1965: Contribution à l'étude de la specificité chez les principales Coccinelles aphidiphages des Alpes-Maritimes et des Basses-Alpes. Entomophaga, 10: 159—178.
Iperti G. 1966: Comportement naturel des Coccinelles aphidiphages du Sud-Est de la France. Leur type de spécificité, leur action prédatrice sur Aphis fabae L. Entomophaga, 11: 203—210.
Iperti G., Brun J. 1969: Role d'une quarantaine pour la multiplication des Coccinellidae coccidiphages destinés à combattre la cochenille du palmier-dattier (Parlatoria Blanchardi Targ.) en Adrar Mauritanien. Entomophaga, 14: 149—157.
Iperti G., Laudého Y. 1968: Intervention bio-écologique en Adrar mauritanien destinée à lutter contre la cochenille du palmier dattier: Parlatoria Blanchardi Targ. (Coccoidea-Diaspididae). Fruits, 23: 543—552.
Iperti G., Laudého Y. 1969: Les entomophages de Parlatoria Blanchardi Targ. dans les palmeraies de l'Adrar mauritanien. I. Études biologiques préliminaires. Perspectives d'acclimatation de nouveaux prédateurs Coccinellidae. Ann. Zool. Écol. anim., 1: 17—30.
Iperti G., Laudého Y., Brun J., Choppin de Janvry E. 1970: Les entomophages de Parlatoria abalrdinch Targ. dans les palmeraies de l'Adrar mauritanien. III. Introduction, acclimatation et efficacité d'un nouveau prédateur Coccinellidae: Chilocorus bipustulatus L. (Souche d'Iran). Ann. Zool. Écol. anim., 2:617—638.
Iperti G., Waerebeke van D. 1968: Description, biologie et importance d'une nouvelle espèce d'Allantonematidae (Nématode), parasite des coccinelles aphidiphages: Parasitylenchus Coccinellinae, n. sp. Entomophaga, 13: 107—119.
Iwao S. 1954: On the distributions of Epilachna sparsa orientalis Dieke and E. vigintioctomaculata Motschulsky at the boundary of their geographical distributions (I). Oyo-Kontyu, 9: 135—141 (in Japanese, Engl. summ.).
Iwao S., Machida A. 1961: Further experiments on the host-plant preference in a phytophagous ladybeetle, Epilachna pustulosa Kono. Insect Ecol., 9: 9—16 (in Japanese, Engl. summ.)
Iwata K. 1932: On the biology of two large lady-birds in Japan. Trans. Kansai ent. Soc., 3: 13—26.

Iwata K. 1965: Supplement on the biology of two large lady-birds in Japan. Acta coleopter., 2: 57—68.
Johnson C. G. 1969: Migration and dispersal of insects by flight. 763 pp., Methuen, London.
Johnson R. H. 1907: Economic notes on aphids and coccinellids. Ent. News, Philadelphia, 18: 171—174.
Johnson R. H. 1910: Determinate evolution in the colour-pattern of the lady beetles. 104 pp., Carnegie Inst., Washington.
Jöhnssen A. 1930: Beiträge zur Entwicklungs- und Ernährungsbiologie einheimischer Coccinelliden unter besonderer Berücksichtigung von Coccinella septempunctata L. Z. angew. Ent., 16: 87—158.
Kaddou I. K. 1960: The feeding behaviour of Hippodamia 5-signata (Kirby) larvae. Univ. Calif. Publs Ent., 16: 181—230.
Kamiya H. 1965: Comparative morphology of larvae of the Japanese Coccinellidae, with special reference to the tribal phylogeny of the family (Coleoptera). Mem. Fac. lib. Fukui Univ. (2), 14: 83—100.
Kanervo V. 1940: Beobachtungen und Versuche zur Ermittlung der Nahrung einiger Coccinelliden Annls ent. fenn., 6: 89—110.
Kanervo V. 1941: Zur Morphologie der präimaginalen Stadien von Calvia 15-guttata F. (Col. Coccinellidae). Annls ent. fenn., 7: 52—60.
Kanervo V. 1946: Studien über die natürlichen Feinde des Erlenblattkäfers, Melasoma aenea L. (Col., Chrysomelidae). Annls zool. Soc. Vanamo, 12: 206 pp. (in Finnish, Germ. summ.).
Kanervo V. 1962: Einfluss der Bekämpfungsmassnahmen im Apfelbau auf die Populationsentwicklung der Obstbaumspinnmilbe (Metatetranychus pilosus C. F.) und ihre natürlichen Feinde in Finnland. 11. Int. Congr. Ent. Vienna 1960, 2: 64—72.
Kapur A. P. 1950: The biology and external morphology of the larvae of Epilachninae. Bull. ent. Res., 41: 161—208.
Kapur A. P. 1954: Mass assemblage of the Coccinellid beetle Epilachna bisquadripunctata (Gyllenhal) in Chota Nagpur. Curr. Sci., 23: 230—231.
Karafiat H., Franz J. 1956: Studien zur Populationsdynamik der Tannenstammlaus Adelges (Dreyfusia) piceae (Ratz.) (Hemipt. Adelgidae). Zool. Jb. (Syst.), 84: 467—504.
Kehat M. 1967a: Survey and distribution of common lady beetles (Col. Coccinellidae) on date palm trees in Israel. Entomophaga, 12: 119—125.
Kehat M. 1967b: Studies on the biology and ecology of Pharoscymnus numidicus (Coccinellidae) an important predator of the date palm scale Parlatoria blanchardi. Ann. Soc. ent. Fr. (N. S.), 3: 1053—1065.
Kehat M. 1968a: The feeding behaviour of Pharoscymnus numidicus (Coccinellidae), predator of the date palm scale Parlatoria blanchardi. Entomologia exp. appl., 11: 30—42.
Kehat M. 1968b: The phenology of Pharoscymnus spp. and Chilocorus bipustulatus L. (Coccinellidae) in date palm plantations in Israel. Ann. Epiphyties, 19: 605—614.
Kehat M., Greenberg S., Gordon D.: Factors causing seasonal decline in Chilocorus bipustulatus L. (Coccinellidae) in citrus groves in Israel (in press).
Kehat M., Swirski E. 1964: Chemical control of the date palm scale, Parlatoria blanchardi, and the effect of some insecticides on the lady beetle Pharoscymnus aff. numicidus Pic. Israel J. agric. Res., 14: 101—110.
McKenzie H. L. 1932: The biology and feeding habits of Hyperaspis lateralis Mulsant (Coleoptera — Coccinellidae). Univ. Calif. Publ. Ent., 6: 9—20.
Kesten U. 1969: Zur Morphologie und Biologie von Anatis ocellata (L.) (Coleoptera, Coccinellidae). Z. angew. Ent., 63: 412—445.
El-Khidir E. 1969: A contribution to the biology of Epilachna chrysomelina F., the melon lady-bird beetle in the Sudan (Col., Coccinellidae). Sudan agric. J., 4: 32—37.
Kieckheffer R. W., Miller E. L. 1967: Trends of populations of aphid predators in South Dakota cereal crops 1963—1965. Ann. ent. Soc. Am., 60: 516—618.
Klausnitzer B. 1965: Zur Biologie der Epilachna argus Geoffr. (Col. Coccinellidae). Ent. Nachr., 9: 87—89.
Klausnitzer B. 1966: Übersicht über die Nahrung der einheimischen Coccinellidae (Col.). Ent. Ber., 1966: 91—102.
Klausnitzer B. 1967: Zur Kenntnis der Beziehungen der Coccinellidae zu Kiefernwäldern (Pinus silvestris L.). Acta ent. bohemoslov., 64: 62—68.
Klausnitzer B. 1968: Zur Biologie von Myrrha octodecimguttata (L.). (Col. Coccinellidae). Ent. Nachr., 12: 102—104.
Klausnitzer B. 1969a: Degeeria luctuosa Mg. (Diptera, Tachinidae) as a parasite of Synharmonia conglobata L. (Coleoptera, Coccinellidae). Ent. Obozr., 48: 500—501 (in Russian, Engl. summ.).

Klausnitzer B. 1969b: Zur Kenntnis der Entomoparasiten mitteleuropäischer Coccinellidae. Abh. Ber. NaturkMus.—ForschStelle, Görlitz, 44: 9: 1—15.
Klausnitzer B. 1969c: Zur Kenntnis der Larve von Lithophilus conatus (Panzer) (Col., Coccinellidae). Ent. Nachr., 13: 33—38.
Klausnitzer B. 1970a: Zur Larvalsystematik der mitteleuropäischen Coccinellidae (Col.). Ent. Abh. Staatl. Mus. Tierk. Dresden, 38: 55—110.
Klausnitzer B. 1970b: Zur Kenntnis der Larven der palaearktischen Brumus-Arten (Col. Coccinellidae). Ent. Nachr., 14: 52—55.
Klausnitzer B. 1971a: Über die verwandtschaftlichen Beziehungen der Lithophilinae und Coccidulini (Col. Coccinellidae). Dt. ent. Z., 18: 145—148
Klausnitzer B. 1971b: Zur Kenntnis der Larven der palaearktischen Arten von Harmonia Muls., Adonia Muls. und Tytthaspis Crotch (Col. Coccinellidae). Ann. Zool. (in press).
Klausnitzer B. 1971c: Zur Kenntnis der Larven der kubanischen Chilocorini (Col. Coccinellidae). Zool. Anz. 186: 224—229.
Klausnitzer B., Bellmann C. 1969: Zum Vorkommen von Coccinellidenlarven (Coleoptera) in Bodenfallen auf Fichtenstandorten. Ent. Nachr., 13: 128—132.
Klingauf F. 1967: Abwehr- und Meideraktionen von Blattläusen (Aphididae) bei Bedrohung durch Räubern und Parasiten. Z. angew. Ent., 60: 269—317.
Knowlton G. F. 1939: Utah Coleoptera 1. Utah Agric. Exp. Stn., Mimeogr. Ser. 200, part 3: 1—25.
Knowlton G. F., Goodarzy K. 1956: The spotted alfalfa aphid in Utah. Utah State Agric. Coll. Mimeogr. Ser. 155: 1—7.
Knowlton G. F., Smith C. F., Harmston F. C. 1938: Pea aphid investigation. Proc. Utah Acad. Sci., 15: 71—80.
Koide T. 1962: Observations on the feeding habit of the larva of Coccinella septempunctata bruckii Mulsant. The feeding behaviour and number of prey fed under different temperatures. Kontyu, 30: 236—241 (in Japanese, Engl. summ.).
Komai T. 1956: Genetics of lady-beetles. Adv. Genet., 8: 155—185.
Korschefsky R. 1931, 1932: Coleopterorum catalogus, pars 118, 120. Coccinellidae I, II. 224 and 435 pp., Berlin.
Korschefsky R. 1934: Platynaspis luteorubra Goeze, ein neuer Larventypus der Coccinelliden. Arb. physiol. angew. Ent. Berl., 1: 278—279.
Kryl'cov A. I. 1956: Geographical variability of lady-birds (Coleoptera, Coccinellidae) in North Kirghisia. Ent. Obozr., 35: 771—781 (in Russian).
Kuenen D. J. 1947: On the ecological significance of two predators of Metatetranychus ulmi Koch (Acari, Tetranychidae). Tijdschr. Ent., 88 (1945): 303—312.
Kurir A. 1964: Erstmaliges Massenauftreten der Koniferenlaus Cinaropsis pilicornis Hartig auf Fichte in Österreich. Zentbl. ges. Forstw., 81: 139—157.
Kuznetsov N. Ya. 1948: Principles of insect physiology. Vol. I, 380 pp., Izd. Akad. Nauk SSSR (in Russian).
Lane C., Rothschild M. 1960: Notes on wasps visiting a mercury vapour trap, together with some observations on their behaviour towards their prey. Entomologist's mon. Mag., 95: 277—279.
Laster M. L., Brazzel J. R. 1968: A comparison of predator populations in cotton under different control programs in Mississippi. J. econ. Ent., 61: 714—719.
Latta R. 1928: The effect of the extreme temperature on Dec. 7, 8, 9, 1927 on hibernating Crioceris asparagi L. and Hippodamia convergens Guer. at Ames, Iowa. Psyche, 35: 229—231.
Laudého Y., Choppin de Janvry E., Iperti G., Brun J. 1970: Intervention bio-écologique contre la cochenille blanche du palmier-dattier (Parlatoria Blanchardi Targ.) (Coccoidea-Diaspididae) en Adrar Mauritanien. Fruits, 25: 147—160.
Laudého Y., Ormières R., Iperti G. 1969: Les entomophages de Parlatoria Blanchardi Targ. dans les palmeraies de l'Adrar mauritanien. II. Étude d'un parasite de Coccinellidae "Gregarina Katherina" Watson. Ann. Zool. Écol. anim., 1: 395—406.
Lees A. D. 1955: The physiology of diapause in arthropods. 151 pp., Cambridge Univ. Press, London.
Legay J.-M. Reggi de M. 1962: Sur un lieu d'hibernation de la Coccinelle Titthaspis sedecimpunctata dans la région Lyonnaise. Bull. mens. Soc. linn. Lyon, 31: 267—269.
Lim Sook Ming 1971: The biology and ecology of some aphidophagous Coccinellidae. PhD Thesis, University of Malaya. Kuala Lumpur.
Lingren P. D., Ridgway R. L. 1967: Toxicity of five insecticides to several insect predators. J. econ. Ent., 60: 1639—1641.
Liu C. L. 1950: Contribution to the knowledge of Chinese Coccinellidae. X. Occurrence of Perilitus

coccinellae (Schrank), a parasite of adult Coccinellidae in North China (Hymenoptera, Braconidae). Ent. News, 61: 207—208.
Loughton B. G. et al. 1963: Spiders and the spruce budworm. Mem. ent. Soc. Can., 31: 249—268.
Lucas A. M. 1969: The effect of population structure on the success of insect introductions. Heredity, 24: 151—157.
Lusis Ya. Ya. 1932: An analysis of the dominance phenomenon in the inheritance of the elytra and pronotum colour in Adalia bipunctata. Trudy Lab. Genet., 9: 135—162 (in Russian, Engl. summ.).
Lusis Ya. Ya. 1947a, b: Some aspects of the population increase in Adalia bipunctata L. 1. Heterozygosity of populations in lethal alleles. 2. The strains without males. Dokl. Akad. Nauk SSSR, 57: 825—828, 951—954 (in Russian).
Lusis Ya. Ya. 1961: On the biological meaning of colour polymorphism of lady-beetle Adalia bipunctata L. Latv. Ent., 4: 2—29 (in Russian, Engl. summ.).
Lusis Ya. Ya. 1971: Experimental data on the three species of genus Calvia (Coleoptera, Coccinellidae) from Central Asia. Latv. Ent., 14: 3—29 (in Russian, Engl. summ.).
Maeta Y. 1969a: Biological studies on the natural enemies of some Coccinellid beetles. I. On' Perilitus coccinellae (Schrank). Kontyu, 37: 147—166.
Maeta Y. 1969b: Some biological studies on the natural enemies of some Coccinellid beetles. II. Pharactrophora sp. Tohoku Konchu Kenkyu. 4: 1—6.
Mani M. S. 1962: Introduction to high altitude entomology. Insect life above the timber-line in the north-west Himalaya. 302 pp., Methuen, London.
Marriner T. F. 1939: Movements of Coccinellidae. Ent. Rec., 51: 104—106.
Matsumoto B., Nishida T. 1966: Predator-prey investigations on taro leafhopper and its egg predator. Tech. Bull. Hawaii agric. Exp. Stn., 64: 1—32.
Minoranskiĭ V. A. 1966: Some data on Aphis fabae (Scop.). Zool. Zh., 45: 1164—1169 (in Russian).
Mogi M. 1969: Predation response of the larvae of Harmonia axyridis Pallas (Coccinellidae) to the different prey density. Jap. J. appl. Ent. Zool., 13: 9—16.
Moreton B. D. 1969: Beneficial insects and mites. 118 pp., Min Agric. Fisher. Food, London.
Moter G. 1959: Untersuchungen zur Biologie von Stethorus punctillum Weise. Doktor-Dissertation, Mathem. — Naturwiss. Fak. Univ. Köln.
Moursi A. A., Kamal M. 1946: Notes on the biology and feeding habits of the introduced beneficial insect Leis conformis Boisd. (Coccinell.). Bull. Soc. Fouad I. Ent., 30: 63—74.
McMullen R. D. 1967a: A field study of diapause in Coccinella novemnotata (Coleoptera: Coccinellidae). Can. Ent., 99: 42—49.
McMullen R. D. 1967b: The effects of photoperiod, temperature and food supply on rate of development and diapause in Coccinella novemnotata. Can. Ent., 99: 578—586.
Müller H. J. 1958: Über den Einfluss der Photoperiode auf Diapause und Körpergrösse der Delphacide Stenocranus minutus Fabr. (Homoptera Auchenorrhyncha). Zool. Anz., 160: 294—312.
Müller H. J. 1965: Probleme der Insektendiapause. Verh. dt. Zool. Ges. Jena, 1965: 192—222.
Müller H. J. 1966: Über mehrjährige Coccinelliden-Fänge auf Ackerbohnen mit hohem Aphis fabae-Besatz. Z. Morph. Ökol. Tiere, 58: 144—161.
Müller H. J. 1970: Formen der Dormanz bei Insekten. Nova Acta Leopoldina, 35: 1—27.
Muma M. H. 1953—1954: Lady beetle predators of citrus (1) aphids, (2) scale insects, (3) white flies, (4) mealybugs, (5) mites. Citrus Mag., 1953 (April): 32—33, (July): 24—25, (November): 12—13, 1954 (April): 16—17, (July): 10—11.
Muma M. H. 1955a: Lady beetles (Coccinellidae: Coleoptera) found on citrus in Florida. Fla Ent., 38: 117—124.
Muma M. H. 1955b: Some ecological studies on the twice stabbed lady beetle, Chilocorus stigma (Say). Ann. ent. Soc. Am., 48: 493—498.
Nadel D., Biron S. 1964: Laboratory studies and controlled mass rearing of Chilocorus bipustulatus L. a citrus scale predator in Israel. Riv. Parassit., 25: 165—206.
Nagui A. M. K. 1969: Einfluss der mechanisch wirkenden Pflanzenschutzmittel Sommer-Weissöl Volck Eté und Polykil H 100 E auf zwei bedeutende Prädatoren von Spinnmilben Stethorus punctillum Weise (Coleoptera, Coccinellidae) und Phytoseiulus riefeli Dosse (Acari, Phytoseiidae). Z. angew. Zool., 56: 261—313.
Nakao S. 1962: A list of insects collected in a citrus grove near Fukuoka City (Ecological studies on the insect community of citrus groves, IV). Kontyu, 30: 50—71 (in Japanese, Engl. summ.).
Nakao S. 1964: The interspecific relations among insects in a citrus grove (Ecological studies on the insect community of citrus groves, V). Kontyu, 32: 490—503 (in Japanese, Engl. summ.).
Newsom L. D. 1967: Consequences of insecticide use on nontarget organisms. A. Rev. Ent., 12: 257—286.

Nielson M. W., Curie W. E. 1960: Biology of the convergent lady beetle when fed a spotted alfalfa aphid diet. J. econ. Ent. 53: 257—259.
Nixon G. E. 1951: The association of ants with aphids and coccids. 36 pp., Commonwealth Inst. Ent., London.
Nohara K. 1962: On the overwintering of Chilocorus kuwanae Silvestri (Coleoptera, Coccinellidae). Sci. Bull. Fac. Agric. Kyushu Univ., 20: 33—39 (in Japanese, Engl. summ.).
Nohara K. 1963: Observations on the activity of Coccinellid beetles in the citrus groves near Hagi, Honshu. Sci. Bull. Fac. Agric. Kyushu Univ., 20: 157—168 (in Japanese, Engl. summ.).
Nohara K., Harada T., Onon T. 1965: On the effect of some insecticides upon the Coccinellid-beetles. Sci. Bull. Fac. Agric. Kyushu Univ., 22: 1—7 (in Japanese, Engl. summ.).
Norris M. J. 1964: Environmental control of sexual maturation in insects. Insect Reproduction, Symposium No. 2, R. Ent. Soc. London, pp. 56—65.
Novák B. 1965: Beitrag zu den Labor-Untersuchungen der Aggregation von Coccinelliden. Konf. Schädl. Hackfrüchte I, Praha 1965: 85—89.
Novák B. 1966: Die Bewirkung der Georeaktionen von Coccinella septempunctata L. durch die Feuchtigkeit. Sb. prací přír. fak. Palack. univ. Olomouc, 22: 147—151 (in Czech, Germ. summ.).
Novák B., Grenarová K. 1967: Coccinelliden an der Grenze des Feld- und Waldbiotops — Hibernationsversuche mit den Imagines führender Arten. Konf. Schädl. Hackfrüchte III, Praha 1967: 49—59. (in Czech, Germ. summ.).
Odum E. P. 1953: Fundamentals of ecology. 384 pp., Saunders, Philadelphia and London.
Ogloblin A. A. 1913: On the biology of coccinellids (Coleoptera, Coccinellidae). Russk. ent. Obozr., 13: 27—43 (in Russian).
Ogloblin A. A. 1924: Le róle du blastoderme extraembryonnaire du Dinocampus terminatus Nees pendant l'état larvaire. Mém. Soc. r. Sci. Bohème (II), 1924: 1—27.
Okada I. 1970: A new method of artificial rearing of coccinellid, Harmonia axyridis Pallas. Heredity, Tokyo, 24: 32—35 (in Japanese).
Okamoto H. 1961: Comparison of ecological characters of the predatory ladybird Coccinella septempunctata bruckii fed on the apple grain aphids, Rhopalosiphum prunifoliae and the cabbage aphids, Brevicoryne brassicae. Jap. J. appl. Ent. Zool., 5: 277—278 (in Japanese).
Paliï V. F. 1960: Causes of fluctuation in numbers of Aphis fabae Scop. in beet-growing regions of central black-soil areas in RSFSR. Zool. Zh., 39: 534—539 (in Russian).
Palmer M. A. 1914: Some notes on life history of lady beetles. Ann. ent. Soc. Am., 7: 213—238.
Pantyukhov G. A. 1965: Influence of temperature and relative humidity on development of Chilocorus renipustulatus Scriba (Col. Coccinellidae). Trudy zool. Inst., Leningrad, 36: 70—85 (in Russian).
Pantyukhov G. A. 1968a: On photoperiodic reaction of Chilocorus renipustulatus Scriba (Coleoptera, Coccinellidae). Ent. Obozr., 47: 376—385 (in Russian, Engl. summ.).
Pantyukhov G. A. 1968b: A study of ecology and physiology of the predatory beetle Chilocorus rubidus Hope (Coleoptera, Coccinellidae). Zool. Zh., 47: 376—386 (in Russian, Engl. summ.).
Peairs L. M. 1914: The relation of temperature to insect development. J. econ. Ent., 7: 174—179.
Peck O. 1963: A catalogue of the Nearctic Chalcidoidea (Insecta: Hymenoptera). Can. Ent. Suppl., 30: 1—1092 (427—428).
Pemberton C. E. 1948: History of the entomology department experiment station, H. S. P. A., 1904—1945. Hawaii. Plrs'Rec., 52: 53—90.
Pienkowski R. L. 1965: The incidence and effect of egg cannibalism in first-instar Coleomegilla maculata lengi (Coleoptera: Coccinellidae). Ann. ent. Soc. Am., 58: 150—153.
Plaut H. N. 1965: On the phenology and control value of Stethorus punctillum Weisse as a predator of Tetranychus cinnabarinus Boisd. in Israel. Entomophaga, 10: 133—137.
Pontin A. J. 1960: Some records of predators and parasites adapted to attack aphids attended by ants. Entomologist's mon. Mag., 95: 154.
Popov P. 1960: Role of Coccinella septempunctata L. in the biological control of aphids. Selskostop. Misul, Sofiya, 1960: 152—153 (in Bulgarian).
Poulton E. B. 1904: A possible explanation of insect swarms on mountain tops. Trans. R. ent. Soc. Lond., 1904: 24—26.
Poulton E. 1936: Assemblies of coccinellid beetles observed in N. Uganda (1927) by Prof. Hale Carpenter and in Bechualand (1935) by Dr W. A. Lamborn. Proc. R. ent. Soc. Lond. (A), 11: 99—100.
Poutiers R. 1930: Sur le comportement du Novius cardinalis vis-à-vis de certain alcaloïdes. C. r. Séanc. Soc. Biol., 103: 1023—1025.
Pradhan S. 1936: The alimentary canal of Epilachna indica (Coccinellidae: Coleoptera), with a discussion on the activity of the midgut epithelium. Jl R. Asiat. Soc. Beng., 2: 127—156.
Pradhan S. 1938: Neuromuscular study of the mouth parts of Coccinella septempunctata with

a comparison of the mouth-parts in carnivorous and herbivorous Coccinellids. Rec. Indu.M s. Calcutta, 40: 341—358.
Pradhan S. 1939: The alimentary canal and pro-epithelial regeneration in Coccinella septempunctata, with comparison of carnivorous and herbivorous Coccinellids. Q. Jl microsc. Sci., 81: 451—478.
Principi M. M., Castellari P. L., Giunchi P. 1967: Observations sur les infestations de pucerons et leurs predateurs et parasites dans des parcelles traitées avec des produits phytiatrique polyvalents ou sélectifs. Entomophaga, Mém. H. S., 3: 103—107.
Pulliainen E. 1963: Preliminary notes on the humidity reactions of Myrrha 18-guttata L. (Col., Coccinellidae). Ann. ent. Fenn., 29: 240—246.
Pulliainen E. 1964: Studies on the humidity and light orientation and the flying activity of Myrrha 18-guttata L. (Col., Coccinellidae). Ann. ent. Fenn., 30: 117—141.
Pulliainen E. 1966: On the hibernation sites of Myrrha octodecimguttata L. (Col., Coccinellidae) on the butts of the pine (Pinus silvestris L.). Ann. ent. Fenn., 32: 99—104.
Putman W. L. 1955: Bionomics of Stethorus punctillum Weise in Ontario. Can. Ent., 87: 9—33.
Putman W. L. 1957: Laboratory studies on the food of some coccinellids (Coleoptera) found in Ontario peach orchards. Can. Ent., 89: 527—579.
Putman W. L. 1964: Occurrence and food of some coccinellids (Coleoptera) in Ontario peach orchards. Can. Ent., 96: 1149—1155.
Putman W. L. 1965: Paper chromatography to detect predation on mites. Can. Ent., 97: 435—441.
Radziesvkaya S. B. 1939: Concerning the hibernation of the ladybird and the struggle with the aphids. Vop. Ékol. Biotsen., 4: 268—275 (in Russian, Engl. summ.).
Read D. B. 1965: The field recognition of the larvae of three common aphid-feeding Coccinellids. N. Z. Ent., 3: 14—17.
Redenz-Rüsch I. 1959: Untersuchungen über die Schädlings- und Nützlingsfauna einer Obstanlage im „Bergischen Land" und deren Beeinflussung durch chemische Bekämpfungsmittel. Höfchenbr. Bayer. PflSchutz-Nachr., 12: 171—258.
Reineck G. 1918: Massenauftreten von Coccinella septempunctata L. Ent. Bl. Biol. Syst. Käfer, 14: 349.
Remington C. L. 1968: The population genetics of insect introduction. A. Rev. Ent., 13: 415—426.
Robertson G. J. 1961: Ovariole numbers in Coleoptera. Can. J. Zool., 39: 245—263.
Robinson A. G. 1951: Annotated list of predators of Tetranychid mites in Manitoba. Rep. ent. Soc. Ont., 82: 33—37.
Rosen D., Gerson U. 1965: Field studies of Chilocorus bipustulatus (L.) on citrus in Israel. Ann. Épiphyties, 16: 71—76.
Rothschild M. 1961: Defensive odours and Müllerian mimicry among insects. Trans. R. ent. Soc. Lond., 113: 101—121.
Rothschild G. H. L. 1966: A study of natural population of Conomelus anceps (Germar) (Homoptera: Delphacidae) including observations on predation using the precipitin test. J. anim. Ecol., 35: 413—434.
Roubaud E. 1930: Suspension évolutive et hivernation larvaire obligatoire provoquée par la chaleur, chez le moustique commun, Culex pipiens L. Les diapauses vraies et les pseudodiapauses chez les insectes. C. R. Acad. Sci., Paris, 190: 324—326.
Rubtsov I. A. 1954: Citrus pests and their natural enemies. 260 pp. Izd. AN SSSR, Moscow-Leningrad (in Russian).
Ruscinsky A. 1933: Über Anhäufungen von Vibidia duodecimguttata Poda. in Wäldern von Palanča (Bessarabien). Bul. Muz. natn. Ist. nat. Chisinau, 5: 162—163.
Ruszkowski A. 1961: Observations on the appearance of lady-birds (Coccinellidae). Polskie Pismo ent. (B), 21—22: 57—69 (in Polish, Engl. summ.).
Sakurai H. 1969: Respiration and glycogen contents in the adult life of the Coccinella septempunctata Mulsant and Epilachna vigintioctopunctata Fabricius (Coleoptera: Coccinellidae). Appl. Ent. Zool., 4: 55—57.
Salt R. W. 1961: Principles of insect cold-hardiness. A. Rev. Ent., 6: 55—74.
Salt R. W. 1964: Trends and needs in the study of insect cold-hardiness. Can. Ent., 96: 400—405.
Sanderson E. D., Peairs L. M. 1913: The relation of temperature to insect life. Tech. Bull. New Hamps. agric. Exp. Stn, 7: 1—125.
Sasaji H. 1967: A revision of the Formosan Coccinellidae (I), the subfamily Sticholotinae, with an establishment of a new tribe (Coleoptera). Etizenia, 25: 1—28.
Sasaji H. 1968a: Description of the Coccinellid larvae of Japan and the Ryukyus (Coleoptera). Mem. Fac. Educ. Fukui Univ., Sec. 2, 18: 93—136.
Sasaji H. 1968b: Phylogeny of the family Coccinellidae (Col.). Etizenia, 35: 1—37.

Sasaji H. 1971: Coccinellidae. Fauna Japonica. 340pp. Acad. Press of Japan.
Savastano L. 1918: Talune notizie sul Novius e l'Icerya riguardanti l'arboricoltore. Boll. R. Staz. sper. Agrum. Fruttic., 32, 2 pp.
Savoïskaya G. I., 1955: Saxaulical lady-bird — Brumus jacobsoni Bar. (Col., Coccinellida). Uchen. Zap. biol. pochv. Fak. Kirgiz. Univ., 5: 319—326 (in Russian).
Savoïskaya G. I. 1957: The experience of composition of key of Coccinellids' larvae. Dokl. VII nauch. konf. Tomsk. Univ., 3: 62—63 (in Russian).
Savoïskaya G. I. 1960a: On morphology and taxonomy of lady-birds larvae (Col., Coccinellidae) from South-East Kazakhstan. Ent. Obozr., 29: 122—133 (in Russian).
Savoïskaya G. I. 1960b: Dot-like lady—bird — Stethorus punctillum Ws. Trudy Inst. Zool. Alma-Ata, 9: 140—144, (in Russian).
Savoïskaya G. I. 1960c: Hibernation sites of some Coccinellids in south eastern Kazakhstan. Zool. Zh., 39: 882—887 (in Russian).
Savoïskaya G. I. 1961: Coccinellidae of the tribe Psylloborini. Uchen. Zap. tomsk. gos. ped. Inst., 19: 92—100 (in Russian).
Savoïskaya G. I. 1962a: Morphology and taxonomy of the larvae of the genus Coccinella L. (Col., Coccinellidae). Trudy Inst. Zashch. Rast., Alma-Ata, 7: 299—315 (in Russian).
Savoïskaya G. I. 1962b: On the Coccinellid genus Coccinula Dobzh. from South-East Kazakhstan. Trudy Inst. Zashch. Rast., Alma-Ata, 8: 323—333 (in Russian).
Savoïskaya G. I. 1962c: Coccinellidae of the tribe Chilocorini (Col., Coccinellidae). Trudy Inst. Zool., Alma-Ata, 17: 189—200 (in Russian).
Savoïskaya G. I. 1964a: Materials on morphology and taxonomy of larvae of the tribe Coccinellini (Col., Coccinellidae). Trudy Inst. Zashch. Rast., Alma-Ata, 8: 310—368 (in Russian).
Savoïskaya G. I. 1964b: On some larvae of the tribe Coelopterini and Hyperaspini, with a description of a new species (Col., Coccinellidae). Trudy Inst. Zashch. Rast., Alma-Ata, 8: 358—370 (in Russian).
Savoïskaya G. I. 1965a: Biology and perspectives of utilisation of coccinellids in the control of aphids in south-eastern Kazakhstan orchards. Trudy Inst. Zashch. Rast., Alma-Ata, 9: 128—156 (in Russian).
Savoïskaya G. I. 1965b: On a new species of Coccinellidae. Vest. sel'.-khoz. Nauki, Alma-Ata, 1: 31—37 (in Russian).
Savoïskaya G. I. 1966: On coccinellids — pests of coccids and aphids. Vest. sel'.-khoz. Nauki, Alma-Ata, 3: 53—56 (in Russian).
Savoïskaya G. I. 1968: Beetles of the genus Exochomus and Platynaspis — predators of the aphids and coccids. Trudy Inst. Zashch. Rast., Alma-Ata, 10: 158—180 (in Russian).
Savoïskaya G. I. 1969a: New genus and species of Coccinellidae from Kazakhstan. Vest. sel'.-khoz. Nauki, Alma-Ata, 6: 37—43 (in Russian).
Savoïskaya G. I. 1969b: On revealing of new taxonomical categories of the coccinellids. Vest. sel'.-khoz. Nauki, Alma-Ata, 9: 101—106 (in Russian).
Savoïskaya G. I. 1970a: Coccinellids of the Alma-Ata reserve. Trudy Alma-Altin. gos. Zapov., 9: 163—187 (in Russian).
Savoïskaya G. I. 1970b: Introduction and acclimatisation of some coccinellids in the Alma-Ata reserve. T u ly Alma-Altin. gos. Zapov., 9: 138—162 (in Russian).
Schiefelbein J. W., Chiang H. C. 1966: Effects of spray of sucrose solution in a corn field on the populations of predatory insects and their prey. Entomophaga, 11: 333—339.
Schilder F. A. 1952/53: Neue Variationsstudien an Coccinelliden. Wiss. Z. Martin-Luther-Univ. Halle-Wittenb., 2: 143—163.
Schilder F. A., Schilder M. 1928: Die Nahrung der Coccinelliden und ihre Beziehung zur Verwandtschaft der Arten. Arb. biol. Reichs Anst. Land-u. Forstw., 16: 213—282.
Schilder F. A., Schilder M. 1951/52: Methoden der Phänoanalyse von Tieren. Wiss. Z. Martin--Luther-Univ. Halle-Wittenb., 1: 81—91.
Schlinger E. I., van den Bosch R., Dietrick E. J. 1959: Biological notes on the predacious earwig Labidura riparia (Pallas) a recent immigrant to California (Dermaptera, Labiduridae). J. econ. Ent., 52: 247—249.
Schlinger E. I., Dietrick E. J. 1960: Biological control of insect pests aided by stripfarming alfalfa in experimental program. Calif. Agric., 14: 8—9, 15.
Schmidt G. 1954: Coccinellidae. In P. Sorauer: Handbuch der Pflanzenkrankheiten. Bd. 2, Teil 2, pp. 99—104.
Schwerdtfeger F. 1963: Autökologie. 461 pp., P. Parey, Hamburg, Berlin.
Semenov Tyanshanskiï 1911: (Discussion.) Russk. ent. Obozr., 11: XIX—XX.
Sem'yanov V. P. 1965a: Fauna, biology and usefulness of coccinellids (Coleoptera, Coccinellidae) in Belorus SSR. Zap. leningr. sel'.-khoz. Inst., 95: 106—120 (in Russian).

Sem'yanov V. P. 1965b: Fauna and distribution in habitats of coccinellids (Coleoptera, Coccinellidae) in the Leningrad area. Ent. Obozr., 44: 315—323 (in Russian).
Sem'yanov V. P. 1965c: Conservation of coccinellids at the chemical treatment of orchards. Zashch. Rast. Vredit. Bolez., 10: 20—21 (in Russian).
Sem'yanov V. P. 1970: Biological properties of Adalia bipunctata L. (Coleoptera, Coccinellidae) in conditions of Leningrad region. Zashch. Rast. Vredit. Bolez., 127: 105—112 (in Russian).
Sethi S. L., Atwal A. S. 1964: Influence of temperature and humidity on the development of different stages of Coccinella septempunctata (Coleoptera: Coccinellidae). Ind. J. agric. Sci., 34: 166—171.
Shade R. E., Hansen H. L., Wilson M. C. 1970: A partial life table of the cereal leaf beetle, Oulema melanopus, in northern Indiana. Ann. ent. Soc. Am., 63: 52—59.
Shands W. A., Holmes R. L., Simpson G. W. 1970: Improved laboratory production of eggs of Coccinella septempunctata. J. econ. Ent., 63: 315—317.
Shands W. A., Shands M. K., Simpson G. W. 1966: Techniques for Massproducing Coccinella septempunctata. J. econ. Ent., 59: 102—103.
Shelford V. E. 1929: Laboratory and field ecology. 545 pp., Williams and Wilkins, Baltimore.
Sherman F. 1938: Massing of convergent ladybeetle at summits of mountains in southeastern United States. J. econ. Ent., 31: 320—322.
Sinha A. K., Krishna S. S. 1969: Feeding behaviour of Epilachna vigintioctopunctata (Coleoptera: Coccinellidae) on Luffa aegyptiaca. Entomologia exp. appl., 12: 268—274.
Skuhravý V. 1959: Die Nahrung der Feldcarabiden. Čas Čs. spol. ent., 56: 1—18 (in Czech, Germ. summ.).
Skuhravý V. 1960: Die Nahrung des Ohrwurms (Forficula auricularia L.) in den Feldkulturen. Čas. Čs. spol. ent., 57: 329—339.
Skuhravý V., Novák K. 1957: Entomofauna des Kartoffelfeldes und ihre Entwicklung. Rozpr. čs. Akad. Věd., 67 (7): 1—50 (in Czech, Germ. summ.).
Skuhravý V. Novák. K. 1966: Die quantitativen Beziehungen zwischen der Schwarzen Bohnenblattlaus (Aphis fabae Scop.) und ihren Räubern an der Zuckerrübe. Z. angew. Ent., 57: 141—166.
Sluss R. R. 1967: Population dynamics of the walnut aphid, Chromaphis juglandicola (Kalt.) in Northern California. Ecology, 48: 41—58.
Sluss R. 1968: Behavioural and anatomical responses of the convergent lady beetle to parasitism by Perilitus coccinellae (Schrank) (Hymenoptera: Braconidae). J. Invertebr. Path., 10: 9—27.
Smee C. 1922: British ladybird beetles. Their control of aphids. Fruit Grow., 53: 675—676, 717—718, 759.
Smirnoff W. A. 1958: An artificial diet for rearing coccinellid beetles. Can. Ent., 90: 563—565.
Smith B. C. 1958: Notes on relative abundance and variation in elytral patterns of some common coccinellids in the Belleville district (Col.: Coccinellidae). Rep. ent. Soc. Ont., 88: 59—60.
Smith B. C. 1960a: A technique for rearing Coccinellid beetles on dry foods, and influence of various pollens on the development of Coleomegilla maculata lengi Timb. (Coleoptera: Coccinellidae). Can. J. Zool., 38: 1047—1049.
Smith B. C. 1960b: Note on parasitism of two coccinellids, Coccinella trifasciata perplexa Muls. and Coleomegilla maculata lengi Timb. (Coleoptera: Coccinellidae) in Ontario. Can. Ent., 92: 652.
Smith B. C. 1961: Results of rearing some coccinellid (Coleoptera: Coccinellidae) larvae on various pollens. Proc. ent. Soc. Ont., 91 (1960): 270—271.
Smith B. C. 1965a: Growth and development of coccinellid larvae on dry foods (Coleoptera, Coccinellidae). Can. Ent., 97: 760—768.
Smith B. C. 1965b: Effects of food on the longevity, fecundity and development of adult coccinellids. Can. Ent., 97: 910—919.
Smith B. C. 1965c: Differences in Anatis mali Auct. and Coleomegilla maculata lengi Timberlake to changes in the quality and quantity of the larval food. Can. Ent., 97: 1159—1166.
Smith B. C. 1971: Effects of various factors on the local distribution and density of coccinellid adults on corn (Coleoptera: Coccinellidae). Can. Ent., 103:1115—1120.
Smith B. D. 1966: Effects of parasites and predators on a natural population of the aphid Acyrthosiphon spartii Koch on broom (Sarothamnus scoparius L.). J. Anim. Ecol., 35: 255—267.
Smith H. S., Armitage H. M. 1931: The biological control of mealybugs attacking citrus. California Agr. Exp. Stn. Bull., 509, 74 pp.
Smith R. F., Reynolds H. T. 1966: Principles, definitions and scope of integrated pest control. Proc. FAO Symp. Integr. Pest Control, Rome 1965, 1: 11—17.
Smith S. G. 1956: Extreme chromosomal polymorphism in a Coccinellid beetle. Experientia, 12: 52—55.

Smith S. G. 1959: The cytogenetic basis of speciation in Coleoptera. 10. Int. Conf. Genet., 1: 444—450.
Smith S. G. 1960: Chromosome numbers of Coleoptera, II. Can. J. Genet. Cytol., 2: 67—88.
Smith S. G. 1965: Heterochromatin, colchicine, and karyotype. Chromosoma (Berl.), 16: 162—165.
Smith S. G. 1966: Natural hybridization in the coccinellid genus Chilocorus. Chromosoma, 18: 380—406.
Sol R. 1961: Über den Eingriff von Insektiziden in das Wechselspiel von Aphis fabae Scop. und einigen ihrer Episiten. Entomophaga, 6: 7—33.
Southwood T. R. E. 1966: Ecological methods, with particular reference to the study of insect populations. 391 pp., Methuen Co. Ltd, London.
Sparks A. N., Chiang H. C., Burkhardt C. C., Fairchild M. L., Weekman G. T. 1966: Evaluation of the influence of predation on corn borer populations. J. econ. Ent., 59: 104—107.
Speyer W. 1934: Die an der Niederelbe in Obstbaumfanggürteln überwinternden Insekten. III. Mitteilung. Coleoptera: Coccinellidae. Z. PflKrankh. PflPath. PflSchutz, 44: 321—330.
Speyer W. 1935: Coccinelliden als Blutlaus-Feinde. NachrBl. dt. PflSchutzdienst, Berl., 15: 83.
Stewart J. W., Whitcomb W. H., Bell K. O. 1967: Estivation studies of the convergent lady beetle in Arkansas. J. econ. Ent., 60: 1730—1735.
Strouhal H. 1926a: Die Larven der Palaearktischen Coccinellini und Psylloborini (Coleopt.). Arch. Naturgesch., 92 (A), 3: 1—63.
Strouhal H. 1926b: Pilzfressende Coccinelliden (Tribus Psylloborini) (Col.). Z. wiss. InsektBiol., 21: 131—143.
Strübing H. 1963: Zum Diapauseproblem in der Gattung Stenocranus (Homoptera, Auchenorrhyncha). Zool Beitr., 9: 1—119.
Sundby R. 1966: A comparative study of the efficiency of three predatory insects — Coccinella septempunctata L. (Coleoptera, Coccinellidae), Chrysopa carnea St. (Neuroptera, Chrysopidae) and Syrphus ribesii L. (Diptera, Syrphidae) at two different temperatures. Entomophaga, 11: 395—404.
Sundby R. A. 1968: Some factors influencing the reproduction and longevity of Coccinella septempunctata L. (Coleoptera, Coccinellidae). Entomophaga, 13: 197—202.
Svihla A. 1952: Two-spotted lady beetles biting man. J. econ. Ent., 45: 134.
Sysoev A. T. 1958: Attempt at conservation of some coccinellid species. In: Biological control of agricultural and forest pests. Izd. Min. sel'.-khoz. Mold. SSR, Kishinev, 63: 44—45 (in Russian).
Szalay-Marzsó L. 1957: Population dynamic examination of Doralis fabae population of a beet field. Ann. Inst. Prot. Plant Hung., 7: 91—99 (in Hungarian, Engl. summ.).
Szalay-Marzsó L. 1958: Populationsdynamische Untersuchungen an Beständen der Rübenblattlaus (Aphis (Doralis) fabae Scop.) in Ungarn, in den Jahren 1955 und 1956. Acta agron. hung., 8: 187—211.
Szumkowski W. 1952: Observations on Coccinellidae. II. Experimental rearing of Coleomegilla on a non-insect diet. 9. Int. Congr. Ent. Amsterdam, 1951, pp. 781—785.
Szumkowski W. 1955: Observaciones sobre la biologia de algunos Coccinellidae (Coleoptera). I. Boln Ent. venez., 11: 1—20.
Szumkowski W. 1961a: Dietas sin insectos vivos para la cria de Coleomegilla maculata Deg. (Coccinellidae, Coleoptera). Agronomía trop. (Venezuela), 10: 149—154.
Szumkowski W. 1961b: Aparicion de un coccinelido predator nuevo para Venezuela. Agronomía trop. (Venezuela), 11: 33—37.
Tadmor U., Applebaum S. W. 1971: Adult diapause in the predaceous coccinellid, Chilocorus bipustulatus: Photoperiodic induction. J. Insect Physiol., 17: 1211—1215.
Takeda S., Hukusima S., Yamada S. 1964: Seasonal abundance in coccinellid beetles. Res. Bull. Fac. Agric. Gifu Univ., 19: 55—63 (in Japanese, Engl. summ.).
Takeda S., Hukusima S., Yamada H. 1965: Some physiological aspects of coccinellid beetles in relation to their tolerance against pesticide treatments. Res. Bull. Fac. Agric. Gifu Univ., 21: 83—93 (in Japanese, Engl. summ.).
Tamaki G., Halfhill J. E. 1968: Bands on peach trees as shelters for predators of the green peach aphid. J. econ. Ent., 61: 707—711.
Tamaki G., Weeks R. E. 1968: Use of chemical defoliants on peach trees in integrated program to suppress populations of green peach aphids. J. econ. Ent., 61: 431—435.
Tanaka M., Maeta Y., 1965: Rearing of some predacious coccinellid beetles by artificial diets. Bull. Hort. Res. Stn. (D) , 63: 17—35 (in Japanese, Engl. summ.).
Tanasijevic N. 1958: Zur Morphologie und Biologie des Luzernemarienkäfers Subcoccinella vigintiquatuorpunctata (Coleoptera: Coccinellidae). Beitr. Ent., 8: 23—78.

Telenga N. A. 1948: Biological method of the insect pest control (predaceous coccinellids and their utilisation in USSR). 120 pp., Izd. AN SSSR, Kiev (in Russian).
Telenga N. A., Bogunova M. V. 1936: The most important predators of coccids and aphids in the Ussuri region of Far East and their utilisation. Zashch. Rast., 1936 (10): 75—87 (in Russian).
Thiele H. U. 1966: Einflüsse der Photoperiode auf die Diapause von Carabiden. Z. angew. Ent., 58: 143—149.
Thomas W. A. 1932: Hibernation of the 13-spotted ladybeetle. J. econ. Ent., 25: 136.
Thompson W. R. 1951: The specificity of host relations in predaceous insects. Can. Ent., 83: 262—269.
Throne A. H. 1935: An unusual occurrence of the convergent ladybeetle. Ecology, 16: 125.
Timberlake P. H. 1918: Notes on some of the immigrant parasitic Hymenoptera of the Hawaiian Islands. Proc. Hawaii. ent. Soc., 3: 399—404.
Timofeeff-Ressovsky H. A. 1941: Temperatur-modifikabilität des Zeichnungsmusters bei verschiedenen Populationen von Epilachna chrysomelina F. Biol. Zbl., 61: 68—84.
Tischler W. 1949: Grundzüge der terrestrischen Tierökologie. Braunschweig.
Turian G. 1969: Coccinelles micromycétophages (Col.). Mitt. Schw. ent. Ges., 42: 52— 57 .
Ul'yanova L. S. 1956: On the possibility of acclimatisation of the Far-East coccinellid Harmonia axyridis Pall. in the conditions of Uzbekistan. Trudy Inst. Zool. Parazit., Tashkent, 6: 111—119 (in Russian).
Uvarov B. P. 1931: Insects and climate. Trans. R. ent. Soc. Lond., 79: 1—247.
Vanderzant E. S., Davich T. B. 1961: Artificial diets for the adult boll weevil and techniques for obtaining eggs. J. econ. Ent., 54: 923—928.
Varley G. C., Gradwell G. R. 1970: Recent advances in insect population dynamics. A. Rev. Ent., 15: 1—24.
Walker M. F. 1961: Some observations on the biology of the ladybird (Coccinellidae) parasite Perilitus coccinellae (Schrank) (Hym., Braconidae), with special reference to host selection and recognition. Entomologist's mon. Mag., 97: 240—244.
Walker M. F. 1962: Degeeria luctuosa (Meig.) (Dipt., Tachinidae). Entomologist's mon. Mag., 98: 20.
Warren L. O., Tadic, M., 1967: Biological observations on Coleomegilla maculata and its role as a predator of the fall webworm. J. econ. Ent., 60: 1492—1496.
Watson J. R., Thompson W. L. 1933: Food habits of Leis conformis Boisd. (Chinese lady-beetle). Fla Ent., 17: 27—29.
Way M. J. 1963: Mutualism between ants and honeydew — producing Homoptera. A. Rev. Ent., 8: 307—344.
Way M. J., Banks C. J. 1958: The control of Aphis fabae Scop. with special reference to biological control of insects which attack annual crops. 10. Int. Congr. Ent., Montreal, 1956, 4: 907—909.
Way M. J., Banks C. J. 1968: Population studies on the active stages of the black bean aphid. Aphis fabae Scop., on its winter host Euonymus europaeus L. Ann. appl. Biol., 62: 177—197.
Way M. J., Smith P. M., Potter C. 1954: Studies on the bean aphid (Aphis fabae Scop.) and its control on field beans. Ann. appl. Biol., 41: 117—131.
Werner F. 1913: Massenansammlung von Coccinella. Z. wiss. Insektbiol., 9: 311.
Westigard P. H., Gentner L. G., Berry D. W. 1968: Present status of biological control of the pear psylla in Southern Oregon. J. econ. Ent., 61: 740—743.
Wiackowski S. K., Dronka K. 1968: Laboratory investigations on the effect of aphicides available in Poland on the most important natural enemies of aphids. Polskie Pismo ent., 38: 159—173.
Wiackowski S. K., Nowacka B. 1968: Laboratory investigations on the effect of insecticides on the larvae and adults of Coccinella septempunctata L. (Col., Coccinellidae). Polskie Pismo ent., 38: 441—452.
Wiesmann R. 1955: Untersuchungen an den Prädatoren der Baumwollschadinsekten in Aegypten im Jahre 1951/52. Acta trop., 12: 222—239.
Wigglesworth V. B. 1953: The principles of insect physiology. 546 pp., Vth ed. with add., Methuen, London.
de Wilde J. 1969: Diapause and seasonal synchronization in the adult Colorado beetle. Symp. Soc. exp. Biol., 23: 263—284.
de Wilde J. 1970: Hormones and insect diapause. Mem. Soc. Endocr., 18: 487—514.
Wille J. 1926—27: Curinus (Orcus) zonatus Muls. (Coccinellidae), ein Feind der Schildläuse an Orangenbäumen. Beiträge zu seiner Morphologie, Biologie und bekämpfungs-technischen Bedeutung. Z. angew. Ent., 12: 357—375.
Williams J. L. 1945: The anatomy of the internal genitalia of some Coleoptera. Proc. ent. Soc. Washington, 47: 73—87.

Wilson F. 1971: Biotic agents of pest control as an important natural resource (the 4th Gooding Memorial Lecture). Cent. Ass. BeeKeepers, Ilford, Essex, 12 pp.
Wylie H. G. 1958: Observations on Aphidecta obliterata (L.) (Coleoptera: Coccinellidae), a predator of conifer-infesting Aphidoidea. Can. Ent., 90: 518—522.
Yadava C., Shaw F. R. 1968: The preference of certain coccinellids for pea aphids, leafhoppers and alfalfa larvae. J. econ. Ent., 61: 1104—1105.
Yakhontov V. V. 1940: Mass migrations and winter aggregations of coccinellids. Tez. Dokl. ekol. Konf. Probl. mass. razmnozh. zhivotn., Kiev, pp. 104—108 (in Russian).
Yakhontov V. V. 1957: A novelty in the biological control of pests (heterosis in Coccinellidae). Zashch. Rast., 1957: 32—33 (in Russian).
Yakhontov V. V. 1960: Utilisation of coccinellids in the control of agricultural pests. 85 pp., Izd. AN Uzbeksk. SSR, Tashkent (in Russian).
Yakhontov V. V. 1962: Seasonal migrations of lady-birds Brumus octosignatus Gebl. and Semiadalia undecimnotata Schneid. in Central Asia. 11. Int. Congr. Ent. Wien, 1960, pp. 21—23.
Yinon U. 1969a: Food consumption of the armored scale lady-beetle, Chilocorus bipustulatus (Coleoptera, Coccinellidae). Entomologia exp. appl., 12: 139—146.
Yinon U. 1969b: The natural enemies of the armored scale lady-beetle, Chilocorus bipustulatus (Col., Coccinellidae). Entomophaga, 14: 321—328.
Yun M. Y., Ruppel R. F. 1964: Toxicity of insecticides to a coccinellid predator of the cereal leaf beetle. J. econ. Ent., 57: 835—837.
Zarapkin S. R. 1930: Über die gerichtete Variabilität der Coccinelliden. I. Allgemeine Einleitung und Analyse der ersten Pigmentierungsetappe bei Coccinella 10-punctata. Z. Morph. Ökol. Tiere, 17: 719—736.
Zarapkin S. R. 1938a, b: Über die gerichtete Variabilität der Coccinelliden. V. Die Reihenfolge der Fleckenentstehung auf den Elytren der Coccinella 10-punctata (Adalia 10-punctata) in der ontogenetischen Entwicklung. VI. Biometrische Analyse der gerichteten Variabilität. Z. Morph. Ökol. Tiere, 34: 565—572, 573—583.
Zaslavskiĭ V. A. 1962: New palaearctic species of Chilocorus (Col., Coccinellidae). Ent. Obozr., 31: 399—401 (in Russian).
Zaslavskiĭ V. A. 1967: Reproductive selfdestruction as the ecological factor (ecological effect of genetic interaction of population). Zh. obshch. Biol., 28: 3—11 (in Russian).
Zaslavskiĭ V. A. 1970: Geographical races of Chilocorus bipustulatus (Coleoptera, Coccinellidae). I. Two types of photoperiodical reaction controlling the imaginal diapause in the northern race. Zool. Zh., 49: 1354—1365 (in Russian, Engl. summ.).
Zaslavskiĭ V. A., Bogdanova T. P. 1965: Properties of imaginal diapause in two Chilocorus species (Coleoptera, Coccinellidae). Trudy zool. Inst., Leningrad, 36: 89—95 (in Russian).
Zelený J. 1969: A biological and toxicological study of Cycloneda limbifer Casey (Coleoptera, Coccinellidae). Acta ent. bohemoslov., 66: 333—344.
Zimmermann K. 1931: Wirkung von Selektion und Temperatur auf die Pigmentierung von Epilachna chrysomelina F. Naturwissenschaften, 19: 768—771.

# Subject index

## Coccinellidae[1])



[1]) The recently valid names are in parentheses [ ]. The synonymy has been elaborated by I. Kovář.

## Aphidoidea[1])



[1]) The synonymy has been elaborated by Dr. J. Holman, Institute of Entomology, Czechoslovak Academy of Sciences, Prague. The recently valid names are in brackets.

## Other prey

Photo 1 Adult of *Anatis ocellata* (photo J. Křeček).

Photo 2 Larva of *Scymnus sp.* (photo J. Křeček).

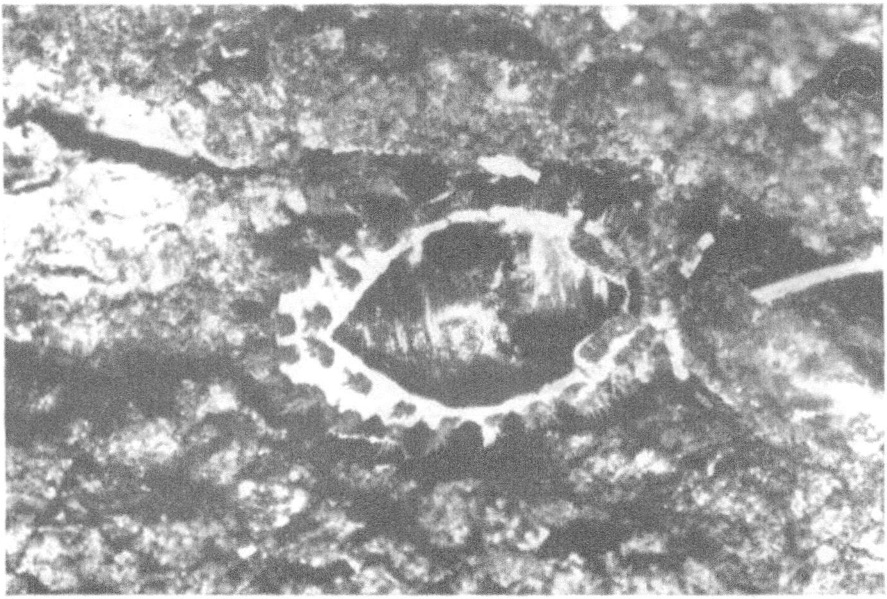

Photo 3 Pupa of *Exochomus quadripustulatus* (photo J. Křeček).

Photo 4 Adult of *Exochomus quadripustulatus* (photo J. Křeček).

Photo 5, 6 *Hippodamia tredecimpunctata* after the adult ecdysis (photo J. Křeček).

Photo 7 *Hippodania tredecimpunctata* after the adult ecdysis (photo J. Křeček).

Photo 8 Adult of *Hippodamia tredecimpunctata* (photo J. Křeček).

Photo 9 Larva of *Hippodamia tredecimpunctata* (photo J. Křeček).

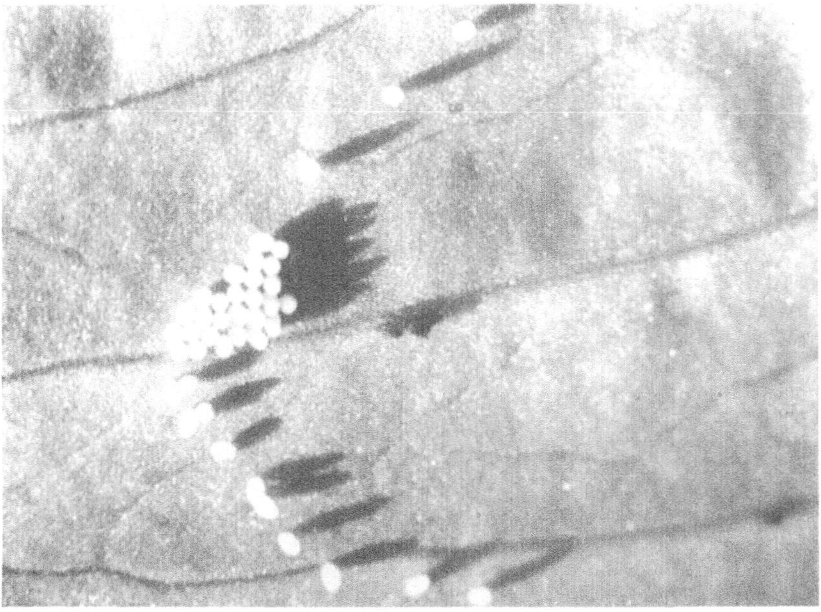

Photo 10 Eggs of *Coccinella septempunctata* (photo I. N. R. A., Antibes).

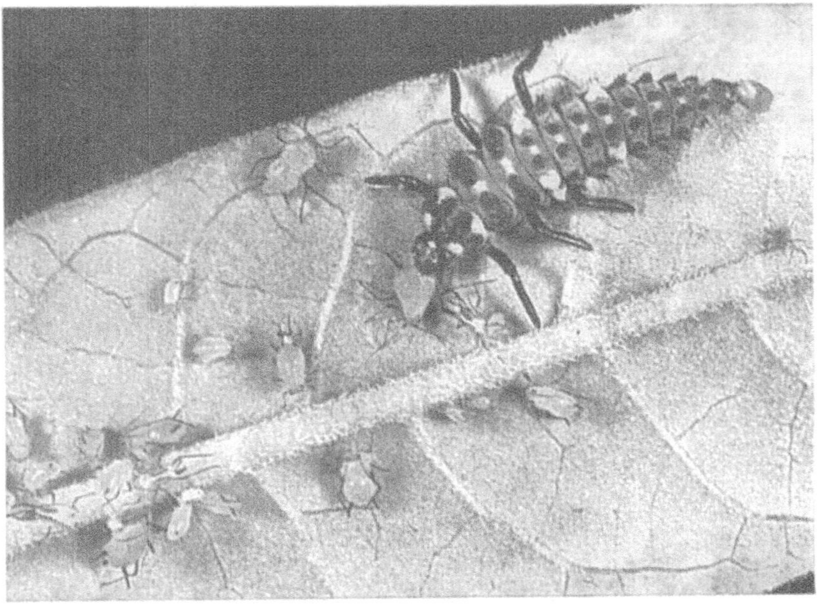

Photo 11 Larva of *Coccinella septempunctata* (photo I. N. R. A., Antibes).

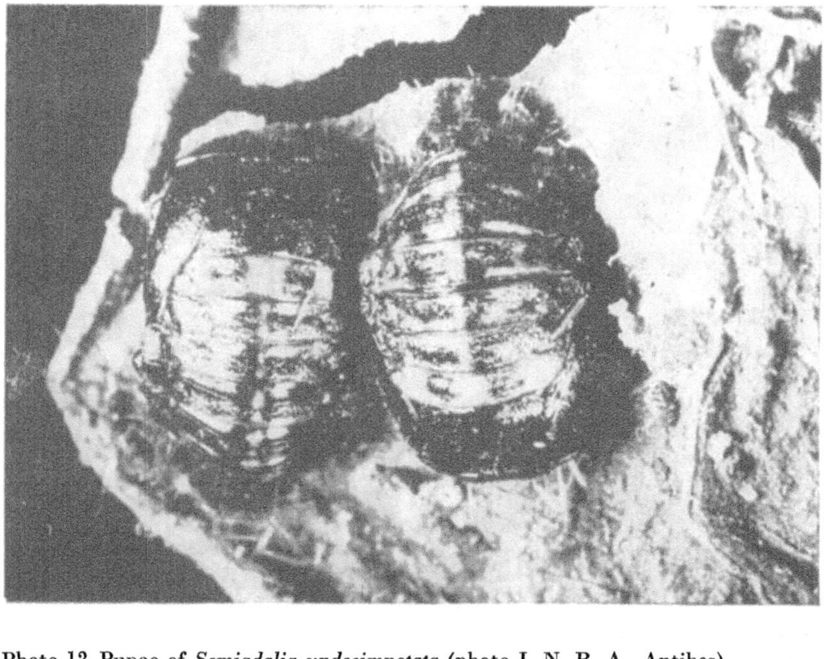

Photo 12 Pupae of *Semiadalia undecimnotata* (photo I. N. R. A., Antibes).

Photo 13 Adult of *Coccinella septempunctata* (photo I. N. R. A., Antibes).

Photo 14 Adult of the mycophagous *Thea vigintiduopunctata* (photo J. Křeček).

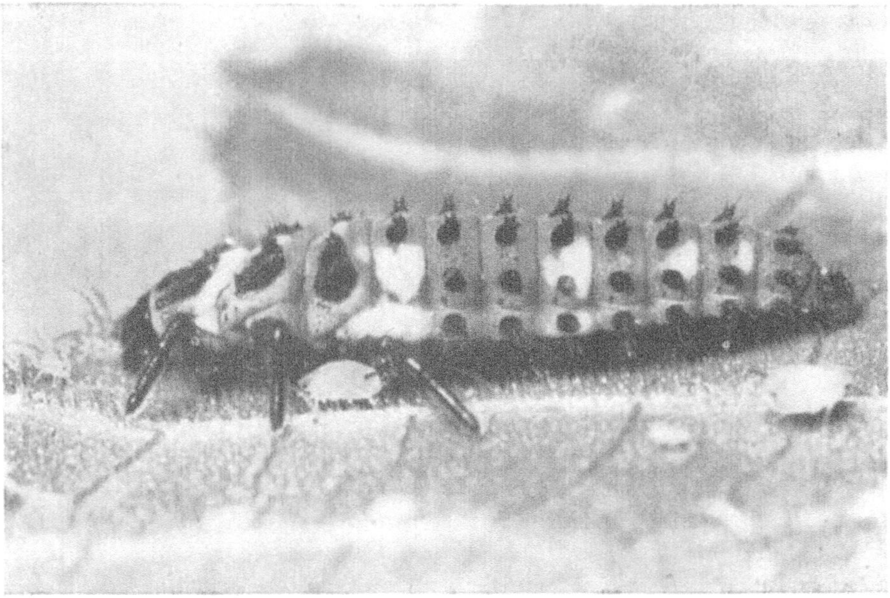

Photo 15 Larva of *Hippodamia sp.* (photo F. E. Skinner).

Photo 16 Adult of *Hippodamia convergens* (photo F. E. Skinner).

Photo 17 Ovaries of a diapausing female of *Semiadalia undecimnotata* (photo J. Kubec).

Photo 18 Ovaries of a reproducing female of *Semiadalia undecimnotata* (photo J. Kubec).

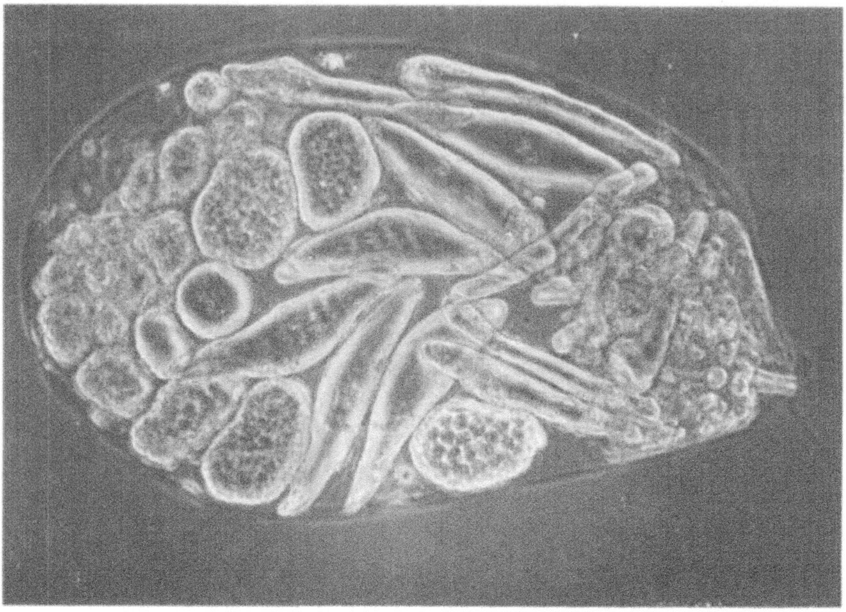

Photo 19 Testicular follicle of *Semiadalia undecimnotata* still active in the first period of dormancy (whole mount, phase contrast, photo V. Landa).

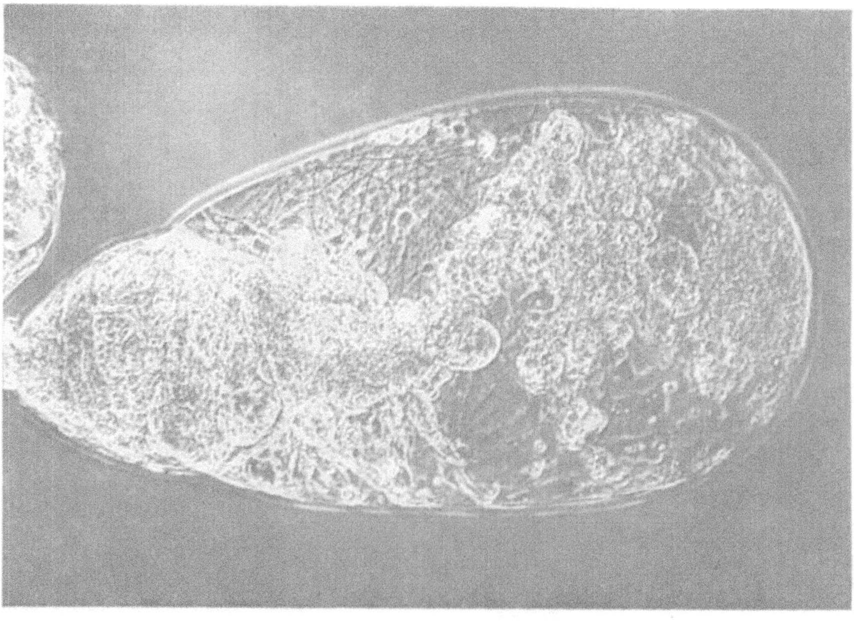

Photo 20 Testicular follicle of *S. undecimnotata* after the arrest of spermatogenesis (early November) (whole mout, phase contrast, photo V. Landa).

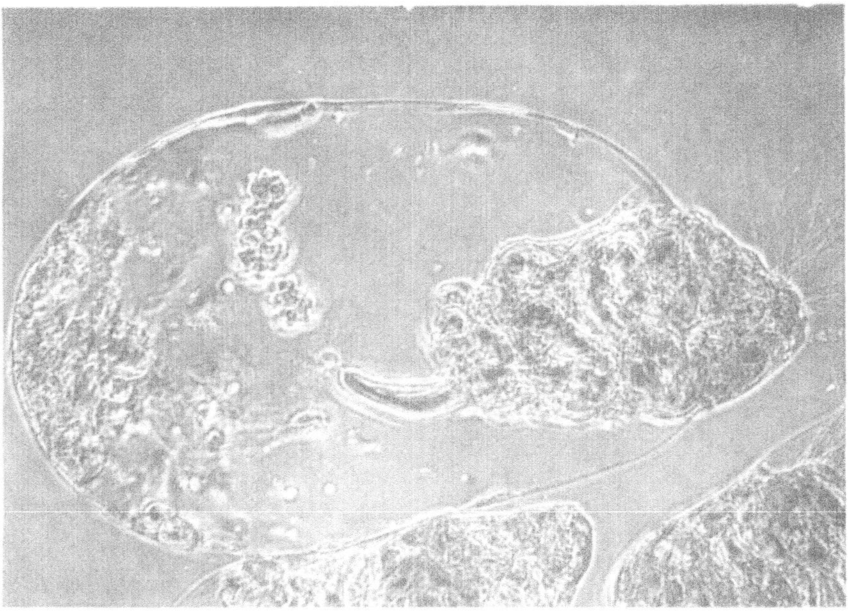

Photo 21 Inactive testicular follicle of *S. undecimnotata* (mid-December) (whole mount, phase contrast, photo V. Landa).

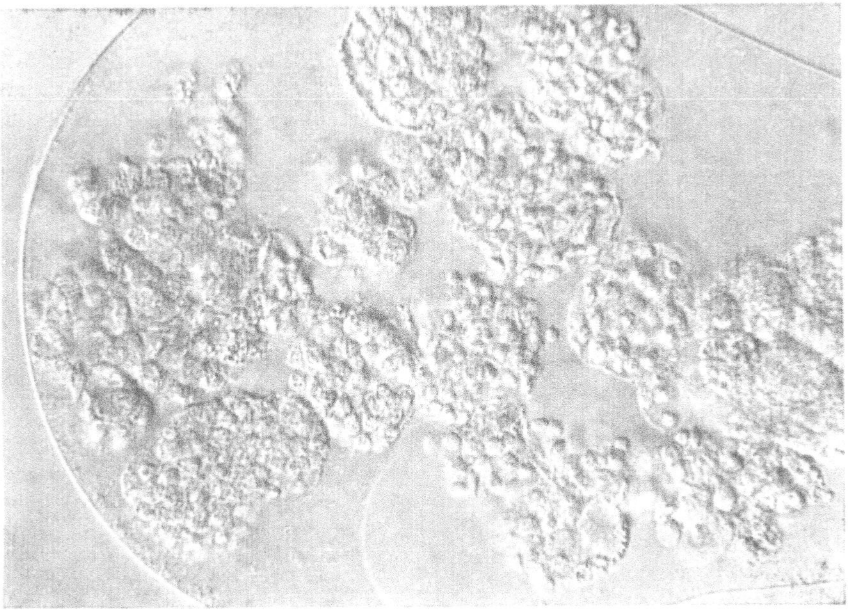

Photo 22 A gradual resumption of spermatogenesis in the testicular follicle of *S. undecimnotata* (whole mount, interference phase contrast, photo V. Landa).

Photo 23 Louny-hills in northern Bohemia (from the left: Raná, Stříbrník, Oblík, Srdov, Brník)

Photo 24 The hill Oblík (in Louny hills) (photo I. Hodek).

with the dormancy sites of *Semiadalia undecimnotata* (photo I. Hodek).

Photo 25 The hill Klapý (with the ruin Hazmburk) in northern Bohemia. (The arrow shows the place with dormancy sites of *S. undecimnotata*. There are no sites at the tower of the ruin.) (Photo I. Hodek).

Photo 26, 27 In the Louny hills the adults of *S. undecimnotata* aggregate in the crevices of rocks or among the low twigs of a shrub and tke rock (photo I. Hodek).

Photo 28 The aggregations of *Semiadalia undecimnotata* on the hill Klapý are in the crevices of high basalt columns (photo I. Hodek).

Photo 29 The limestone mountain Děvín (Pavlov hills) in southern Moravia, with dormancy sites of *S. undecimnotata* right and *Coccinella septempunctata* left (photo L. Peške).

Photo 30, 31 The northern part of Děvín where *C. septempunctata* overwinters in small aggregations on the lower surface of stones (photo L. Peške).

Photo 32 The largest aggregation of *S. undecimnotata* on Děvín can be found among stones or between the post and a stone in a heap of rocks; (from the left A. C. Smith, Belleville, Canada; K. S. Hagen, Berkeley, California, USA; author; G. Iperti, Antibes, France) (photo M. Hodková).

Photo 33 Smallr aggregations of *S. undecimnotata* are situated in the rocks near the peak of Děvín (photo I. Hodek).

**Photo 34 Vršatec mountains in western Slovakia with the dormancy site of *C. septempunctata*(arrow) (photo L. Peške).**

Photo 35 The dormancy site of *C. septempunctata* in grass tussocks on Vršatec (indicated by the arrow on the photo 34) (photo I. Hodek).

Ohoto 36 Zobor hill in the Tribeč mountains (western Slovakia) where several dormancy sises of *S. undecimnotata* are situated (photo I. Hodek).

Photo 37, 38 Places with dormancy sites on the main and northern tops of Zobor (photo I. Hodek, L. Peške).

Photo 39 A large aggregation of *S. undecimnotata* on Nagy-hegy (261 m) (southeastern Slovakia) is situatedin a wide crack of a rock; (no aggregations at the foot of the triangulation tower) (photo I. Hodek).

Photo 40 Cousson mountain (1516 m) with dormancy sites of *S. undecimnotata* in south-eastern France (near Digne, Basses Alpes) (photo I. N. R. A., Antibes).

Photo 41 Courbons mountain (220 m) with dormancy sites of *S. undecimnotata* in south-eastern France (Basses Alpes) (photo I. N. R. A., Antibes).

Photo 42 Saint-Amand hill (732 m) with dormancy sites of *S. undecimnotata* in south-eastern France (Vaucluse); [described by Fabre (1879) as a hibernation quarter of *C. septempunctata*] (photo I. N. R. A,. Antibes).

Photo 43 Dormancy sites of *Adalia spp.* at the foot of a Tyan-shan spruce at c. 1700 m (the Great Alma-Atinka valley, Zailiïskiï Ala-Tau, near Alma Ata, Kazakh. SSR; (lower down the valley, at c. 1300—1400 m, *C. septempunctata* hibernates below stones in the dry river bed) (photo I. Hodek).

Photo 44 Dormancy sites of *Brumus octosignatus* (amongst vegetation in the left foreground) and *Coccinella septempunctata* (the hill in the middle background) in the Chatkalskii khrebet (Su Kok region near Tashkent, Uzbek SSR) (photo I. Hodek).

Photo 45 Dormancy site of *C. septempunctata* in the Chatkalskii khrebet (Su Kok region near Tashkent, Uzbek SSR) (photo I. Hodek).

**Photo 46** The mountain Teshik Tash with the dormancy sites of *C. septempunctata* (Chatkalskiï khrebet, near Taschkent, Uzbek SSR) (photo I. Hodek).

Photo 47 The onset of spring dispersion of *S. undecimnotata* from a rock crevice (photo I. Hodek).

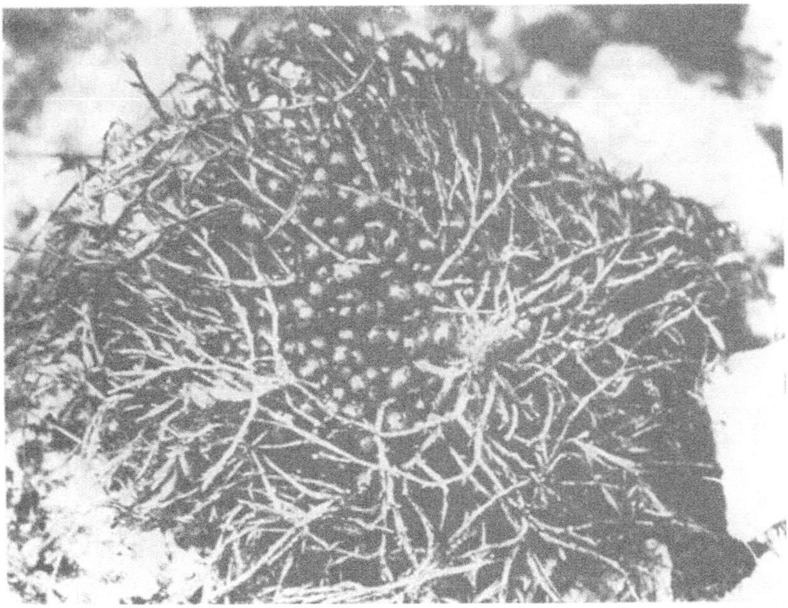

Photo 48 Aggregation of *S. undecimnotata* in vegetation (photo I. N. R. A., Antibes).

Photo 49, 50 Artificial dormancy site divised by G. Iperti (photo I. N. R. A., Antibes).

Photo 51 Adult of *Cycloneda sanguinea* eating an aphid (photo F. E. Skinner).

Photo 52 Adults of *Adalia bipunctata* (photo J. Křeček).

Photo 53, 54 The spring dispersion of *Semiadalia undecimnotata* from an aggregation in Louny-hills, Bohemia (photo I. Hodek).

Photo 55, 56 A large aggregation of *Semiadalia undecimnotata* on a wooden wall protected by stones in Pavlov-hills, Moravia (photo H. J. Müller).

Photo 57, 58 Large aggregations of *Hippodamia convergens* on vegetation ni Sierra Nevada, California (photo B. Matsumoto).

GPSR Compliance

*The European Union's (EU) General Product Safety Regulation (GPSR) is a set of rules that requires consumer products to be safe and our obligations to ensure this.*

*If you have any concerns about our products, you can contact us on ProductSafety@springernature.com*

In case Publisher is established outside the EU, the EU authorized representative is:

Springer Nature Customer Service Center GmbH
Europaplatz 3
69115 Heidelberg, Germany

Batch number: 10755466

Printed by Printforce, the Netherlands